Sustainability Science and Engineering

VOLUME 2

Sustainability Science and Engineering

Series Editors: M.A. Abraham and R. Sheldon

In this series:

Volume 1: Sustainability Science and Engineering: Defining Principles
Edited by M.A. Abraham

Volume 2: Sustainability Science and Engineering: Water Recyclying versus Desalination
Edited by Isabel C. Escobar and Andrea I. Schäfer

IWA Publishing
London • New York

Co-published by IWA Publishing, Alliance House, 12 Caxton Street, London SW1H 0QS, UK
Tel. +44 (0) 20 7654 5500, Fax +44 (0) 20 7654 5555
publications@iwap.co.uk
www.iwapublishing.com
ISBN 1843393409
ISBN13 9781843393405

Sustainability Science and Engineering

Sustainable Water for the Future: Water Recycling versus Desalination

VOLUME 2

Edited by

ISABEL C. ESCOBAR
The University of Toledo, Ohio, USA

ANDREA I. SCHÄFER
The University of Edinburgh, Edinburgh, UK

ELSEVIER

Amsterdam • Boston • Heidelberg • London • New York • Oxford
Paris • San Diego • San Francisco • Singapore • Sydney • Tokyo

Elsevier
Radarweg 29, PO Box 211, 1000 AE Amsterdam, The Netherlands
Linacre House, Jordan Hill, Oxford OX2 8DP, UK

First edition 2010

British Library Cataloguing in Publication Data
A catalogue record for this book is available from the British Library

Library of Congress Cataloging-in-Publication Data
A catalog record for this book is available from the Library of Congress

ISBN: 978-0-444-53115-5
ISSN: 1871-2711

Printed and bound in The Netherlands

10 11 12 13 14   10 9 8 7 6 5 4 3 2 1

Isabel C. Escobar dedicates this book to her mother and father, and to the memory of her grandmother and aunt.

Andrea I. Schäfer dedicates this book to Riko Hannes, born during the production of the book.

With this book, Isabel and Andrea honor Professor Menachem Elimelech's intellectual contribution to research along with his legacy of supporting and mentoring the careers of so many women in the field. Last, their friendship is what made this book possible.

CONTENTS

PART 2: Desalination Technologies

PART 3: Water Recycling Technologies

Color Plate Section to be found in the back of this book

LIST OF CONTRIBUTORS

Souhail R. Al-Abed
National Risk Management Research Laboratory, US Environmental Protection Agency, 26W. Martin Luther King Dr., Cincinnati, OH 45268, USA

Gary Amy
UNESCO-IHE Institute for Water Education, Delft, The Netherlands

Pierre Bérubé
Department of Civil Engineering, University of British Columbia, Vancouver, BC V6T 1Z4, Canada

Rick G. Bond
Black and Veatch

Hyeok Choi
Department of Civil Engineering, University of Texas at Arlington, 416 Yates Drive, Arlington, TX 76019-0308, USA

Yoram Cohen
University of California, Los Angeles, CA 90095-1592, USA

Brett Digman
Chemical and Environmental Engineering Department, The University of Toledo, Toledo, OH 43606-3390, USA

Dionysios D. Dionysiou
Department of Civil and Environmental Engineering, University of Cincinnati, Cincinnati, OH 45221-0071, USA

Sara Dolnicar
Marketing Research Innovation Centre, School of Management and Marketing, University of Wollongong, Northfields Ave., 2522 Wollongong, NSW, Australia

Isabel C. Escobar
Chemical and Environmental Engineering Department, The University of Toledo, 2801 West Bancroft Street, MS 305, Toledo, OH 43606-3390, USA

Christopher J. Gabelich
Palos Verdes, CA 90275, USA

Colleen Gorey
Chemical and Environmental Engineering Department, The University of Toledo, Toledo, OH 43606-3390, USA

Tilak Gullinkala
Chemical and Environmental Engineering Department, The University of Toledo, Toledo, OH 43606-3390, USA

Richard Hausman
Chemical and Environmental Engineering Department, The University of Toledo, Toledo, OH 43606-3390, USA

Anna Hurlimann
Faculty of Architecture Building and Planning, The University of Melbourne, Melbourne 3010, Australia

Maria D. Kennedy
UNESCO-IHE Institute for Water Education, Delft, The Netherlands

Sabine Lattemann
Institute for Chemistry and Biology of the Marine Environment (ICBM), University of Oldenburg, Oldenburg, Germany

Joseph F. Leising
Applied Research and Development Center, Southern Nevada Water Authority, Las Vegas, NV, USA

Panagiotis Lianos
Department of Engineering Science, University of Patras, 26500 Patras, Greece

Sybrand Metz
Wetsus, Centre of Excellence for Sustainable Water Technology, P.O. Box 1113, 8900 CC Leeuwarden, The Netherlands

William Mitch
Department of Chemical Engineering, Yale University, New Haven, CT 06520, USA

Albert Munoz
Environmental Engineering, University of Wollongong, NSW 2125, Australia

Kitty Nijmeijer
Membrane Technology Group, Institute of Mechanics, Processes and Control Twente (IMPACT), University of Twente, P.O. Box 217, 7500 AE Enschede, The Netherlands

Martin Reinhard
Department of Civil and Environmental Engineering, Stanford University, Stanford, CA 94305-4020, USA

Bryce S. Richards
School of Engineering and Physical Sciences, Heriot-Watt University, Edinburgh EH14 4AS, UK

Andrea I. Schäfer
School of Engineering, University of Edinburgh, Edinburgh EH9 3JL, UK

Jan C. Schippers
UNESCO-IHE Institute for Water Education, Delft, The Netherlands

Shane A. Snyder
Applied Research and Development Center, Southern Nevada Water Authority, Las Vegas, NV, USA

Benjamin D. Stanford
Applied Research and Development Center, Southern Nevada Water Authority, Las Vegas, NV, USA

Elias Stathatos
Department of Electrical Engineering, Technological-Educational Institute of Patras, 26334 Patras, Greece

H. Strathmann
Institute of Chemical Engineering, University of Stuttgart, Stuttgart, Germany

Emma Teuten
School of Engineering, University of Edinburgh, Edinburgh EH9 3JL, UK

B. Van der Bruggen
K.U. Leuven, Department of Chemical Engineering, Section Applied Physical Chemistry and Environmental Technology, W. de Croylaan 46, B-3001 Heverlee (Leuven), Belgium

Sophie Walewijk
Department of Civil and Environmental Engineering, Stanford University, Stanford, CA 94305-4020, USA

Pei Xu
Colorado School of Mines, Golden, CO 80401-1887, USA

FOREWORD

Sustainability means "development that meets the needs of the present without compromising the ability of future generations to meet their own needs" (Brundtland Commission report). While originally envisioned as an expansion on concerns of pollution prevention, sustainability more directly addresses the issue of resource allocation. It is a holistic approach to the use of resources: use what you need, but return it to nature in a form in which it can then be used by others. Be considerate of other species; as stewards of the planet have concern for the well-being of the land, the trees, and the animals. In the grandest sense, sustainability is a return to the ideals of our ancestors and their goal of protecting the world around them.

But sustainability also recognizes the need for society to develop, for technology to flourish, and for humanity to expand. And this means the consumption of resources and the disposal of our wastes. The resources needed to support a population of nearly 7 billion are tremendous, placing an enormous strain on the ability of the planet to provide for the current generation, much less to look to the protection of future generations. Arnulf Grubler, during his keynote address to the conference that kicked off this book series, and included in the book *Sustainability Science and Engineering*, explained this as the first paradox: "We need green engineers to solve the problems created by the success of engineering." In other words, technological development that has made the current generation the wealthiest in history (as measured in terms of average life expectancy, leisure time, and similar) has been achieved at the expense of huge resource consumption. To continue this growth requires another way of looking at development, another way of consuming our resources.

Sustainability recognizes that this tremendous wealth is not distributed equitably across the planet. While people in developed nations enjoy the fruits of these developments, those in lesser developed countries continue to suffer in substantial poverty. However, our environmental challenges are global, and those in lesser developed nations do not have the means to address these challenges. Thus, it is incumbent upon scientists and engineers to continue to develop new technologies that make better use of our scarce resources, so that resources can be distributed more equitably and the world's population can continue to flourish.

The current book looks at the issues of water sustainability. Water is one of the primary resources needed to sustain life, and the availability of naturally occurring fresh water is declining. If we are to achieve a sustainable water system, we need to make better use of existing water supplies, we need to identify new ways to create drinking water from nonpotable sources, and we need to develop technologies that reuse the water that we have already extracted. And while solutions to the water challenge exist, their application across a broad population in wealthier nations poses an economic constraint, while their application in remote locations in which limited resources and energy exist poses a more practical challenge to resource allocation. Improved technologies that can be applied without great cost or consumption of energy resources are needed to achieve a sustainable water supply.

Membrane technologies are a potential sustainable solution, if they can be applied with limited materials consumption and a small energy footprint. Understanding these processes and developing these technologies provides a possible pathway to a sustainable future. The current book explores these concepts, evaluates alternatives, and describes opportunities. It is a promising vision of the future, built around a particular technology whose time has perhaps come.

Martin Abraham
Youngstown State University
Youngstown, OH

ACKNOWLEDGMENTS

Isabel C. Escobar would like to acknowledge the help and support of Martin T. Geithmann, who spent many hours helping her edit figures. I also thank the work and commitment of all of the authors who submitted chapters according to our requests and dealt with our numerous comments. My students, Tilak, Colleen, Rick, and Brett, were crucial in the writing of our chapter. However, knowledge shared and learned from all my past and present students was essential to my knowledge and abilities. Encouragement was unending from my friends, especially Maria R. Coleman, Cyndee Gruden, and Nancy Ruiz. The patience and support from my department, college, and university allowed me to pursuit the goal of developing a book. Without research funding, knowledge would have never been available to write and edit this book, so I thank the past and present funding agencies, the U.S. National Science Foundation, U.S. Environmental Protection Agency, U.S. Department of Interior – Bureau of Reclamation, and Office of Naval Research. Lastly, I thank Martin Abraham for giving me the idea (and the push) to write this book, and Andrea for agreeing to co-edit it with me.

Andrea I. Schäfer would like to thank Isabel for her invitation to contribute to this book and taking the lion's share of the workload. Isabel is a very special peer and most wonderful to work with. A special thank you also to Bryce and our children Moana, Tane, and Riko, for putting up with the many late nights of Mama stuck to a laptop screen. Let's hope that the impacts of our work will make our world a better place. My research group, in the past and the present, has contributed to the knowledge we have accumulated through hard work and much enjoyable fun events. I am looking forward to many more years of working with energetic young researchers and seeing them fly off to their own careers.

Isabel and Andrea also would like to thank the constant encouragement and unending patience of their Elsevier editor, Derek Coleman.

PART One

Water Situation

CHAPTER 1

An Overview of the Global Water Situation

Isabel C. Escobar*

Chemical and Environmental Engineering Department, The University of Toledo, 2801 West Bancroft Street, MS 305, Toledo, OH 43606-3390, USA

According to the International Water Management Institute (IWMI), one in three people globally endures some form of water scarcity, one-quarter of the world's population lives in areas where water is physically scarce, and over one billion people live where water is economically scarce, or places where "water is available in rivers and aquifers, but the infrastructure is lacking to make this water available to people." Fig. 1 shows a United Nations history of global water scarcity since 1950, and a projection to 2025. The World Water Institute states that water scarcity is not a factor of absolute quantity; rather, it is a relative concept comparing the availability of water to actual use. In the United States and Europe, the average individual uses between 200 and 600 liters of water per day (UN – Coping with Water Scarcity, 2007 World Water Day, http://www.unwater.org/wwd07), compared to the 20 liters deemed to be the minimum daily requirement for drinking, washing, cooking, and sanitation. A significant cause of water scarcity is agriculture since crop production requires up to 70 times more water than is used in drinking and other domestic purposes. IWMI approximates that each calorie consumed as food requires about 1 liter of water to produce. Such unsustainable consumption levels have led to localized areas of water scarcity and significantly altered freshwater ecosystems.

Existing water supplies may be limited in quantity or quality for meeting the increasing demands from population growth and industry expansion. In many arid and semi-arid areas, providing the large volume of water required for industrial, agricultural, recreational, and potable applications is especially difficult. So, searching for "new" water sources is a task for researchers in the water works field. Municipal wastewater, which comprises between 75% and 80% of consumed water in most cities, is one of the most reliable sources of

*Corresponding Author.
E-mail address: Isabel.Escobar@utoledo.edu

Sustainability Science and Engineering, Volume 2 ISSN 1871-2711, DOI 10.1016/S1871-2711(09)00201-3

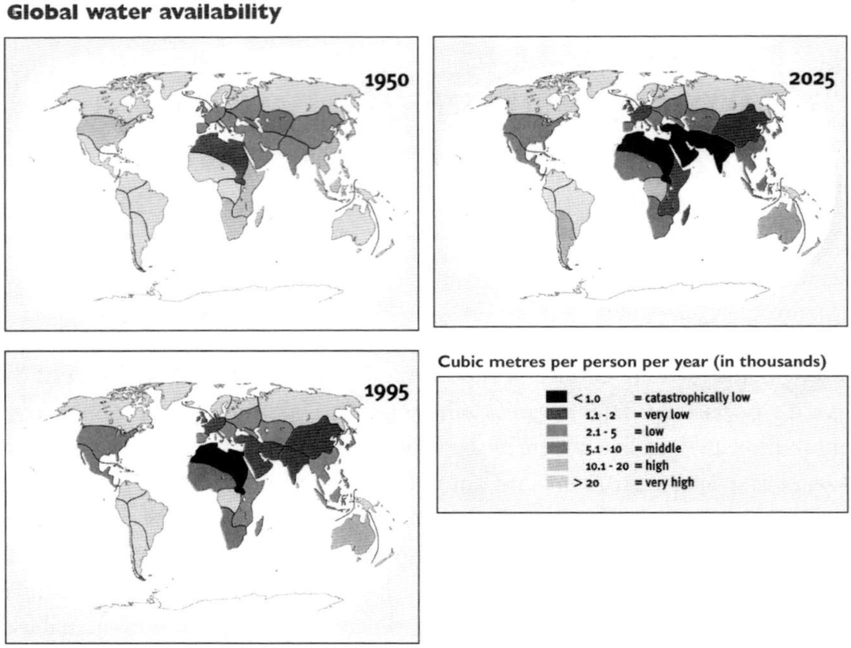

Figure 1 A history of global water scarcity since 1950, and a projection to 2025. *Sources:* http://www.un.org and http://www.un.org/events/water/images/WaterYear-Graph.jpg

water, since its volume varies little throughout the year. Through suitable treatment, reclaimed wastewater can meet various water quality requirements for potential wastewater reuse/recycle [1]. Recycled water can be used in numerous applications to satisfy many water demands such as agricultural and landscape irrigation, industrial processes, toilet flushing, or replenishing a groundwater basin, depending on the level of treatment. Usually, treatment includes four stages: primary, secondary, tertiary or advanced, and disinfection. Among the many available treatment alternatives, membrane filtration technologies are attractive treatment options since they can meet stringent regulation standards.

Membrane processes are now a proven and reliable method of providing high-quality, cost-effective water. Membrane technologies have immediate applications to treatment of fresh, brackish, and seawaters, as well as wastewater reclamation. With innovative module design and engineering, micro- and ultrafiltrations have become effective and economical for drinking water production, particularly for removal of microorganisms. Membrane bioreactors are being developed for municipal and industrial water recycling. Various membrane processes are also used to remove contaminants from industrial wastewaters. The use of membrane technologies for aqueous separations has become very popular over the past 20 years. Successful use of membranes was first seen with desalination of brackish water and seawater. However, improvements in materials and manufacturing technology, mechanical configuration, and cleaning have expanded membrane technology to the treatment of waters of varying quality. Communities are searching for alternatives to conventional treatment for the production of high-quality effluents, and membrane technologies are emerging as treatment of choice for communities, as such technologies become better understood and widely available [2].

The United Nations declared 2005–2015 the "Water for Life" Decade because water is crucial for sustainable development. The goals of the "Water for Life" Decade are to reduce by half the proportion of people without access to safe drinking water, to stop unsustainable exploitation of water resources, to aim to develop integrated water resource management and water efficiency plans, and to halve the proportion of people who do not have access to basic sanitation. Providing safe, clean water in a sustainable fashion is the focus of this book as it covers the fundamental and practical concepts and issues dealing with the application of different technologies for sustainable water treatment. It describes and compares the effectiveness of desalination versus water recycling for long-term sustainable water use.

REFERENCES

[1] I.C. Escobar, Membrane developed systems for water and wastewater treatment, Environmental Progress 24(4) (2005) 355–357.

[2] I.C. Escobar, S. Ritchie, Foreword: Selected water/wastewater membrane-related presentations from the North American, Environmental Progress 27(2) (2008) 169–172.

CHAPTER 2

Global Desalination Situation

Sabine Lattemann[1,*], Maria D. Kennedy[2], Jan C. Schippers[2] *and* Gary Amy[2]

[1]Institute for Chemistry and Biology of the Marine Environment (ICBM), University of Oldenburg, Oldenburg, Germany
[2]UNESCO – IHE Institute for Water Education, Delft, The Netherlands

Contents

1. INTRODUCTION

Seawater and brackish groundwater have become the most important sources of drinking water in a few arid countries of the Middle East, such as Kuwait or the United Arab Emirates, which depend heavily on desalination. Many industrialized and developing regions, however, have

*Corresponding author:
E-mail address: sabine.lattemann@uni-oldenburg.de

Sustainability Science and Engineering, Volume 2 ISSN 1871-2711, DOI 10.1016/S1871-2711(09)00202-5

recently also started to use desalination as a way to supplement and diversify their water supply options. Until a few years ago, desalination plants were limited to the water-scarce but oil-rich countries of the Middle East and North Africa, and to some tropical and subtropical islands. Today, desalinated water has become a commodity for many countries in order to satisfy their growing demand for water.

For the "pioneering" countries, the driving factors were often a lack of surface and groundwater resources, coupled with sufficient natural or financial resources to engage in energy-intensive and costly desalination projects. For the newly emerging desalination markets, driving factors are more diverse, including economic and demographic growth, prolonged droughts, climate change, or declining conventional water resources in terms of quality and quantity due to overuse, pollution, or salinization. Moreover, as conventional water production costs have been rising in many parts of the world and the costs of desalination – particularly seawater desalination – have been declining over the years, desalination also becomes economically more attractive and competitive (Fig. 1).

This chapter gives a short account of the historical development of desalination technologies, an overview on the presently installed worldwide desalination capacity, distinguishing between different raw water sources, processes and use types. It furthermore discusses regional and future trends, driving factors such as cost and energy demand, as well as concerns, such as the environmental impacts of the discharges into the sea.

2. HISTORICAL DEVELOPMENT

The extraction of salt from salty water by means of natural evaporation has been practiced for a long time, dating from the time when salt, not water, was the precious commodity [1]. Advanced technologies that mimic natural processes such as evaporation or osmosis in order to extract the water have only been developed in modern times. Basic desalting processes were first used on naval ships in the 17th to 19th centuries. The island of Curaçao in the Netherlands Antilles was the first location to make a major commitment to desalination in 1928, followed by a major seawater desalination plant built in what is now Saudi Arabia in 1938 [1,2].

A major step in the development of desalination technologies came during World War II in order to supply water to military establishments in arid areas. After the war, the potential of desalination was recognized and more research work was continued in various countries. The American government, through creation and funding of the Office of Saline Water

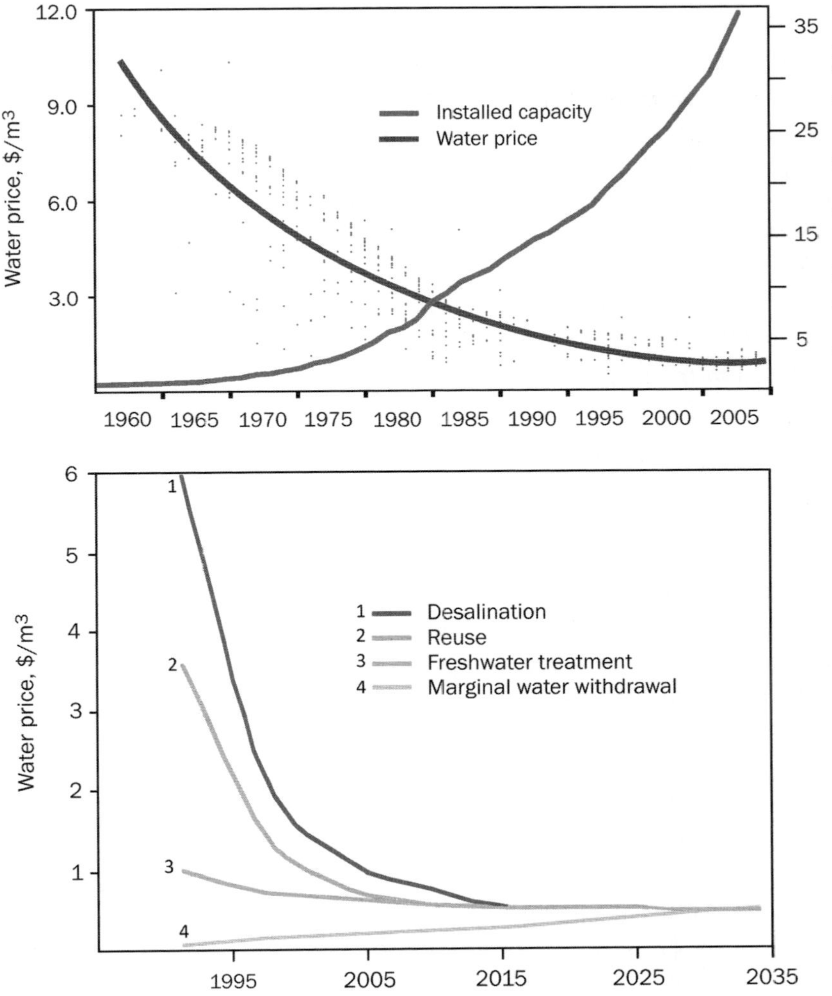

Figure 1 Water resource cost trends in the desalination market in US$ per cubic meter. Top: Total installed capacity and water price development. Bottom: Differentiated between water source type. Adapted from Ref. [6].

(OSW) in the early 1950s and its successor organization, the Office of Water Research and Technology (OWRT) in 1974, helped to provide much of the basic research and development of the different desalting technologies [2,3].

Many of the early projects focused on thermal processes. Significant work was completed on construction materials, heat transfer surfaces, and corrosion, which was instrumental in assisting the design and construction of the first large distillation systems in the Middle East [2]. The multieffect distillation (MED) process has been used in industry for a long time, traditionally for the production of sugar and salt. Some of the early distillation plants also used the MED process; however, the multistage flash (MSF) process that was developed in the 1950s continually displaced the MED process due to a higher resistance against scaling. A revived interest in MED can be observed since the 1980s due to a lower operating temperature and energy demand of the process [3].

During the late 1950s, the first asymmetric membrane for desalination was developed by Loeb and Sourirajan, which consisted of cellulose acetate polymer [4]. The electrodialysis (ED) process, which was commercially introduced in the early 1960s, moves salts selectively through a membrane driven by an electrical potential. It was the first cost-effective way to desalt brackish water and spurred a considerable interest in using desalting technologies for municipal water supply, especially in the United States. ED is exclusively applied to low brackish and fresh water desalination, since the energy consumption for seawater treatment would be far too high. Other milestones included the commercialization of reverse osmosis (RO), a pressure-driven membrane process, in the early 1970s [3], followed by the development of a more robust composite aromatic polyamide spiral wound membrane in the 1980s [4].

Although a wide variety of membrane materials and module configurations have been developed over the years, including hollow fine fibers from cellulosic or noncellulosic materials, composite aromatic polyamide membranes in spiral wound configuration are almost exclusively used in modern RO plants today. While cellulose acetate seawater membranes had a specific permeate flux of $0.5\,\mathrm{L/(m^2\,h\,bar)}$ and a salt rejection of 98.8% in the 1970s, the latest polyamide seawater membranes have a specific flux of more than $1.2\,\mathrm{L/(m^2\,h\,bar)}$ and a salt rejection of 99.8%. The improvement in specific flux translates into a significant reduction of the specific energy demand of the RO process [4]. Another significant power and cost reduction stems from the development of energy recovery devices, which result in a total energy demand of $3\text{–}4\,\mathrm{kWh/m^3}$ of permeate water using state-of-the-art technology.

To conclude, it took about 50 years to evolve from the first land-based distillations plants into a fully developed industry in the 1980s. By the 1990s, the use of desalting technologies for municipal water supplies had become commonplace [3]. Today, municipalities are the main end users of desalinated water and the market continues to grow exponentially, with a doubling of the installed capacity expected from 2006 to 2015. RO has emerged as the most important desalination process today (Section 3, [5,6]). In 1969, the world's largest RO system in operation was a $380\,\mathrm{m^3/day}$

brackish water plant in Dallas, Texas [7]. Today, the largest seawater RO plant produces 330,000 m^3 of water per day, the equivalent of 132 Olympic-size swimming pools, using 27,000 membrane elements with an active surface area of about 99 ha (or 200 football fields), which need to be replaced every 3–7 years. Currently, the membrane market is estimated to have current sales in excess of US$ 500 million per year, and an annual market growth at about 16% annually [7].

3. GLOBAL INSTALLED DESALINATION CAPACITY

The worldwide installed desalination capacity is increasing at a rapid pace. The latest figures from the 20th International Desalination Association (IDA) Worldwide Desalting Plant Inventory [5] indicate that the production capacity of all desalination plants worldwide was around 44.1 million cubic meters per day (Mm^3/day) by the end of 2006. This figure includes all facilities listed in the inventory that treat seawater, brackish water, river water, wastewater, brine, and pure water, which are either in construction, online, or presumed online.

3.1 Projected Growth of the Desalination Market

The worldwide installed capacity grew at a compound average rate of 12% a year over the past 5 years, and the rate of capacity growth is expected to increase even further. Based upon country-by-country analysis involving desalination projects and official data on water supply and demand from agencies around the world, it is projected that the installed capacity will presumably reach 64 Mm^3/day by 2010 and 98 Mm^3/day by 2015 (Fig. 2) [6].

3.2 Global capacity by source water type

Much of the expected growth of the desalination market will take place in the seawater sector, although brackish water and wastewater desalination processes will presumably become more important in the future. Only 5% of the total volume of 44.1 Mm^3/day presently comes from wastewater sources, 19% is produced from brackish water sources, and 63% from seawater sources (primary data from Ref. [5]). Desalination of seawater is hence the dominant desalination process and accounts for a worldwide water production of 27.9 Mm^3/day (Fig. 3, top). For illustration, this is a volume comparable to the average discharge of the Seine River at Paris (average flow of 28.3 Mm^3/day).

3.3 Global capacity by process

All source water types included, RO is the prevalent desalination process. It accounts for slightly more than half (51% or 22.4 Mm^3/day) of the global capacity (Fig. 3, second row). Forty percent or 17.7 Mm^3/day of the global

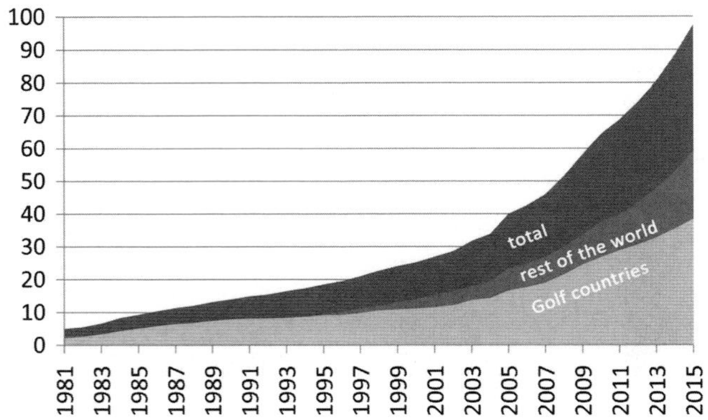

Figure 2 Projected growth of the desalination market (including seawater, brackish water, river water, wastewater, brine, and pure water desalination processes). The installed capacity was 44.1 Mm^3/day in 2007 [5] and is expected to more than double by 2015. At that time, 38 Mm^3/day will presumably be installed in the Gulf region and 59 Mm^3/day in the rest of the world [6].

production of desalinated water comes from distillation plants, either using the MSF or the MED process, with relative market shares of 32% (14 Mm^3/day) and 8% (3.7 Mm^3/day), respectively. Other minor desalination processes include the membrane-based nanofiltration (NF) and ED process with about 4% market share each (2 and 1.6 Mm^3/day, respectively).

The picture changes if one distinguishes between the different source water types. Thermal desalination processes account for 61% (17.2 Mm^3/day) of the production in all desalination plants that use *seawater* as raw water source, of which 50% is produced in MSF plants. Only 35% of the water comes from RO seawater desalination plants. On the contrary, RO accounts for 84% (6.9 Mm^3/day) and 79% (1.7 Mm^3/day) of the production in brackish water and in wastewater applications, respectively, whereas distillation processes play a negligible role in brackish water (<2%, 0.1 Mm^3/day) and a minor role (13%, 0.3 Mm^3/day) in wastewater desalination (primary data from Ref. [5]).

3.4 Global capacity by use type

All source water types included, desalinated water is mainly used for municipal and industrial purposes: 70% (31 Mm^3/day) of the globally desalinated water is used by municipalities and 21% (9 Mm^3/day) by industries (Fig. 3, third row). Other end users include the power generation industry (4%), irrigation (2%), military (1%), and tourism (1%).

Again, the picture is different if one distinguishes between the different source water types. Municipalities are also the main end users of desalinated

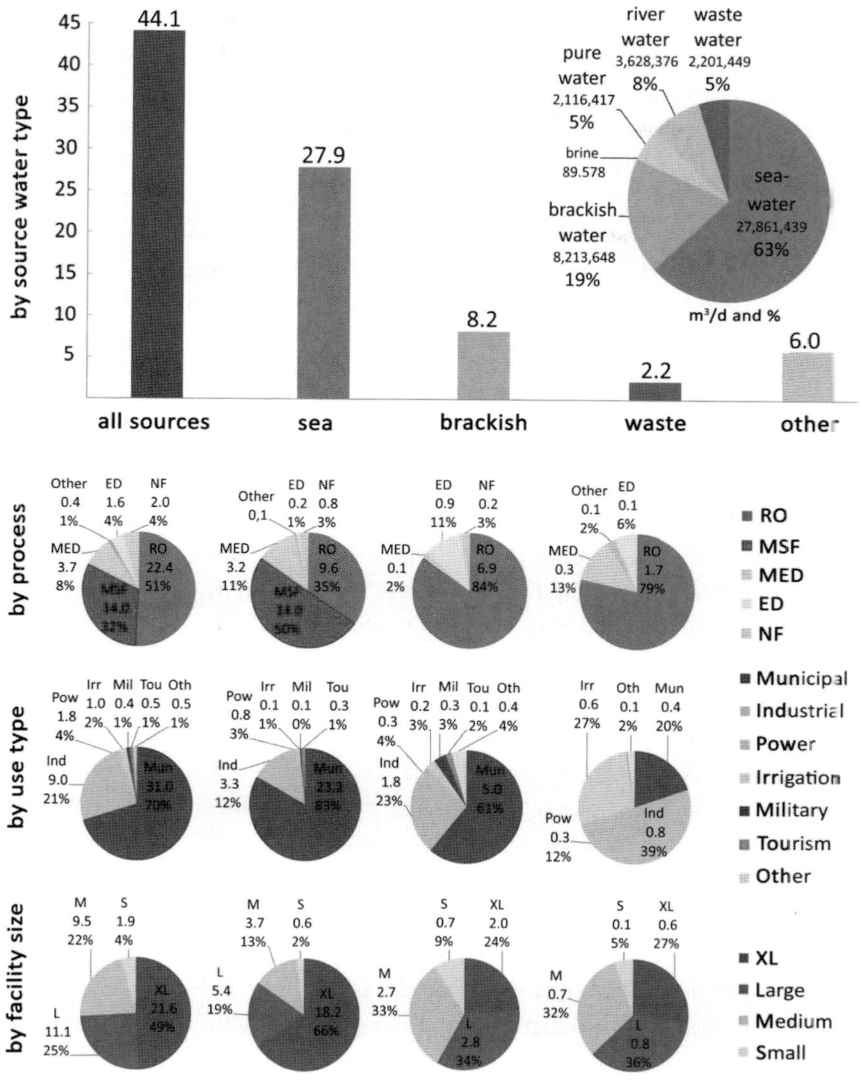

Figure 3 Global desalination capacities (in Mm^3/day and %) by source water type (top row), by process and source water type (second row), by use type and source water type (third row). Data analysis based on primary data from Ref. [5]. Abbreviations: RO, reverse osmosis; MSF, multistage flash distillation; MED, multieffect distillation; NF, nanofiltration; ED, electrodialysis; $XL \geq 50{,}000\ m^3/day > L \geq 10{,}000\ m^3/day > M \geq 1{,}000\ m^3/day > S$ (see plate 1 in color plate section at the end of this book).

sea and brackish water and account for 83% (23.2 Mm^3/day) and 61% (5 Mm^3/day) of the production, respectively, and 20% of the production of repurified wastewater. As one moves from seawater to brackish water and wastewater, the share of municipal use decreases, while the share of industrial use increases. The latter accounts for 12% of the production from seawater, 23% of the production from brackish water sources, and is the primary user of repurified wastewater: 39% (0.8 Mm^3/day) is used for industrial purposes plus an additional 12% is used by the power industry. Irrigation is only the second most important use of repurified waste water with a share of 27% (0.6 Mm^3/day) after industrial use (39% of the wastewater) (primary data from Ref. [5]).

3.5 Global capacity by plant size

Forty-nine percent of the desalinated water is produced by very large facilities with production capacities of 50,000 m^3/day or more ("XL-sized" plants, Fig. 3, last row). The share of production in very large facilities is even higher in the *seawater* sector, where 66% (18.2 Mm^3/day) of the water is produced in only 122 industrial-sized plants. On the other end of the scale, about 1660 small seawater desalination facilities with production capacities of less than 1000 m^3/day account for only 2% (0.6 Mm^3/day) of the production. The plant size distribution is a bit more homogeneous in the *brackish* (BW) and *wastewater* (WW) sectors, where 24% (BW) and 27% (WW) of the water is produced in XL-sized plants, where large plants account for 34% (BW) and 36% (WW) of the production, medium plants for 33% (BW) and 32% (WW) of the production, and small plants for 9% (BW) and 5% (WW) of the production, respectively (primary data from Ref. [5]).

To conclude, most of the desalinated water today is produced in industrial-sized facilities. These include the large thermal distillation plants in the Middle East with production capacities up to 1.6 Mm^3/day. Outside the Middle East region, seawater reverse osmosis (SWRO) is the dominant process that finds application. Majority of the SWRO plants (59%) are small ($<1000\ m^3$/day) and account for only 5% of the worldwide production of 9.4 Mm^3/day, while 2% or 42 large facilities ($\geq 50{,}000\ m^3$/day) account for almost (45%) half the worldwide production. The largest RO plant currently produces 330,000 m^3/day and a few RO projects up to 500,000 m^3/day are being planned.

3.6 Costs and energy demand of desalination processes

The rising costs of conventional water production (Fig. 1) that are observed in some parts of the world are caused by increasing technical expenditure and costs for treating water from conventional sources and for transporting water over long distances. The water production costs have risen – depending on the country, supply, demand, and technology – to US$ 1–1.5 m^{-3}. At the same time, the cost of drinking water from desalination has been decreasing, in some places even below the cost of conventional water production. The causes included improved design and technology, especially of RO processes, the adaptation of facilities to local demand, or the use of cheapest energy sources [8].

The average investment cost required for engineering, procuring, and constructing a MSF distillation plant is US$ 1235 per cubic meter and day installed capacity. The capital costs for MED and SWRO plants are lower, with an average of US$ 916 and US$ 641 per cubic meter and day installed capacity, respectively. The average production costs of desalinated seawater are in the range of US$ 0.45–0.60 m^{-3} (Fig. 4) [6]. This includes the replacement of parts and membranes, chemicals for pretreatment of the intake water, plant cleaning and posttreatment of the product water, labor costs, and – as the most important cost factor – energy demand. With

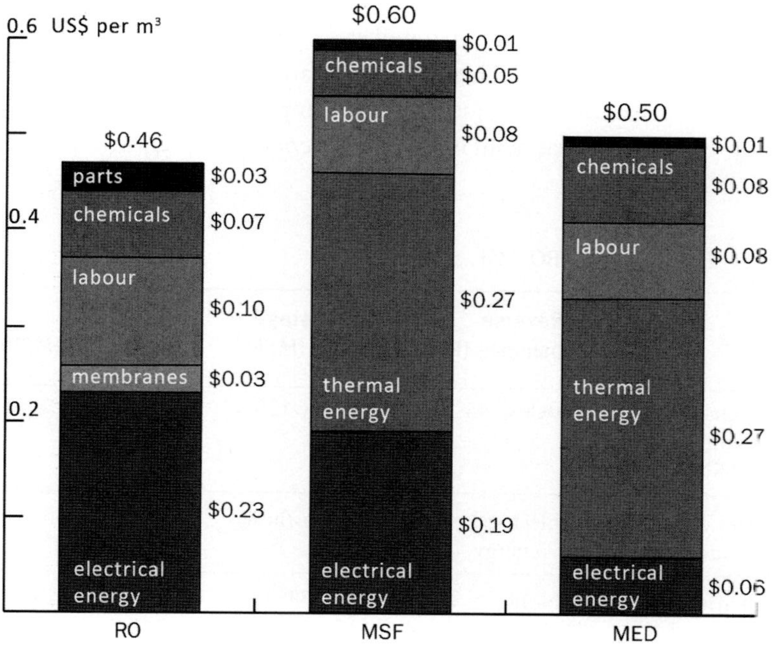

Figure 4 Relative operation costs in US$ of the main desalination processes. Adapted from Ref. [6].

current energy prices on the increase, desalination may again become a more costly water supply option in the future.

The amount of energy needed for water production is process-dependant: MSF plants, having a maximum operating temperature of 120 °C, require about 250–330 MJ/m^3 of thermal and 3.5 kWh of electrical energy for the production of 1 m^3 of water. MED plants, which operate at temperatures below 70 °C, require 145–390 MJ/m^3 of thermal and 1.5 kWh of electrical energy per cubic meter of water. Seawater RO requires less energy than distillation processes. The energy demand of SWRO plants depends on the process design and equipment used (Table 1). The use of low-energy membrane elements, variable frequency pumps, and pressure exchangers can significantly reduce the specific energy demand of a plant. While older plants without energy saving equipment may still require about 5 kWh/m^3, modern plants usually achieve a total energy demand of 3–4 kWh/m^3. For example, the Spanish National Hydrological Plan assumes a total energy value of 4 kWh/m^3 under the assumption that plants are equipped with state-of-the-art technologies [9]. The Affordable Desalination Collaboration operated a demonstration plant in California over 2 years using state-of-the-art, off-the-shelf technology and set a world record in specific energy consumption of 1.58 kWh/m^3 with a low-energy membrane operated at 42% recovery, but at the expense of permeate water quality. The energy demand therefore also depends on the required permeate water quality. For example, employing a second RO pass for boron removal will increase the energy demand. The specific

Table 1 Energy data of RO, MSF, and MED

	Reverse osmosis (RO)	Multistage flash (MSF)	Multieffect distillation (MED)
Operating temperature [2,8]	Below 45 °C	Below 120 °C	Below 70 °C
Main energy source [8]	Electrical energy	Steam (heat)	Steam (heat)
Thermal energy demand [8]	None	250–330 kJ/kg	145–390 kJ/kg
Electrical energy demand [2]	2.5–7 kWh/m^3	3–5 kWh/m^3	1.5–2.5 kWh/m^3

energy demand for SWRO plants usually increases with recovery, but the total energy demand decreases with the recovery rate as less feedwater must be pumped and treated to obtain the same volume of permeate at a higher recovery. Therefore, it is important to analyze the desalination process as a whole, and not just the SWRO-specific energy demand. At the most affordable point for a 190,000 m^3/day plant, total treatment energy in the range of 2.75–2.98 kWh/m^3 was demonstrated [10].

4. REGIONAL DESALINATION SITUATION

Forty-eight percent (21.0 Mm^3/day) of the global desalination production takes place in the Middle East, mainly in the Gulf country states (19.3 Mm^3/day). Nineteen percent of the desalinated water is produced in the Americas (8.2 Mm^3/day), 14% in the Asia-Pacific region (6.2 Mm^3/day), 14% (6.0 Mm^3/day) in Europe, and 6% in Africa (2.8 Mm^3/day; Fig. 5, primary data from Ref. [5]).

Except for one region, seawater desalination is the prevalent process. Sixty-one percent (17.1 Mm^3/day) of the global seawater desalination capacity is located in only six GCC countries, including Saudi Arabia, the United Arab Emirates, Kuwait, Bahrain, Qatar, and Oman. Another 11% (2.9 Mm^3/day) of the global seawater desalination capacity is located in Southern Europe and 7% (2.0 Mm^3/day) in North Africa. Three enclosed sea areas therefore account for the lion's share of the global seawater desalination capacity – the Gulf, the Red Sea, and the Mediterranean Sea.

North America is the only region where brackish water desalination is the dominating process. The production capacity is 3.0 Mm^3/day, which represents more than one-third (36%) of the global brackish water desalination capacity. Twenty-one percent of the production from brackish water sources takes place in the GCC states (1.7 Mm^3/day) and 13% (1.1 Mm^3/day) in Southern Europe.

Wastewater purification is also primarily practiced in North America (22% or 0.49 Mm^3/day), closely followed by East Asia (21% or 0.46 Mm^3/day) and the GCC country states (19% or 0.42 Mm^3/day). Each of these three regions accounts for roughly one-fifth of the global wastewater treatment capacity, followed by Japan, Korea, and Taiwan (12%) and Southern Europe (10%).

In the following, emphasis will be given to the Gulf, the Red Sea, and the Mediterranean Sea and the installed *seawater* desalination capacity in these sea regions.

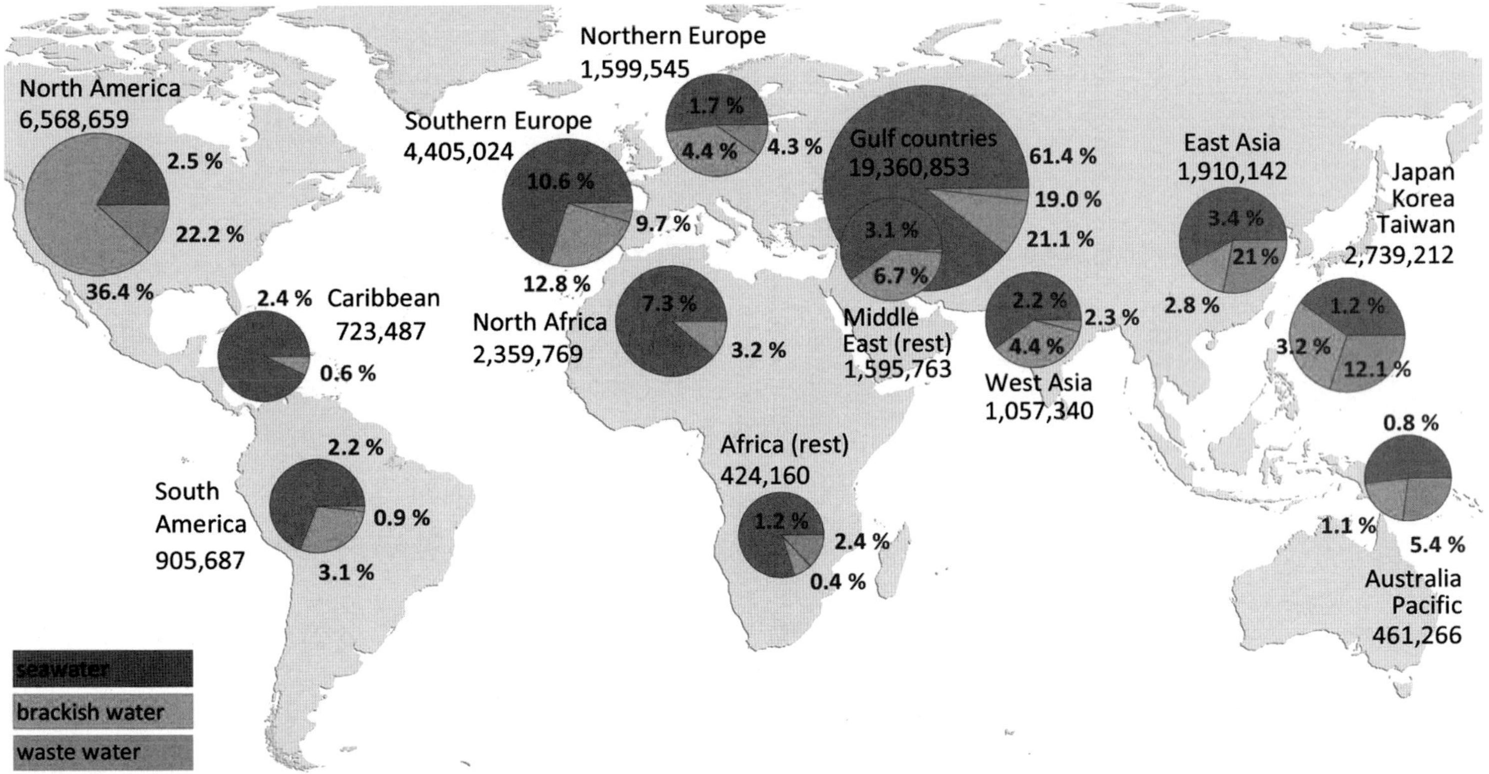

Figure 5 Global desalination capacities in cubic meters per day. For example, the installed capacity in Southern Europe is 4,405,024 m^3/day. This figure includes all source water types. Seawater desalination accounts for most of the production in Southern Europe, brackish water for about one-fourth of the production and wastewater desalination plays a relatively minor role (pie diagram). The figures next to the pie diagram give the contribution to the *global* production, that is, the seawater desalination capacity in Southern Europe represents 10.6% of the *global seawater* desalination capacity. The brackish water capacity – though it is less than half the seawater desalination capacity in Southern Europe – represents 12.8% of the *global brackish*

4.1 The Gulf region

In terms of *sea areas*, the largest number of *seawater* desalination plants can be found in the Gulf with a total desalination capacity of approximately 12.1 Mm^3/day – or a little less than half (44%)[1] of the worldwide daily production (Fig. 6). The main producer in the Gulf (and worldwide) is Saudi Arabia with 25% of the worldwide seawater desalination capacity, of which 11% are located on the Gulf shore and 12% on the Red Sea coast (2% unaccounted for), followed by the United Arab Emirates (23%) and Kuwait (6%).

Thermal desalination processes dominate in the Gulf region (about 94% of all production), as water and electricity are often generated by large cogeneration plants that use low value steam and electricity from power plants as a heat source for desalination. Most of the water (81%) in the Gulf is produced by the MSF distillation process. Minor processes are MED distillation and RO, which account for 13% and 6% of the production, respectively (primary data from Ref. [5]).

4.2 The Red Sea

In the Red Sea region, desalination plants have a combined production capacity of 3.6 Mm^3/day (13% of the worldwide capacity, Fig. 7). Similar to the Gulf, most of the water is produced by large cogeneration plants mainly on the Saudi Arabian coast in the locations of Yanbu, Rabigh, Jeddah, Assir, and Shoaiba, where the world's largest desalination complex with a capacity of 1.6 Mm^3/day is located. Saudi Arabia accounts for more than 92% of the desalinated water production from the Red Sea, with 2.6 Mm^3/day (78%) produced by thermal plants. Egypt, the second largest producer of desalinated water in the region, accounts for only 7% of the production from the Red Sea, with 90% (0.2 Mm^3/day) coming from smaller RO plants on the Sinai Peninsula and in the tourist resorts along the Red Sea coast.

4.3 The Mediterranean Sea

In the Mediterranean, the total water production from seawater is about 4.0 Mm^3/day (14% of the worldwide capacity, Fig. 8). Spain, with about 8% of the worldwide desalination capacity, is the largest producer of desalinated water in the region with an installed capacity of 2.2 Mm^3/day. About 65% (1.4 Mm^3/day) of the Spanish capacities are located on the

[1]The figure of 44% includes only those plants located on the shores of the Gulf. In contrast to the figure of 61%,which is given for the GCC states above, the figure of 44% does not include plants in Oman and on the Red Sea coast of Saudi Arabia, but it does include plants in Iran.

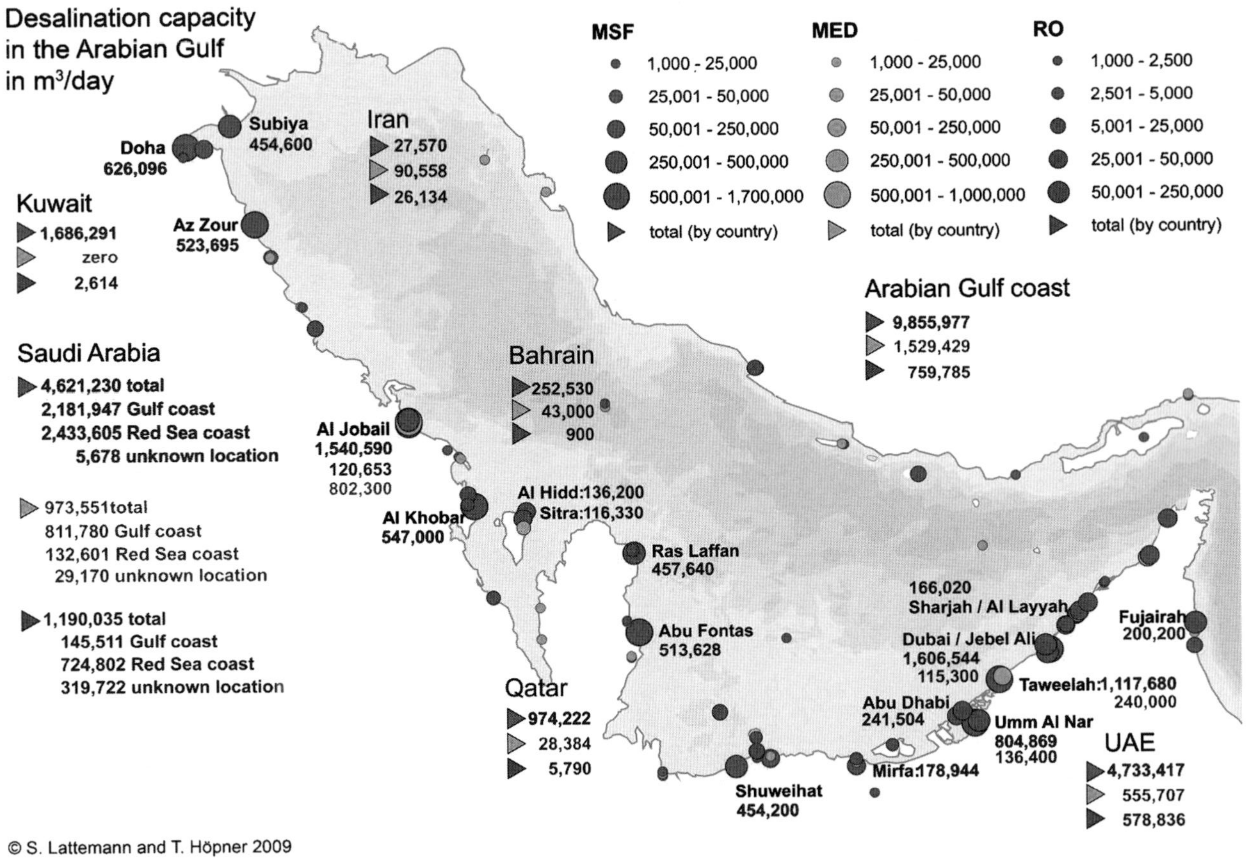

Figure 6 Cumulative MSF, MED, and RO capacities in the Arabian Gulf in cubic meters per day by site location (dots) and by country (triangles). The map shows all sites with an installed capacity $\geq 1000\ m^3$/day and displays sites with a capacity $\geq 100{,}000\ m^3$/day by name and capacity. The map was first published in Ref. [17,28] and updated using raw data from Ref. [5] (see plate 3 in color plate section at the end of this book).

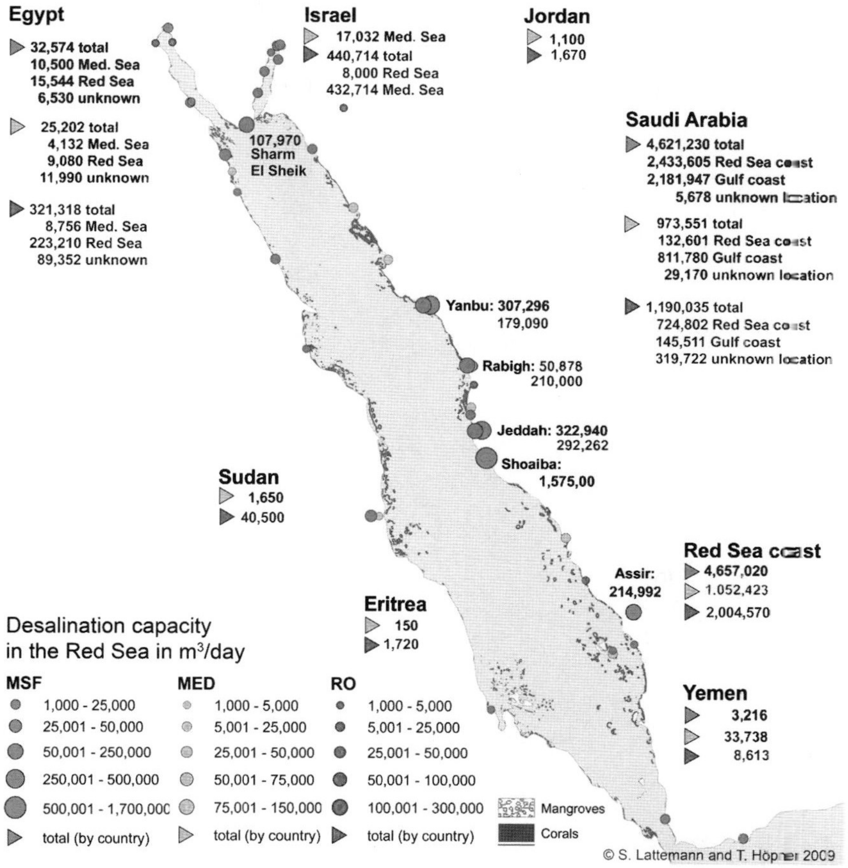

Figure 7 Cumulative MSF, MED, and RO capacities in the Red Sea in cubic meters per day by site location (dots) and by country (triangles). The map shows all sites with an installed capacity $\geq 1000\,m^3$/day and displays sites with a capacity $\geq 100{,}000\,m^3$/day by name and capacity. The map was first published in Ref. [17,28] and updated using raw data from Ref. [5] (see plate 4 in color plate section at the end of this book).

Mediterranean coast and the Balearic Islands, and 25% on the Canary Islands. The Spanish A.G.U.A. program[2] will further augment water supply on the Mediterranean coast by increasing the desalination capacity to over $2.7\,Mm^3$/day until 2010. While thermal processes are dominating in the Gulf and Red Sea, 70% of the Mediterranean and 99% of the Spanish

[2]The program "Actuaciones para la Gestión y la Utilización del Agua" was introduced by the Spanish government in 2004 following the decision not to divert the Ebro river to Southern Spain. The package of measures includes desalination but also water saving and efficiency of use and water reuse.

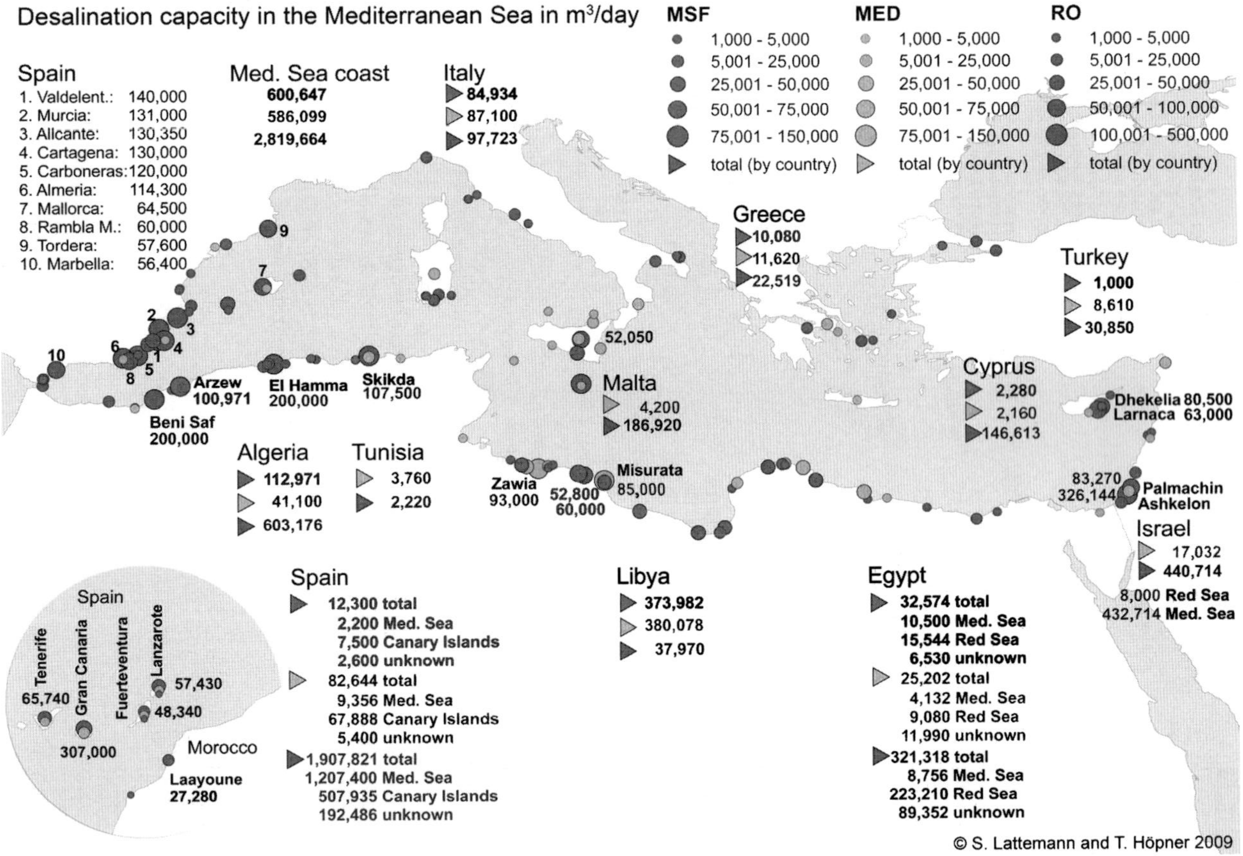

Figure 8 Cumulative MSF, MED and RO capacities in the Mediterranean Sea in cubic meters per day by site location (dots) and by country (triangles). The map shows all sites with an installed capacity $\geq 1000\,m^3$/day and displays sites with a capacity $\geq 50{,}000\,m^3$/day by name and capacity. The map was first published in Ref. [17,28] and updated using raw data from Ref. [5] (see plate 5 in color plate section at the end of this

production on the Mediterranean coast is produced by the process of SWRO.

Larger numbers of *distillation* plants are only found along the coasts of Libya and Algeria in North Africa, and also in Italy. However, new plants in these countries are also often SWRO plants. A tremendous expansion of capacities is currently taking place in Algeria, North Africa's fastest growing desalination market, where the first large SWRO plant (200,000 m^3/day) was opened in February 2008 [11]. It is the first in a series of other projects with capacities between 50,000 and 500,000 m^3/day, which will increase the country's desalination capacity to 4 Mm^3/day by 2020 [2].

On the Mediterranean coast of Israel, two large SWRO are currently in operation, the Ashkelon plant with a capacity of 330,000 m^3/day – the world's largest SWRO project to date – and the Palmachin plant (83,000 m^3/day). Desalination presently accounts for approximately 8% of Israel's water supply. According to original plans, this would have been increased to more than 30% (1.8 Mm^3/day) by 2020 [12]. In 2008, however, the Israeli government approved a new, even more ambitious emergency program to address the country's growing water shortage, which will raise the target for desalinated water production to 1.6 Mm^3/day by 2013 and to 2.1 Mm^3/day by 2020, which may also reach 2.7 Mm^3/day depending on water demand and other alternatives [13]. Several large SWRO desalination plants with capacities up to 274,000 m^3/day are currently being planned along Israel's Mediterranean coast [14]. Furthermore, it is planned to sharply increase the use of the country's brackish water resources, from presently around 16,500 m^3/day to somewhere between 220,000 and 274,000 m^3/day [15]. Other measures include more water efficient practices, fixed water quotas, greater enforcement of water restrictions, and upgrading wastewater treatment capacities in order to increase recycling of wastewater from 75% at present to 95% in 5 years [13].

4.4 Other regions

While seawater desalination is already a well-established technology in the above-mentioned sea regions, the era of large-scale desalination projects is about to start in other parts of the world, such as California, Australia or China, just to name a few.

In California, a potential for 15–20 new desalination projects is expected until 2030 with a combined production of 1.7 Mm^3/day (Fig. 9). The two most advanced and largest projects are the 200,000 m^3/day facilities in Carlsbad and Huntington Beach, which will presumably start operation in 2009 [5].

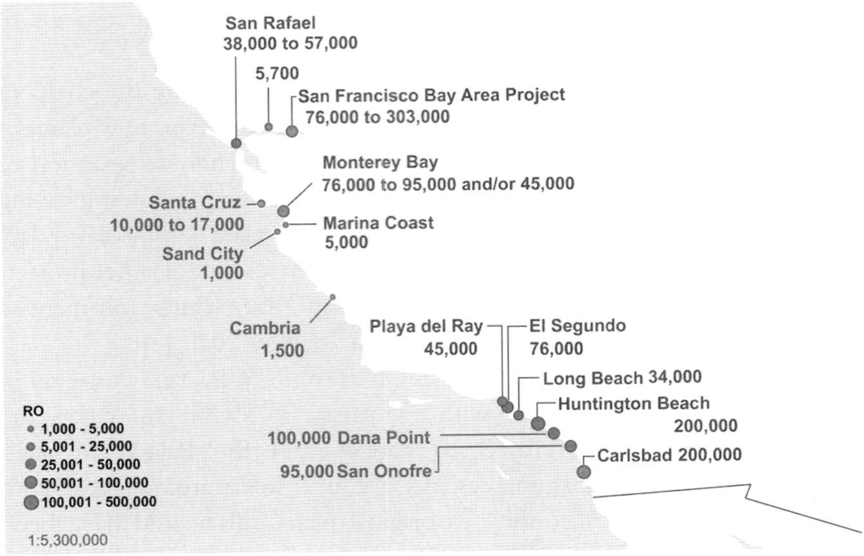

Figure 9 Seawater desalination projects in California (green: in operation or construction, blue: in planning). Adapted from Ref. [25].

In Australia (Fig. 10), the first large SWRO plant with a capacity of 144,000 m^3/day became operational in Perth in 2006. Another project currently under construction is the Sydney plant with an initial capacity of 250,000 m^3/day, which can, if necessary, be expanded to 500,000 m^3/day. Further projects include the Melbourne, Brisbane, and South East Queensland plants, with projected capacities up to 400,000 m^3/day each, and projects in Adelaide, the Upper Spencer Gulf, and a second plant near Perth, with capacities between 120,000 and 140,000 m^3/day each.

A third impressive example is China. The country is expected to dramatically expand its desalination capacity and might establish itself as another important market in the near future. In order to alleviate expected severe water shortages, China's desalination capacity may be increased 100-fold by 2020 – i.e. from presently around 366,000 m^3/d to 36 Mm^3/d. Besides desalination of seawater, wastewater treatment is a serious option under consideration [2].

5. ENVIRONMENTAL CONCERNS OF SEAWATER DESALINATION

The desalination industry has undergone many gradual changes since its beginnings in the early 1960s. Today, the trend is towards large,

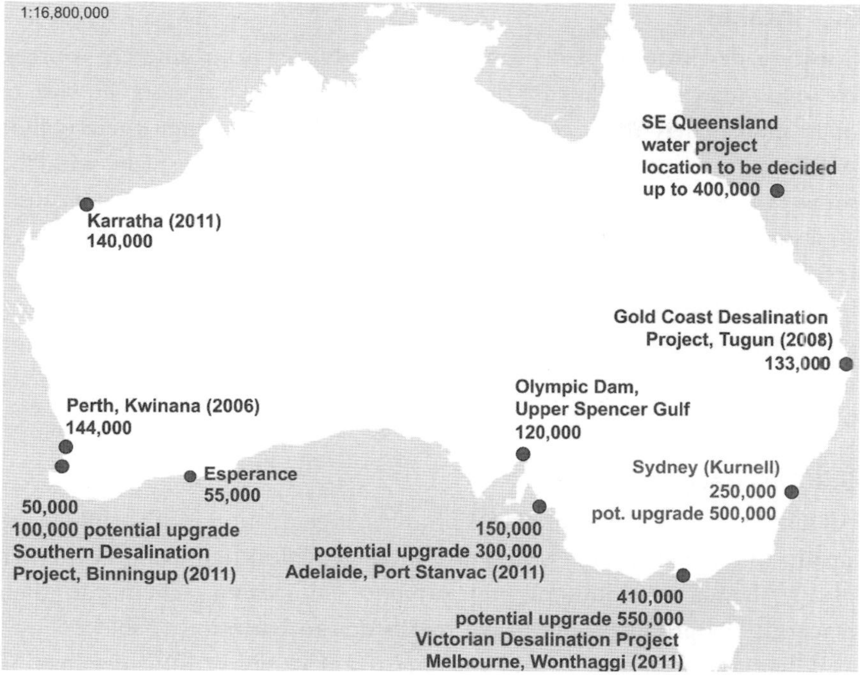

Figure 10 Seawater desalination projects in Australia (green: in operation or construction, blue: in planning). Based on Refs. [14,26].

industrial-sized facilities with production capacities in the range of 100,000 m^3/day or more. The implementation of large desalination facilities is no longer limited to a few water-scarce but oil-rich countries of the Middle East. Desalinated water has become a commodity that amends and diversifies conventional water supplies in many parts of the world. Due to the growing desalination activity in many sea regions and the growing number of large facilities, concerns over potentially negative impacts of the technology on the environment are being raised. The main environmental concerns of desalination activity revolve around the emissions of greenhouse gases and air pollutants, the concentrate and chemical discharges into the sea, the use of large quantities of seawater for cooling purposes and as feedwater, causing the impingement and entrainment of marine organisms, and construction-related impacts on the coastal and near-shore habitats. A brief overview of the main concerns is given in the following sections. More details can be found in recent literature surveys (e.g., [16–18]).

5.1 Intakes

Seawater desalination plants can receive feedwater from different sources, but open seawater intakes are the most common intake option. The use of open intakes may result in losses of aquatic organisms when these collide with intake screens (impingement) or are drawn into the plant with the source water (entrainment). The construction of the intake structure causes an initial disturbance of the seabed, which may result in the resuspension of sediments, nutrients, or pollutants that may affect water quality. After installation, the structures can affect water exchange and sediment transport, act as artificial reefs, or may interfere with shipping routes or other maritime uses. Alternatives are beachwell intakes and infiltration galleries, which are placed below the seabed.

5.2 Discharges

All seawater desalination processes produce large quantities of a saline waste stream (the concentrate), which may be increased in temperature (thermal plants), contain residues of pretreatment and cleaning chemicals, their reaction (by-)products, and heavy metals due to corrosion (Table 2).

Chemical pretreatment and cleaning is a necessity in most desalination plants, which typically includes the treatment against biofouling, scaling, foaming, and corrosion in thermal plants, and against biofouling, suspended solids, and scale deposits in membrane plants. The chemical residues and by-products are typically washed into the sea along with the concentrate. The concentrate of distillation plants is increased in temperature and salinity and typically contains residual chlorine and chlorination by-products, antiscalant, and antifoaming agents and certain heavy metals such as copper or nickel. The concentrate of SWRO plants is increased in salinity and typically also contains antiscalants, but residual chlorine is removed by dechlorination with sodium bisulfite to protect the RO membranes from oxidation. The concentrate of SWRO plants does not contain antifoam agents or significant levels of metals from corrosion, but it is often used to dilute other intermittent waste streams such as high-turbidity backwash waters from media filters that contain natural solids and coagulants several times per day or chemical cleaning solutions several times per years. To conclude, the discharge is a mix of these different pollutants, which may have potentially synergistic effects on marine life, such as for example the synergistic effect of chlorine residues and increased temperature which is well documented [19]. The discharge volume depends on the process recovery rate and the size of the facility. Also, the composition and concentration of residual pollutants from the pretreatment process is process- and plant-specific.

Table 2 Effluent properties of RO, MSF, and MED distillation plants, assuming conventional process design [16,27]

Physical parameters	Reverse osmosis (RO)[a]	Multistage flash (MSF)[b]	Multieffect distillation (MED)[b]
Salinity (*S*) (depending on ambient salinity and recovery rate)	• SWRO: 65–85 g/L • BWRO: 1–25 g/L	• Cooling water: ambient salinity (e.g., 40 g/L) • Brine: 60–70 g/L • Combined: 45–50 g/L	• Cooling water: ambient salinity • Brine: 60–70 g/L • Combined: 50–60 g/L
Temperature (*T*)	• If subsurface intakes used: may be below ambient T due to a lower T of the source water • If open intakes used: close to ambient • If water is received from cooling water discharges of power plants: may be above ambient	• Brine: 3–5 °C above ambient • Cooling water: 8–12 °C above ambient • Combined: ~5–10 °C above ambient	• Brine: 5–25 °C above ambient • Cooling water: 8–12 °C above ambient, up to 20 °C possible • Combined: ~10–20 °C above ambient
Plume density (ρ)	• Lower than ambient (negatively buoyant plume)	• Plume can be positively, neutrally, or negatively buoyant depending on the process design and mixing with cooling water before discharge, typically positively buoyant	
Dissolved oxygen (DO)	• If subsurface intakes used: may be below ambient DO due to a lower DO of the source water • If open intakes used and if oxygen scavengers for dechlorination are not overdosed: close to ambient	• Brine: below ambient because of deaeration and use of oxygen scavengers • Cooling water: close to ambient (minor effects on DO because of changes in temperature) • Combined: mixing of brine with cooling water increases the DO content of the combined effluent close to ambient; turbulent mixing allows oxygen take-up from air	

Table 2 *(Continued)*

Chemical parameters	Reverse osmosis (RO)[a]	Multistage flash (MSF)[b]	Multieffect distillation (MED)[b]
Biofouling control additives and by-products			
Oxidants	• Typically dosage of 1–2 ppm to the feed water in all plants operating on open seawater		
• Mainly chlorine • Chlorine dioxide used in some plants	• Oxidants typically removed to prevent membrane damage, using sodium bisulfite (two to four times higher dosage than oxidizing agent dose)	• Discharge level is about 10–25% of dosage due to chlorine demand of the seawater • Both the brine and the cooling water contain residual chlorine • Chlorine typically not removed by a dechlorination step inside the plant	
Halogenated organic by-products such as trihalomethanes (THMs)	• Use of chlorine dioxide reduces the risk of by-product formation		
	• May form during chlorination, but levels are assumed to be low due to dechlorination	• Chlorination of seawater results in varying composition and concentrations of halogenated (chlorinated and brominated) organic by products, mainly THMs such as bromoform	
Removal of turbidity (suspended solids)			
Coagulants • Dosage 1–30 mg/L • Often iron (III) salts	• If filter backwash is discharged to surface waters: may cause turbidity and sedimentation in the discharge site and iron salts may cause effluent coloration (red brines)	• Treatment not applied	• Treatment not applied
Coagulant aids (e.g., polyacrylamide) • Dosage 0.1–5Img/L • e.g. polyacrylamide			

Scale control additives (used in all desalination processes, can be a blend of several different antiscalants in combination with acid treatment)			
Polymeric antiscalants (e.g., polymaleic acids) and phosphonates • Dosage: 1–2 ppm	• Mainly used in RO	• Antiscalant only present in the brine, but not in the cooling water	
	• Dosage/discharge concentration below toxic levels to invertebrate and fish species; some products are classified as being harmful to algae, presumably due to a nutrient inhibition effect • Slow degradation (some products classified as 'inherently' biodegradable) with presumably increased residence times in surface waters		
Phosphates • Dosage: 2 ppm	• Still used at a limited scale	• Not stable at high temperature (blends of polymeric antiscalants and phosphonates preferred)	
	• May cause eutrophication near outlets, as easily hydrolyzed to orthophosphate, a major nutrient for primary producers		
Acid (H_2SO_4) • Dosage: 30–100 ppm	• Lowers the pH from around 8.3 (natural pH of seawater) to pH 6–7 • Effective against calcium carbonate scales but not against sulfate scale, therefore more effective in seawater RO and MED processes where calcium carbonate is the main scale forming species • The acidity is quickly consumed by the natural alkalinity of seawater, so that the pH quickly returns to normal		
Foam control additives			
Antifoaming agents (e.g., polyglycol)	• Treatment not applied	• Typically low dosage (0.1 ppm) below harmful levels • Used in all distillation processes, but primarily in MSF • Antifoam only present in the brine, but not in the cooling water	
Corrosion			
Heavy metals	• Metallic equipment made from corrosion-resistant stainless steel • Concentrate may contain low levels of iron, chromium,	• Metallic equipment made from carbon steel, stainless steel, copper nickel alloys	• Metallic equipment made from carbon and stainless steel, aluminum and aluminum brass, titanium, or copper nickel alloys

Table 2 *(Continued)*

Chemical parameters	Reverse osmosis (RO)[a]	Multistage flash (MSF)[b]	Multieffect distillation (MED)[b]
	nickel, molybdenum if low-quality steel is used	• Concentrate may contain iron and copper, copper levels can be an environmental concern	• Lower corrosion rates than in MSF • No data on brine contamination available
Corrosion prevention	• Not necessary besides choice of materials	• As the feed water is deaerated, the brine is also deaerated before mixing with cooling water, which is not deaerated • In MSF, the feed water (but not the cooling water) may also be treated with oxygen scavengers (e.g., sodium bisulfite), which may also remove residual chlorine	
Cleaning solutions (only present if cleaning solutions are discharged to surface waters)			
Cleaning chemicals (used intermittently)	Alkaline (pH 11–12) or acidic (pH 2–3) solutions with additives, e.g.: – Detergents (e.g., dodecylsulfate) – Complexing agents (e.g., EDTA) – Oxidants (e.g., sodium perborate) – Biocides (e.g., formaldehyde)	Acidic (low pH) washing solution which may containing corrosion inhibitors such as benzotriazole derivates	

[a] No use of cooling water in the process, but RO plants may receive their intake water from cooling water discharges.
[b] Assuming that the two waste streams from the desalination process are combined, that is, the brine is diluted with major amounts of cooling water from the desalination process; further dilution with cooling water from power plants may occur but is not considered here.

Negative effects on the marine environment can occur especially when wastewater discharges and pollutant loads coincide with sensitive ecosystems. The impacts of a desalination plant on the marine environment depend on both the physical and chemical properties of the reject streams and the hydrographical and biological features of the receiving environment. The concentrate of SWRO plants is negatively buoyant due to higher than ambient salinity values, with the potential of plume sinking and seafloor spreading. The concentrate of distillation plants can be negatively, positively, or neutrally buoyant, depending on the salinity and temperature values and the amount of cooling water co-discharge, which results from the desalination process itself and co-located power plants. It is most likely positively buoyant due to large cooling water flows with a higher than ambient temperature. The concentrate of SWRO and distillation plants therefore affects different realms in the marine environment. Seafloor spreading may negatively affect benthic ecosystems such as seagrass meadows or macroalgae stands and associated benthic species such as sea urchins or shrimps, whereas neutrally or positively buoyant plumes spread in the water column and could affect nektonic species such as fish, turtles, or mammals. As these are mobile species, they can be assumed to avoid the discharge site, which could result in a loss of habitat, such as foraging, resting or reproduction areas, for the affected species. Enclosed and shallow sites with abundant marine life can generally be assumed to be more sensitive to desalination plant discharges than exposed, high-energy, open-sea locations [20], which are more capable to dilute and disperse the discharges. Environmental baseline studies thus provide important information for project planning and site selection, while monitoring during construction and operation is useful for compliance and effect monitoring Although the number of publications discussing the *potential* for negative environmental impacts of effluents from desalination facilities has been steadily increasing over the last years, a surprising paucity of useful *experimental* data, either from laboratory tests or from field monitoring still exists. Therefore, a considerable amount of uncertainty still exists about the environmental impacts of desalination [2].

5.3 Energy demand

Desalination of seawater consumes significant amounts of energy (Table 1), either directly in the form of steam (distillation processes) or indirectly through electricity use from the electricity grid. Energy supply is consequently an important factor in the planning of new facilities. The main environmental concern associated with energy demand, both directly and indirectly, is the emission of air pollutants. Air quality may be affected by emissions of greenhouse gases (mainly CO_2), acid rain gases (NO_x, SO_x), fine particulate matter (PM), and other air pollutants that are produced when fossil fuels are used for electricity/steam generation. The production of greenhouse gases is relevant in the context of national and international efforts to limit these emissions to minimize the impacts of climate change. Significant local impacts may further occur if emissions conflict with applicable air quality standards or management plans, contribute substantially to other existing or projected air emissions (cumulative impacts) in the vicinity and expose the resident population to increased pollutant concentrations [18]. Concerns may also arise due to more indirect impacts, such as the cooling water requirements of power plants or the increasing risk for accidents associated with the transport of fuels. When existing power plant capacities are increased or new plants constructed in order to provide additional electricity for desalination, these indirect impacts will likely be intensified.

As the treatment and distribution of water from conventional sources and by conventional processes also requires energy, it is necessary to consider both the *total* energy increase caused by desalination processes and the *relative* increase compared to other water supply options.

Reference values are often used to put the energy demand of desalination into perspective, which may influence how we perceive and evaluate the significance of energy demand and associated environmental impacts, for instance by comparing it to energy demand on a local, regional, or national level or to other energy consumers. Some examples [21]:

- On the Canary Islands, desalination accounts for 14% of all energy demands [22].
- The SWRO plant of Carboneras (capacity of 120,000 m^3/day) on the Mediterranean coast of Spain consumes about one-third of the province's electrical energy [23].
- The Spanish Agua program shall increase desalination capacity on the Mediterranean coast of Spain from 1.1 Mm^3/day (2005) to over 2.7 Mm^3/day (2010). This will require additional 11 GWh/day of electricity assuming an energy demand of 4 kWh/m^3 of desalinated water as foreseen in the Spanish National Hydrological Plan [9] and will cause a 1.4% increase over 2005 *national* electricity generation levels (805 GWh/d or 294 TWh in 2005 [24]). It would result in additional CO_2 emissions of 5475 tons/day, which represents a 0.6 % increase in *national* CO_2 emissions compared to pre-2005 levels of 326 million tons CO_2 in 2004.
- For California, it is estimated that the currently proposed desalination plants with a total capacity of 1.7 Mm^3/day would increase the *water-related energy use* by 5% over 2001 levels assuming an average energy use

of 3.4 kWh/m^3 [1]. The total water-related energy use was 48,012 GWh in 2001, representing 19% of the total energy use in California [25]. In another source [26], an average energy use of 2.9 kWh/m^3 is assumed to produce the 1.7 Mm^3/day by desalination in 2030, which is realistic as further energy savings are to be expected in the future. Desalination would thus increase the water related-energy use by 1800 GWh/year or about 4% over 2001 levels.

- The Sydney desalination plant with an initial capacity of 250,000 m^3/day is expected to result in a 1.2% increase of New South Wales' electricity demand if upgraded to a capacity of 500,000 m^3/day [27]. The Perth SWRO plant in Western Australia is responsible for about 0.67% of the energy demand in the region (at peak power consumption of 3574 MW in summer), compared to 30% as used for air-conditioning in Perth [28].
- In Kuwait, co-generation plants produce 443 Mm^3 of desalinated water (90% of the national water supply) and 42,257 GWh of electricity per year, using 462 million GJ of energy, which is 54% of the national fuel use. About 10% of the national fuel use and the national emissions are thus attributed to the production of desalinated water and 43% to electricity generation. As the plants use mainly heavy oil (78%) and crude oil (20%), air pollution from cogeneration plants is significant and amounts to 7 million tons of CO_2, 0.13 tons of SO_2, and 0.02 tons of NO_X per year for water production, and 30 million tons CO_2, 0.54 tons SO_2 and 0.06 NO_X per year for electricity production. 62% of the total fuel energy (290 M-GJ) are rejected to the atmosphere (46 M-GJ) and to the sea (243 M-GJ) as cooling water. 60% of the cooling water discharges are attributed to the power plants and 40% to the MSF plants [29].

To conclude, desalination can be a significant energy consumer in some parts of the world, which depend heavily on desalinated water. As seen in the aforementioned examples, desalination accounts for 14% of the energy demand on the Canary Islands or for 10% of the national fuel use in Kuwait. On the mainland of Spain, however, desalination accounts for only about 1.4% of national electricity generation, and this value would even be lower if the energy use of desalination would be compared to the total Spanish energy demand taking emissions for example from transportation or heating into account. The value of 1.4% is similar in magnitude to the reference values given for Sydney (0.6% of the regional electricity demand for a single 250,000 m^3/day facility) and Perth (0.67% of the regional peak energy demand for a single 140,000 m^3/day facility). Taking these latter values into consideration, energy use seems to be a minor energy consumer on a regional or national level in industrialized regions. However, environmental impact assessments may still find energy use to be a significant factor, which may entail some form of impact mitigation. For example, the projects in Sydney and Perth compensate the electricity demand by renewable energy projects.

5.4 Impact mitigation measures

A widely recognized and accepted approach for investigating, evaluating, and mitigating impacts of development projects on the environment is the environmental impact assessment (EIA). To date, only a handful of EIA studies have been carried out for desalination plants and made publicly available. In some cases, the investigations are carried out under immense time constraints. For instance, only 4 months were set aside for an EIA study for a 200,000 m^3/day SWRO plant in Algeria [21]. This shows that environmental concerns can be of secondary importance when a ready supply of freshwater is urgently needed. The opposite is also true: comprehensive environmental studies are currently being carried out for the large SWRO projects in Australia, and environmental concerns are the major hurdle in the permitting process of new projects in California, where the planning and permitting process of the first large plant took more than 10 years.

A central element of all EIA studies is the comparison of alternatives, such as alternative project sites or technologies in order to identify the option with the least environmental footprint. Especially the selection of a suitable project site for a new desalination project can be a very effective way of minimizing and preventing impacts on the environment. Furthermore, several technical options can be implemented to mitigate the environmental effects of the waste discharges. For example, advanced diffuser systems can achieve a maximum dilution with a minimum salinity increase of one unit above background levels in the sea. Negative impacts from chemicals can be minimized by treatment before discharge, by substitution of hazardous substances, and by implementing alternative nonchemical treatment options. For instance, backwash waters from pretreatment filters can be dewatered and deposited on land, or membrane cleaning solutions can be treated on-site in special treatment facilities or discharged to a sanitary sewer system [16].

The use of alternative pretreatment methods may be considered where feasible. Prefiltration with ultrafiltration (UF) or microfiltration (MF) membranes may reduce the need for chemical pretreatment. The UF/MF membranes usually require chemically enhanced backwash and periodic cleaning. The process is therefore not entirely "chemical-free," but an advantage of intermittent cleaning over continuous pretreatment is that wastewaters are produced in smaller volumes and can be treated effectively.

A nonchemical treatment option is irradiation of the intake water with UV-light at 200–300 nm wavelengths for disinfection. A major advantage of UV-light is that storage, handling, and disposal of toxic chemicals are avoided; however, UV-irradiation has not been found to be an effective pretreatment for large desalination plants to date.

Air pollutant emissions can be minimized by increasing the energy efficiency of the desalination process. For instance, use of energy recovery devices allow for a reduction of the specific energy demand in seawater RO plants to 2–3 kWh/m^3, which may be decreased further in the future. Furthermore, air emissions can be controlled at the source – the power plant – as emissions depend on the fuel source (e.g., gas, coal), the technology and efficiency of the power plant, as well as on any exhaust purification equipment installed (e.g., scrubbers capturing sulfur emissions). When electricity is taken from the electricity grid, the composition of the energy mix must furthermore be taken into account when estimating the indirect air emissions of a single desalination project.

Finally, the potential for renewable energy use (solar, wind, geothermal, biomass) may be investigated to minimize impacts on air quality and climate. This may be in the form of desalination systems directly driven by renewable energy, or as an indirect compensation measures such as the installation and use of renewable energy in other localities or for other activities. For instance, the large SWRO projects in Perth and Sydney, Australia, compensate for their energy demand through wind farm projects.

6. SUMMARY AND CONCLUSION

In a nutshell, 63% of the worldwide (44.1 Mm^3/day) desalination capacity is produced from seawater sources. Of this water, 61% is produced by thermal processes. The MSF distillation process is almost exclusively used for the desalination of seawater in the Gulf countries. The RO process is the second most important process for treating seawater on a global scale, but it is the first choice in many industrialized and developing countries that are now starting to consider seawater desalination. Eighty-three percent of the treated seawater is for municipal use. Sixty-six percent of the *seawater* desalination capacity is attributed to industrial scale facilities, with production capacities in single MSF distillation plants up to 1.6 Mm^3/day, while proposed capacities for single SWRO plants approach 500,000 m^3/day. Seventy-nine percent of the global seawater desalination capacity is located in the Middle East, North Africa, and Southern Europe, with 71% being located in the Gulf, the Red Sea, and the Mediterranean Sea. The enclosed nature of these sea areas makes them especially susceptible to any form of pollution, and desalination plants have been classified as a main contributor to land-based pollution in the Gulf and Red Sea [22,23].

Only 19% of the global desalination capacity is presently produced from brackish water sources and 5% from wastewater sources, with 84% of the brackish water and 79% of the wastewater being treated by RO. This share increases to 98% and 85%, respectively, if one includes the other membrane-based processes, that is, NF and ED, as well.

Although brackish water and wastewater treatments offer a great future potential, desalination of seawater will remain the dominant process for some time. This is mainly because Saudi Arabia and the United Arab Emirates will continue to be the largest desalination markets in the foreseeable future, where seawater desalination plays a prominent role. MSF distillation will therefore continue to be the main desalination process, but will presumably lose further market shares to MED and RO. While thermal cogeneration facilities predominate in the oil-rich countries of the Middle East, which produce both electricity and water, RO is usually the preferred process where cheap fossil energy or waste heat is not available, due to its lower energy demand. Consequently, most countries outside the Middle East choose RO for seawater desalination.

As the need for desalination accelerates in many parts of the world, the problem spreads from water scarcity to energy use and airborne emissions [9,24], and from overused polluted freshwater bodies to the marine environment. Due to the environmental concerns associated with the desalination of seawater, this option should therefore only be considered after other alternatives have been tapped to the full potential, such as water saving and water reuse. Examples such as Spain or Israel (cf. Section 4.3), however, show that desalination developments are often only one aspect of a whole package of water management measures, and not necessarily the first and only choice to satisfy the ever-growing demand for water and to reduce the burdens of drought. To negate the need for desalination in countries such as Israel or Spain would also mean that societies in the North would have to make concessions, as much of what we eat and wear is grown in sunny but water-scarce regions.

The question is not if desalination will provide the ultimate solution to the world's water problems. In the end, decisions about desalination developments revolve around complex evaluations of local circumstances and needs, economics, financing, environmental and social impacts, and available alternatives [1]. The question is rather which mitigation measures are necessary to reduce the environmental burden of desalination to acceptable levels. Many useful ideas have been put forward in recent literature to minimize the environmental footprint of desalination. The best project design, however, can only be identified in project- and site-specific studies. A catalogue of best available techniques (BAT) and best environmental practices (BEP) may be useful in guiding practitioners, consultants, and decision makers in their choices when undertaking new desalination projects. Furthermore, there is need for ongoing research and demonstration projects to gain experience, knowledge, and trust in new environmentally friendly technologies, as well as political incentives through policies or financial support to implement state-of-the-art technologies. Some of these measures will increase the price of desalinated water production; however, technological advances will most likely result in a lower energy consumption and production cost of desalinated water in the future. Sustainable desalination is not a utopia, but requires a commitment to providing water at a reasonable price, which includes not only the construction and operating costs, but also the costs to mitigate environmental impacts, including the costs for environmental studies, advanced technology, or compensation measures.

In the end, some advantages of wastewater desalination over seawater desalination should be highlighted. Water reuse is practiced in many parts of the world, but the use of desalination technologies in water reuse has been limited so far. The world's largest desalination facility treating waste water with an output capacity of 310,000 m^3/day is located in Sulaibiya, Kuwait. It uses ultrafiltration followed by reverse osmosis to treat secondary effluent waste water. The main advantage of treating waste water is that it is cheaper and the energy demand is lower than for seawater RO. An expansion of waste water desalination is therefore expected in the future. Second, most of the waste is already where it is most needed, that is, near urban areas. Even if the decision is made not to use the purified wastewater for direct potable use (though from a technical point of view, the product can comply with WHO standards), it can be used for industrial use or landscaping activities in urban areas. And third, wastewater and some of its contaminants, including nutrients, metals, or micropollutants such as pharmaceutical and personal care products, are still a burden for many rivers, estuaries, and coastal seas. Purifying and reusing wastewater does not only produce a new source of water supply, but can eliminate a waste product if the waste stream from the desalination process, which is about 15% of the original waste water volume, is treated instead of discharged. Zero liquid discharge (ZLD) technologies could be used for this purpose. While some media vilify reclaimed wastewater by negative headlines, public education programs using terms such as "new" or "purified" water can help to establish a positive attitude.

REFERENCES

[1] H. Cooley, P. Gleick, G. Wolff, Desalination, With a Grain of Salt. A California Perspective, Pacific Institute, California, 2006.
[2] NRC, Desalination: A National Perspective, Committee on Advancing Desalination Technology, Water Science and Technology Board, National Research Council (NRC) of the U.S. National Academies, 2008.
[3] O.K. Buros, The ABCs of Desalting, 2nd ed., FMC Corporation, International Desalination Association (IDA), Massachusetts and Saline Water Conversion Corporation (SWCC), Saudi Arabia, 1999.
[4] M. Wilf, L. Awerbuch, C. Bartels, M. Mickley, G. Pearce, N. Voutchkov, The Guidebook to Membrane Desalination Technology, Balaban Desalination Publications, L'Aquila, Italy, 2007, 524 pp.
[5] IDA and GWI., IDA Worldwide Desalting Plant Inventory Report No. 20 in MS Excel Format, Media Analytics Ltd., Oxford, UK, 2008.
[6] GWI., Desalination Markets 2007. A Global Industry Forecast (CD ROM), Global Water Intelligence, Media Analytics Ltd., Oxford, UK, 2007, www.globalwater intel.com
[7] T. Pankratz, Membrane growth takes off, in: T. Pankratz (Ed.), Water Desalination Report, Media Analytics Ltd., Oxford, UK, 2008, in cooperation with Global Water Intelligence, Houston, TX.
[8] K. Wangnick, IDA Worldwide Desalting Plant Inventory Report No. 18, Wangnick Consulting, Germany, 2004.
[9] G.L.M. von Medeazza, "Direct" and socially-induced environmental impacts of desalination, Desalination 185(1–3) (2005) 57–70.
[10] J. MacHarg, T. Seacord, B. Sessions, ADC baseline tests reveal trends in membrane performance, The International Desalination and Water Reuse Quarterly 18(2) (2008) 30–39.
[11] T. Pankratz, Hamma SWRO on-line, in: T. Pankratz (Ed.), Water Desalination Report, Media Analytics Ltd., Oxford, UK, 2008, in cooperation with Global Water Intelligence, Houston, TX.
[12] Y. Dreizin, A. Tenne, D. Hoffman, Integrating large scale seawater desalination plants within Israel's water supply system, Desalination 220 (2008) 132–149.
[13] Media Analytics Ltd., Israel unveils drought-busting water strategy, in: Global Water Intelligence, Media Analytics Ltd., Oxford, UK, 2008.
[14] GWI, DesalData.com by Global Water Intelligence, 2008.
[15] Media Analytics Ltd., Israel's brackish water challenge, in: Global Water Intelligence, Media Analytics Ltd., Oxford, UK, 2008.
[16] S. Lattemann, T. Höpner, Environmental impact and impact assessment of seawater desalination, Desalination 220 (2008) 1–15.
[17] S. Lattemann, T. Höpner, Seawater Desalination – Impacts of Brine and Chemical Discharges on the Marine Environment, Balaban Desalination Publications, L'Aquila, Italy, 2003, 142 pp.
[18] UNEP., Desalination Resource and Guidance Manual for Environmental Impact Assessment, United Nations Environment Programme, Nairobi, Kenya, 2008.
[19] C.J.L. Taylor, The effects of biological fouling control at coastal and estuarine power stations, Marine Pollution Bulletin 53 (2006) 30–48.

[20] T. Höpner, J. Windelberg, Elements of environmental impact studies on coastal desalination plants, Desalination 108 (1996) 11–18.
[21] C. Mooij, Hamma water desalination plant: planning and funding, Desalination 203 (1–3) (2007) 107–118.
[22] UNEP, Overview on land-based sources and activities affecting the marine environment in the ROPME sea area, UNEP Regional Seas Reports and Studies No. 168, UNEP/GPA Coordination Office, The Hague, The Netherlands and Regional Organization for the Protection of the Marine Environment (ROPME), Safat, Kuwait, 1999.
[23] UNEP, Assessment of land-based sources and activities affecting the marine environment in the Red Sea and Gulf of Aden, UNEP Regional Seas Reports and Studies No. 166, UNEP/GPA Coordination Office, The Hague, The Netherlands, 1997.
[24] S. Dolnicar, A.I. Schäfer, Public perception of desalinated water versus recycled water in Australia, AWWA Desalination Symposium, 2006.
[25] N. Voutchkov, Advances and Challenges of Seawater Desalination in California, IDA World Congress on Desalination and Water Reuse, Maspalomas, Gran Canaria, 2007.
[26] IDA., IDA Worldwide Desalting Plant Inventory Report No. 19 in MS Excel Format, Media Analytics Ltd., Oxford, UK, 2006.
[27] WHO., Desalination for Safe Water Supply, Guidance for the Health and Environmental Aspects Applicable to Desalination, World Health Organization, Geneva, Switzerland, 2007.
[28] S. Lattemann, Protecting the marine environment, in: G. Micale, A. Cipollina, L. Rizzuti (Eds.), Seawater Desalination, Springer-Verlag, GmbH, Berlin, Heidelberg, 2009.

CHAPTER 3

The Global Water Recycling Situation

B. Van der Bruggen*
K.U. Leuven, Department of Chemical Engineering, Section Applied Physical Chemistry and Environmental Technology, W. de Croylaan 46, B-3001 Heverlee (Leuven), Belgium

Contents

1. INTRODUCTION

The awareness that water recycling is the only possible answer to the world's growing water needs is ever increasing. Governments are developing policies of incentives and/or permits to stimulate water recycling in an industrial context. Possible tools that can be deployed are: increasing taxes on wastewater discharge, requiring the development of wastewater treatment techniques that result in enhanced removal of a wide range of contaminants, and linking permits to progressive use of alternative water sources. The *industry* itself is working actively on water reuse projects, mainly for economic reasons, but also from the perspective of environmental responsibility. The benefits of water reuse for the industry include [1]: a supplemental and reliable water source to augment or replace existing freshwater supplies; reduction of the net amount of water consumed; and

*Corresponding author.
E-mail address: bart.vanderbruggen@cit.kuleuven.be

Sustainability Science and Engineering, Volume 2 ISSN 1871-2711, DOI 10.1016/S1871-2711(09)00203-7

reduction of wastewater generation and associated costs of wastewater treatment. Cooling water and process water recycling accounts for about 30% of all reuse applications. Because human contact is minimal or can be avoided, secondary treatment followed by disinfection is often sufficient. There is a large variation of water quality in industrial applications, with the most advanced applications requiring extensive additional treatment, but for some applications, tertiary treatment consists of only filtration and disinfection. These applications have clear and direct advantages for the industry; points of attention are monitoring and acceptance of customers, rather than technical treatment issues.

The *academic community* is the driver for progress and stimulates new projects by investigating feasibility on a small scale and suggesting new processes and approaches. Many examples of such studies can be found in the (recent) scientific literature for many industries [2–6]. These laboratory-scale applications have grown into large-scale projects in many cases.

The *population* is often ignored, because they operate on a small scale and do not have the expertise. Nevertheless, population is the one and only wastewater producer: directly through daily activities, but also indirectly by participating in the global economy, in which water (and, consequently, wastewater generation) plays a central role. More people means more freshwater, which means more wastewater: this is the real challenge for reuse. The overall water balance should remain under control, even though populations keep growing. Population growth will eventually prove to be the central water-related problem. Even more urgent than the energy issue, water scarcity will be the limitation for further growth of established economies and development of new economies. In water-stressed regions, water conflicts are already appearing, although sometimes still hidden.

At this point, it is not yet considered feasible for families to reuse own wastewater, although this is in fact common practice in the developing world, where it is often used for irrigation. Therefore, *agriculture* is also a stakeholder in water reuse. Irrigation seems to be a simple solution; 60% of all reuse applications are to be found here. Another 10% is used for irrigation of parks, sport fields, etc. and for groundwater recharge by percolation. Groundwater recharge requires tertiary and quaternary treatment methods for removal of organic material, micropollutants, heavy metals, and for disinfection. Agricultural applications often use lower standards, with not much attention for possible diffusion of contaminants on the land, even though aspects of crop contamination are taken into account. Water quality standards for irrigation using recycled water are feasible with relatively few extensions to existing treatment plants. Typical standards [7] for agricultural application (food crops commercially processed, surface irrigation of orchards and vineyards) are:

- pH 6–9
- Biological oxygen demand (BOD) $\leq$ 30 mg/L
- Suspended solids $\leq$ 30 mg/L
- fecal coli $\leq$ 200/100 mL
- 1 mg/L rest-Cl_2 minimum

Monitoring includes weekly measurement of pH and BOD, daily measurement of suspended solids and coliforms, and continuous measurement of rest-Cl_2. However, it is increasingly understood that more extensive tertiary and quaternary treatments are necessary in agricultural applications as well to prevent uncontrolled diffusion of contaminants such as pharmaceuticals and endocrine disruptors.

This is a clear and global trend that appears in industrial, agricultural, and municipal context. Water recycling applications on a large and even on a small scale today are well considered technologically and provide high-quality effluents [8]. This chapter aims at giving a tentative overview of current water recycling applications worldwide based on available information.

2. SHORT HISTORY OF REUSE APPLICATIONS

Water recycling is a hot topic today, but it was not invented in the 20th or 21st century. Agricultural applications have been practiced in Ancient Egypt and China, where wastewater was used for irrigation. And in general, it should be recognized that in fact all water is reused through the eternal water cycle. In Rome, water supply problems were solved by constructing aqueducts, which allowed a permanent influent of freshwater. Since the excess of used and unused water had to be removed, sewage was necessary as well. The extreme dilution obtained in this way resulted in remarkably few problems with the effluent quality, which was to be discharged but did not pose severe problems in surface waters to be used as water source by villages located downstream. The aqueducts fell into disuse, and since wastewater treatment systems were nonexistent in the Middle Ages, severe problems arose with surface water quality in large cities. The use of surface water was in fact no less than direct reuse of wastewater without any treatment, apart from some dilution. A typical example from the Middle Ages is London, where population growth was booming from 40,000 in the 12th century to 1,000,000 in 1666, and the Thames was used both as a sewer and for water supply. An even more difficult case was Amsterdam where the Amstel River was dammed to prevent intrusion of the saline water of the Flevo Lake. The city's ring of canals was soon highly polluted by discharge of wastewater. It is evident that this caused epidemic of cholera and other waterborne diseases. It was not until the 16th century that measures were taken, which included prohibiting throwing of dead animals into the canals. These insights were crucial for public health, but also implied that freshwater and wastewater became two separated circuits. The "natural" link between wastewater and freshwater disappeared until the 20th century, where it was argued that in view of water scarcity and population growth, it may be more useful to restore this link, but improve purification and control. This induced the idea of water reuse, and was the starting point for many new projects and applications.

The first attempts toward water reuse were undertaken in the USA in the 1940s when chlorinated domestic wastewater was used in the steel industry [1]. However, it was not until the last quarter of the 20th century that water reuse appeared on the international agenda, at first in industrialized countries such as America and Europe. Between 1930 and 1970, the volume of reused water in Sweden increased by a factor of six. In 1951 in Japan, a program for recycling of the purified water of the Mikawashima Wastewater Treatment Plant in Tokyo started, which was to be reused as process water for a paper mill. In this case, the quality of the purified wastewater was higher than the quality of any other available water source. The fast economic growth in Japan in this period resulted in a strong competition between industry and agriculture for available water sources, which made water recycling even more important. Today, 80% of industrial process water in Japan is already reused.

3. WATER RECYCLING TODAY

Water recycling projects nowadays can be found in all parts of the world. Water recycling is particularly practiced in world regions suffering water scarcity, such as the Middle East, Australia, or southwest USA, or in regions with severe restrictions on disposal of treated wastewater effluents, such as Florida, coastal or inland areas of France and Italy, and densely populated European countries such as the UK and Germany [9]. In China, the average fraction of reused water in 1989 (in 82 important cities) was 56%, with a maximum of 93% [1]. However, a recent survey on municipal wastewater reclamation [10] mainly identified large water reuse projects in Japan (over 1800), the USA (over 800), the EU (over 200), and Australia (over 450). In the Mediterranean and Middle East area, around 100 sites were identified, whereas 50 sites were found in Latin America and 20 in sub-Saharan Africa. Large parts of Asia were not included in the survey. In particular in China, it can be assumed that many water reclamation sites can be identified, based on the numbers given above. Small projects, defined as below 0.5 Mm^3/y reclaimed water for unrestricted use or 2.5 Mm^3/y for restricted use, were also not included in this survey. This limitation may possibly influence the results in terms of fields of application, since it is known that large-scale projects are mostly used for landscape and agricultural irrigation, whereas small-scale projects often have urban, recreational, or environmental uses [10]. This is important, for example in the case of Japan where many relatively small-scale projects can be found [11], in contrast to the USA where water reclamation is mainly dominated by medium- to large-scale projects [10]. The total volume of reused water in the USA at the time of this study was estimated at 6.5 million cubic meters per day.

Among the objectives for water recycling, various fields of application can be distinguished. Bixio et al. [10] identified five categories, i.e., (1) agricultural irrigation, (2) urban, recreational, and environmental uses, including aquifer recharge, (3) process water for industry, (4) direct and indirect potable water production, and (5) combinations of these categories. In what follows, an overview of water recycling projects throughout the world will be given, with a slightly different approach, based on the driving force for implementation of water recycling projects. This includes water scarcity and drinking water supply, irrigation using reclaimed water, source protection, overpopulation, and environmental protection. In the different parts of the world that will be discussed, examples of these driving forces can be found. Finally, some of the remaining challenges will be highlighted.

4. WATER RECYCLING IN THE USA

In the USA, water recycling projects can be found for various purposes, ranging from irrigation and gray water supply to indirect potable reuse projects. Logically, the southern states are the most active in this area, with California and Florida being the most notable forerunners. In California, the use of reclaimed wastewater for the irrigation of corn, barley, lucerne, cotton, and pastures began in Bakersfield in 1912, followed by other projects in the 1920s [12]. In 1970, water reclamation was formally encouraged in the California State Water Code (Asano, 1998). Public health laws were progressively developed, leading to the publication of the so-called Purple Book [13], which is a collection of guidelines, rules, and standards that was later used elsewhere as a basis for regulations. Currently, wastewater recycling projects in California are booming, with ca. 600 GL of recycled water being used across over 4800 locations from 234 wastewater treatment plants [12]. An overview of different uses of recycled water in California is given in Fig. 1. The largest volumes are used for agricultural and landscape irrigation; other applications (industrial reuse, groundwater recharge, seawater barrier, recreation and wildlife, and others) use smaller volumes. In Los Angeles County, reclaimed secondary treated wastewater, followed by dual media filtration and chlorination, is supplied to the Whittier Narrows Groundwater Basin after surface spreading, since 1962, leading to ca. 23% of the potable water being indirectly recycled water [12].

The Irvine Ranch Water District has had separate water lines since the late 1960s to provide raw water and reclaimed water for irrigation [12]. Since 1991, recycled water has been used for high-rise buildings; ca. 20% of the water supply is now reclaimed water.

In the West Basin Municipal Water District and Orange County Water District, reverse osmosis is used to purify secondary treated wastewater; the treated water is injected into coastal aquifers to mitigate salt water intrusion from the ocean. This has been practiced since the 1950s in the West Basin Municipal Water District, and since 1976 in the Orange County Water District. The latter is an example of wastewater recycling in view of source protection; 75% of the water in Orange County comes from groundwater sources; the remaining 25% is imported from Colorado River. Because an

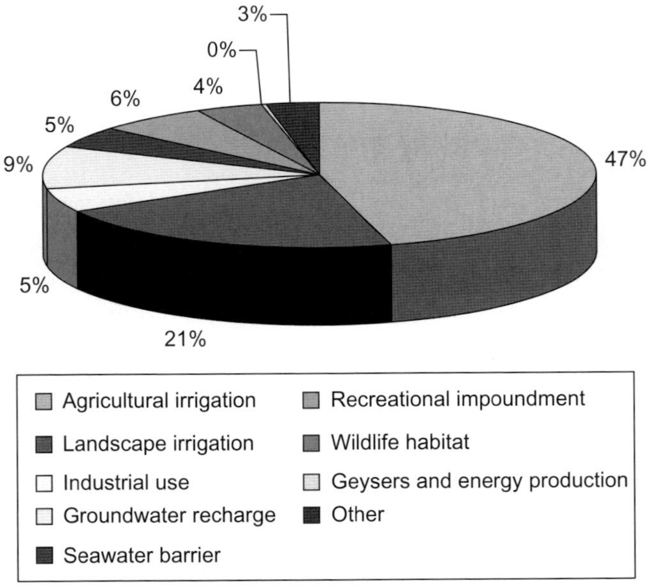

Figure 1 Relative volumes of recycled water used in California for different purposes.

increase in the volume of imported water is not to be expected, and the population growth will lead to more water demand but also to more wastewater generation, wastewater recycling by injection into groundwater layers was a logical option to enhance the capacity of these layers. More extraction without recharge would lead to intrusion of saline water from the ocean; recharge protects the groundwater supplies and at the same time increases the volume that can be extracted.

The initial "Water Factory 21" plant consisted of a complex system of physicochemical processes. The influent was secondary effluent treated with activated sludge. The total capacity of the plant was 57,000 m^3/day. The first step of the process was a combined flocculation and softening by addition of CaO (350–400 mg/L). The pH increases to 11, so that in addition to $CaCO_3$, heavy metals also precipitate. In this way, the concentrations of heavy metals are immediately below the standards for drinking water. Only during periods of high influent concentrations, exceeding values were observed for cadmium and chromium. Flocculation with polymeric flocculants removes suspended solids, leading to a reduction of 90% in turbidity and a chemical oxygen demand (COD) removal of 50%. There is also a disinfection effect: for coliforms and viruses, a removal of ca. 98% is obtained. Further treatment consisted of filtration and activated carbon adsorption, followed by reverse osmosis on a partial stream.

The process scheme for "Water Factory 21" was recently modernized into the Groundwater Replenishment System; in the new process scheme Fig. 2) precipitation, filtration, and activated carbon adsorption is replaced by microfiltration. The permeate from microfiltration contains less suspended solids and less microorganisms, which reduces problems in the reverse osmosis units. Pretreatment prior to microfiltration consists of sieving (opening 1 mm) to remove particles, and the addition of sodium hypochlorite for disinfection. The microfiltration unit is a hollow fiber module with an outside-in flow direction. A light vacuum at the permeate side of the membranes is applied to provide the necessary driving force. Cleaning of the microfiltration membranes is done by a combined air–water backwash, and a chemical cleaning (e.g., base combined with surfactants). A buffer tank decouples flows in the microfiltration and reverse osmosis units. In the reverse osmosis unit, ca. 90% of salts is removed, to a level of 100 mg/l, along with removal of organic solutes. The permeate yield is about 85%. Sodium hexametaphosfate is added as a scaling inhibitor, and chlorine is dosed to avoid biofouling. Sulfuric acid is dosed to obtain a pH of 5.5 in order to avoid scaling. There are three steps with in total 42 membrane modules. Each module consists of six spiral wound composite

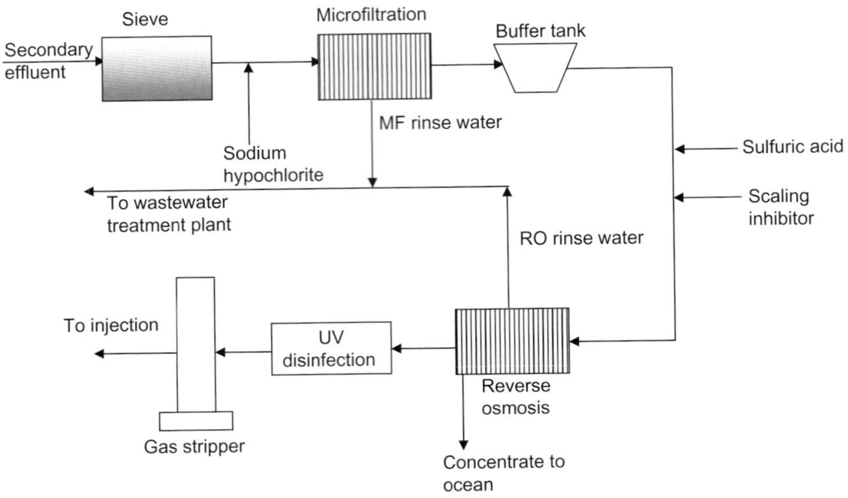

Figure 2 Water reclamation in the Orange County Water District's Groundwater Replenishment System.

polyamide membranes and is 7 m long, 8 in. in diameter. The feed water flows first in parallel through the first step, which consists of 24 modules. The concentrate from this step goes to the second step, consisting of 12 modules; the concentrate from the second step goes to the six remaining modules in the third step. The concentrate (ca. 15%) is discharged into the ocean. The permeate is further treated by UV disinfection and air stripping to remove CO_2 (due to addition of sulfuric acid); CaO is added for stabilization. Water Factory 21 has led to regulations to govern future indirect potable reuse projects involving groundwater recharge by the California Department of Health Services, which include:

- no more than 20% of injected water should return to the potable system over 5 years;
- no more than 50% of surface spread water should return to the potable system over 5 years;
- reclaimed water should remain underground for 12 months for direct injection;
- reclaimed water should remain underground for 6 months for surface spreading;
- the only feasible reclamation technologies are reverse osmosis and carbon adsorption.

Water Factory 21 and the Groundwater Replenishment System is a successful project well known over the entire world and may induce other wastewater recycling projects as well. However, the experience in San

Diego is the other side of the medal: after establishing a water recycling demonstration facility in 1983 using a sequence of processes quite similar to what was used in Water Factory 21, it tried to establish a 80 ML/day plant for groundwater recharge, and this was a failure due to strong opposition from San Diego citizens. The "toilet-to-tap" cartoons that were shown (which in fact suggested that no treatment at all was applied to the wastewater) provoked such a strong reaction that eventually the plan was shelved.

Florida, the fourth most populous state in the USA, has the second most water recycling projects after California. Out of 64 counties, ten have over 80% reuse [12], but highly populated counties such as Miami-Dade have only about 5–6%. The average is ca. 39%. In comparison to California, less reclaimed water is used for agricultural irrigation (19% vs. 47%), but much more for landscape irrigation (44% vs. 21%). Furthermore, more water is used for industrial purposes (15% vs. 5%), whereas groundwater recharge is similar (16% vs. 15%). Some of the most prominent water recycling projects in Florida include [12]; Mantovani et al., 2001):

- the CONSERV II project, in which groundwater is recharged by recycled water in Orange County and Orlando and subsequently used for irrigation of 11500 acres of citrus, eight nurseries, a tree farm, and a fernery;
- a 700 ha farm using recycled water (70 ML/day) in Tallahassee (discussed in more detail below);
- the dual distribution system in St Petersburg (discussed in more detail below);
- a 1240-acre wetland in Orlando for recreational purposes, using 5.1 MGD of recycled water from Ironbridge Sewage Treatment Plant;
- groundwater recharge into the Floridian Aquifer in Gainesville using 7.1 MGD of recycled water from the Kanapaha Sewage Treatment Plant;
- the Walt Disney World Resort Complex where recycled water from Reedy Creek Utilities is used for five golf courses, highway medians, a water park, and a tree park (horticulture);
- the Apricot project ("A Prototype Realistic Innovative Community of Today") in Altamonte Springs, comprising a dual water supply system throughout the city with recycled water being used for household irrigation and car washing, which can also be used for vegetable growing if the produce is peeled, cooked, or thermally processed before consumption;
- energy applications: Curtis Stanton Energy Center and Tampa's McKay Bay Refuse-to-Energy Centre.

Tallahassee is a city with 130,000 inhabitants, having a moderate climate with rainy summers and mild, rainy winters (Viessman & Hammer, 2005).

The average rainfall is 1500 mm. Urban wastewater contains mainly organic pollutants; no industrial wastewaters are mixed with the urban wastewater. Previously, the effluents were discharged in surface water after secondary treatment. This caused eutrophication on a large scale. Therefore, it was decided to stop discharges to surface water and instead use the wastewater for irrigation. A first plant (Thomas P. Smith Wastewater Treatment Facility) has a capacity of 104,000 m^3/day and treats ca. 53,000 m^3/day (average). A second plant (Lake Bradford Road Wastewater Treatment Plant) has a capacity of 17,000 m^3/day and treats 11,000 m^3/day (average). For the largest plant, the BOD, total nitrogen, and fecal coliforms are on average 5 mg/L, 12 mg/L, and 4 per 100 mL. For the smaller plant, this would be 8 mg/L, 25 mg/L, and 18 per 100 mL. The effluent is collected and stored in four stabilization ponds with a total volume of 45,500 m^3 Fig. 3a). From there, the effluent is further pumped to the irrigation site, where a pond is available with a capacity of 530,000 m^3 (incl. emergency basins). The effluent irrigates a total surface area of ca. 1000 ha, organized in "irrigation circles" Fig. 3b. A typical irrigation circle has a surface area of 54 ha; irrigation occurs from a central pivot. Because the crops are organized

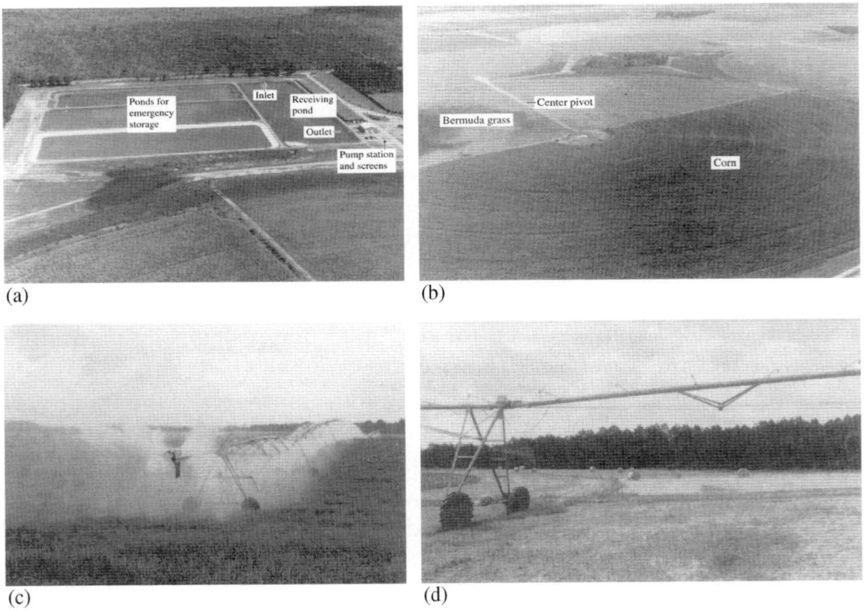

Figure 3 Wastewater recycling in Tallahassee, Florida (Viesmann & Hammer, 2005): (a) storage ponds, (b) irrigation circle with a radius of 400 m, (c) center pivot sprinkling bermuda grass, (d) irrigation tower.

in circles, the fields can be irrigated with a single system. Sprinkling occurs through a circulating pipeline with a total length of 400 m Fig. 3c and d, supported by braces, truss angles, rods, and rotating wheels. Half of the total surface area is used for bermuda grass used on the spot as animal feed (cows). Cows graze on the fenced parts of the irrigation circles; between the fences, the rotating supports for the pipeline are moving.

The primary goal of the farm is not what it produces (cattle, corn, soybeans), but rather the safe discharge of wastewater. The farm's success is a side effect, although welcome. Groundwater is at a depth of 3–6 m; because the soil mainly consists of fine sand, groundwater is supplemented by irrigation water. Therefore, a monitoring program was set up to measure nitrate concentrations on site and around; it was found that on site some places have a slightly increased nitrate concentration, but this is not visible at the site's boundaries at a depth of 40 m.

St. Petersburg is located on a peninsula between Tampa Bay and the Golf of Mexico. No groundwater is available; drinking water is extracted from sources up to 100 km away from the city (Wiessman & Hammer, 2005). The population increase resulted in an increased competition for the use of water, so that water management became a very important matter. Meanwhile, already in 1972 legal regulations for the effluent discharged to Tampa Bay came into force. Enhanced purification was required, with a standard for BOD of 5 mg/L, suspended solids 5 mg/L, total nitrogen 3 mg/L, and phosphorus 1 mg/L. It was decided to extend the existing secondary treatment with sand filtration and extended chlorination (with increased contact time, ca. 40 min based on average daily flow) in view of unlimited use for landscape irrigation and urban reuse. The plant started in 1978 and discharges to Tampa Bay stopped in 1987. The total wastewater purification capacity of St. Petersburg is about 260,000 m^3/day, distributed over four sites. After storage in a reservoir, the water is distributed through pipe network. During wet periods, when not much water is used for landscape irrigation, the excess effluent is injected into a saline water body at a depth of 300 m. Water exceeding the standards is injected here as well.

Standards for the effluent are conform the legislation of Florida, and include: 20 mg/L BOD as yearly average, 5 mg/L suspended solids in any sample (daily sampling), fecal coliforms below detection limit in 75% of the samples during a 30-day period (daily sampling), and no sample above 25 per 100 mL; minimum chlorine residual of 1 mg/L after 15 min contact time at peak hourly flow. The four plants meet these standards; for residual chlorine, a value of 4 mg/L is used, much higher than required. Problems so far are related to the chloride concentration of the effluent. If this is too high, damage to plants is observed. Therefore, effluent with a chloride concentration above 600 mg/L is rejected. The distribution system has 9000 points of use; 95% of these are for residential use. The other applications are parks, schools, and golf courses. The pressure on the distribution system is sufficiently high, allowing the hydrants to provide secondary fire service.

Economic advantages in this case are limited cost of tertiary wastewater treatment (in this case, less than required for discharge in surface water) and the possibility to use water containing organic material at low cost. However, the most important advantage is the fact that no new investments in the drinking water network were required. The population of St. Petersburg and the commercial activity have constantly increased since 1970, yet drinking water consumption remains at a stable level. The reason for this is that a part of the consumption is replaced by recycled water. The price of recycled water for residential use was \$0.30 per 1000 gal (3785 m^3), or \$11 per month for unlimited use. This shows that prices are very low compared to drinking water.

In various other states in the USA, recycling projects can be found as well:

- In Virginia: the UOSA or Upper Occoquan Sewage Authority, using wastewater as a source of indirect potable supply to the 40 GL Occoquan Reservoir starting from 1978, providing water for up to one million people in Northern Virginia. Typically, 15% of the water in the reservoir is reclaimed water [12].
- In Nevada, the Las Vegas Water District traditionally uses water from Lake Meed (Hoover Dam), but due to increased demand, the production volumes could only be met by recycling 180 GL/year from the sewage treatment plant back to the lake [14];
- In Louisiana, bill H.B. 2016, issued in 2003, mandates a reclaimed water program to be used for irrigation at parks, cemeteries, golf courses, and highway landscaped areas.
- In El Paso, Texas, the Fred Hervey Water Reclamation Plant provides water (ca. 7.6 MGD) for groundwater recharge through injection in the Hueco Bolson aquifer, and, to a lesser extent, for cooling in electricity generation and for irrigation of a golf course. This plant uses biological activated carbon, biological denitrification, lime precipitation and ozonation, in addition to screening and conventional primary sedimentation.

Many other recycling projects can be found elsewhere in the USA and Canada, but most of these are of a much smaller scale.

5. WATER RECYCLING IN ASIA

In Asia, Japan and Singapore are the most prominent water recyclers. Japan has a long tradition in reuse; ca. 150 GL of water is recycled annually in

Japan. Water recycling is a necessity in urban environments, where the existing potable water infrastructure is unable to cope with the increasing building density. The piping in large buildings in cities like Tokyo and Fukuoka is fit for using primarily recycling water. Furthermore, toilets have an arrangement in which the water filling the cistern can first be used for washing hands; this is a direct form of reuse. Domestic supply is available at a price 16% below that of potable water. Landscape and agricultural irrigation is much less important Fig. 4).

In Osaka, a "21st Century Master Plan" was developed that aims to develop 30% recycling by 2013, and 100% by 2030 [12]. Since 2002, the Nagisa plant on the left bank of the Yodo River processes 115 ML/day of which 10% is recycled (within the plant, for landscaping, for heat exchanging in the air conditioning system of the City General Welfare Centre, for fire mains and toilet flushing).

Yokohama has a 70,000-seat International Stadium, the largest in Japan; 5 ML/day of treated wastewater is used in the facilities as a heat source for a heat pump, for flushing toilets, for sprinklers, and for artificial streams in parks neighboring the stadium.

Elsewhere in Asia, the NEWater project in Singapore [15] is the most visible. Singapore depends for a large part on Malaysia for freshwater, because it has a very limited catchment area. This is an undesired and insecure position. Therefore, other options have been explored since the 1970s, which has led to a gradual development of the current large-scale water recycling project. Starting from secondary effluent, a multiple barrier approach is used: a dual membrane filtration step consisting of microfiltration/ultrafiltration and

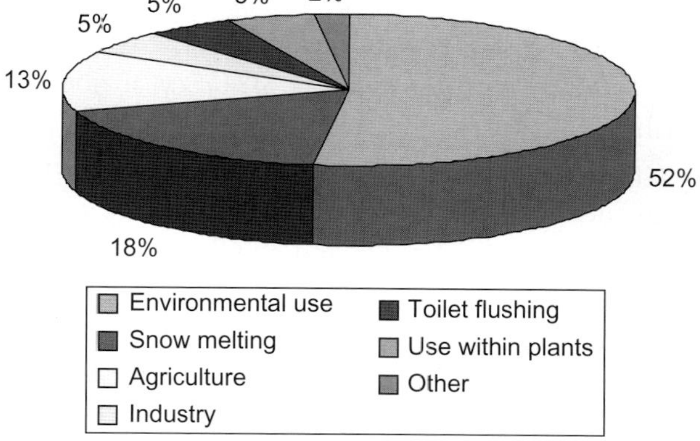

Figure 4 Relative volumes of recycled water used in Japan for different purposes.

reverse osmosis, followed by UV disinfection. NEWater is primarily aimed at nonpotable use, for example, as process water for wafer fabrication plants (as a source for ultrapure water). This is not controversial, because the recycled water is used in an industrial context and in a sequence of further treatment processes with the final result being water without any impurities. NEWater is also used for other industries and as cooling water in air conditioning systems. Since 2003, a small flow of NEWater has been introduced to existing raw water reservoirs for indirect potable reuse. This is similar to the approach in some applications in the USA. However, in this case, it was well understood that information and education are a prerequisite for the success of the project. A Web site was developed (www.pub.gov.sg/NEWater) and an information campaign was held, which contributed to the success of this project.

Elsewhere in Southeast Asia, water is seldom reused after purification, mainly because water sources are often not scarce. Exceptions are India and Vietnam, although in most cases treatment is quite limited. This is partly also true for China, where recycling is more common, but the effluent does often not meet standards for, e.g., agricultural reuse. The situation in China, however, is rapidly improving in terms of quality and quantity.

A similar situation is found in West Asia and the Middle East, where raw effluent is used for irrigation in Yemen, Syria, Lebanon, Palestine Territories, Egypt, and Iran [12]. Notable exceptions are Israel, Jordan, and Kuwait. In Jordan, 95% of the total volume of 74 Mm^3/year is reused, mainly (80%) for (restricted) irrigation in the Jordan Valley, after discharge to the Zerqa River and storage in the King Talal Dam. The remaining 20% is used on site as process water. In Kuwait, 25% of its agriculture and green areas are irrigated using 52 Mm^3/year of treated wastewater. Groundwater recharge is also practiced in Kuwait [12]. In Israel, 20% of its water supply came from recycled water already in 1994 [12]. Large projects include the Dan region where 95 GL/year from Tel Aviv is used for recharge of the coastal aquifer and for irrigation after 2 months storage. In Haifa, 32 GL/year goes to the Kishon complex, which consists of two deep stabilization reservoirs [16]. Currently, 65% of the connected sewage in Israel is reused for irrigation purposes [17]; unrestricted agricultural irrigation is envisaged for future projects as well as public park irrigation, industrial reuse, and aquifer salinity reduction.

6. WATER RECYCLING IN EUROPE

A good overview of water recycling in Europe is given by Bixio et al. (2006) as a result of the EU project AQUAREC. From this, it could be concluded that the water sector in Europe is in a transitionary phase with unique opportunities for improvement of water management, including water reuse. Nevertheless, the success of water reuse depends on many factors such as local communities and companies as well as on centralized rules and regulations; differences between EU member states often complicate the implementation of individual projects. This leads to significantly different levels of implementation in European countries. Guidelines and regulations are often not clear (enough), which results in uncertainties about what is (not) allowed. In addition, the use of recycled water is not the same in all countries (Bixio et al., 2006): in southern Europe, the primary use is agricultural irrigation (44%) and for urban or environmental applications (37%), whereas in northern Europe, the focus is more on industrial use (33%) along with urban and environmental applications (51%). However, this is only a tendency; individual countries may show large deviations from this picture, as was shown by Bixio et al. (2006). Italy is an example where regulations impede water recycling: wastewater recycling is only allowed for agricultural purposes, and on condition that an increase of crop production can be achieved by using recycled wastewater [18]. In a similar way, wastewater reuse is not developed in Poland and Hungary, where not more than a few percent of the total generated wastewater is used, mainly for (small) irrigation projects. Water recycling in the Czech Republic is also still underdeveloped, but a recent study [19] showed the potential for agricultural reuse in the Znojmo area, and, more importantly, industrial reuse as a replacement of surface water for cooling purposes.

European countries have always been very innovative, in particular countries like the Netherlands, where centuries of experience are available. The "third pipe" system was experimented with in Utrecht, denoted as "household water," but this was abandoned because of cross-contamination problems. In Belgium, the recycling project carried out by the local drinking water company IWVA in Koksijde has received much attention; this project involved treatment of 2.5 GL of wastewater by microfiltration and reverse osmosis and subsequent storage of the treated water in an aquifer in the dunes. After a residence time of 1–2 months, the water is distributed as drinking water. No quality problems have been encountered, and customer satisfaction is high. Elsewhere in Belgium, a membrane bioreactor (MBR) is used in Schilde to provide water for unrestricted irrigation. Long-term effluent results for a broad range of water reuse parameters demonstrate the suitability of MBR to meet standards for unrestricted irrigation (Bixio et al., 2006).

In the UK, water recycling came only slowly into practice, with indirect reuse with abstraction points downstream having some proportion of treated wastewater. For example, treated wastewater is discharged into the river Chelmer, and subsequently used for recharging the Hanningfield reservoir in Essex [16]. The Millenium Dome on the bank on the river Thames in Greenwich was, in fact, the first significant recycling project in the UK. Around 500 m^3/day of water, partly from the hand basins in the toilets, partly rainwater and mixed with groundwater, is recycled and used on site for toilet flushing.

In the northern part of Europe (Scandinavian countries and Germany) people have a very high environmental awareness, which translates into many small-scale projects that can be designated as "zero discharge," involving not only wastewater recycling but also minimization of solid waste generation and energy consumption. These "ecological villages" are prototypes and could be seen as experiments in view of a more sustainable future society.

7. WATER RECYCLING IN AUSTRALIA

It could be assumed that due to successful examples, the idea of water recycling would be picked up easily in other regions with similar water supply problems, in the Middle East, Asia, northern Africa, the southwest USA, and Australia. However, public perception is still problematic, impeding new initiatives. This was proved again in 2006 in Toowoomba, Queensland, Australia, where a poll to mix treated wastewater with raw surface water for drinking water production had a negative outcome. Nevertheless, other examples of water recycling can be found throughout Australia, and the debate on water recycling is probably nowhere as intense as in Australia. Recent droughts played a role in the awareness that water sources are limited [20]. A good overview of water recycling projects in the Australian context is given by Radcliffe [12].

In New South Wales, Sydney Water operates 27 sewage treatment systems collecting ca. 1300 ML/day, with 29.9 ML/day processed for reuse. In Rouse Hill, a suburb of Sydney, a third pipe scheme has been installed with treated water from the sewage treatment plant of Rouse Hill, similar to St Petersburg. Around 4.4 ML/day is treated using ozonation, microfiltration, and chlorination Fig. 5). Ozone is used because of the stronger oxidizing effect; nevertheless, microfiltration and chlorination were thought essential as well. Taps contained recycled water are purple and labeled "Not For Drinking." The cost of recycled water (2003) is 28 AUS$c/$m^3$ or 17 EURc/m^3, compared to 98 AUS$c/$m^3$ or 58 EURc/m^3 for potable water. The perception of customers is very positive because it is

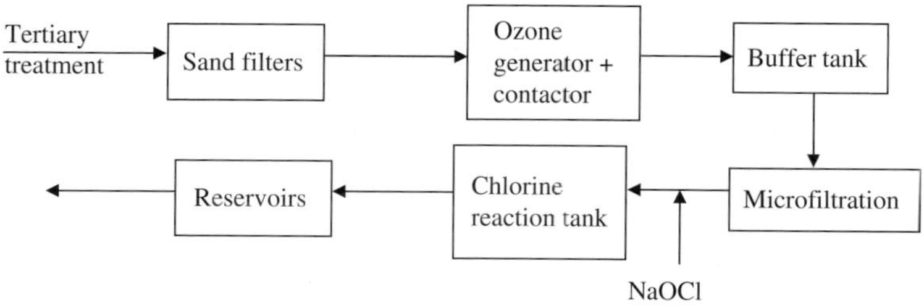

Figure 5 Process scheme for water recycling in Rouse Hill, New South Wales, Australia.

understood that this project is important for a sustainable future. In addition, the economic benefits are appreciated. The consequence was an increase of 20% compared to other regions, whereas ca. 35% of the total consumption consisted of recycled water.

A similar project was carried out in the Olympic Park Sewage Treatment Plant (capacity 2.2 ML/day), constructed for the Olympic Games of 2000 in Sydney. In this case, treatment consists of microfiltration and reverse osmosis, followed by chlorination. The water is distributed at a cost of 83 AUS\$c/m^3 in Newington, a nearby suburb. The production cost, however, is higher at AUS\$ 1,60/m^3, similar to Rouse Hill (AUS\$ 3–4/m^3).

Various other examples can be found in the Sydney area [12]. A typical application for agricultural irrigation is Gerringong-Gerroa, where secondary treatment and tertiary treatment is used comprising a biological reactor, clarification, sand filtration, ozonation, biologically activated carbon, microfiltration, and UV filtration [21]. Other examples can be found in Picton (2.4 ML/day for 134 ha) and in Richmond. Furthermore, 1 ML of tertiary treated water in St. Mary is used for irrigation of a golf club. In Wollongong, the Illawarra Wastewater Strategy will involve production of high-quality recycled water (using microfiltration and reverse osmosis) for a steel factory [22].

In the Newcastle area, the picture is similar. A diagram of the various uses of recycled water (data 2003) is given in Fig. 6, and shows that the share of industrial use is larger than elsewhere. In other parts of New South Wales, numerous small-scale projects can be found.

In Victoria, 174 sewage treatment plants are in operation, involving 15 coastal discharges. Most of the plants are inland and have a high recycling ratio; however, the total volume is relatively low. In Melbourne, only 2% of 295 GL/year of effluent is reused. The total volume for Victoria is ca. 40 GL/year.

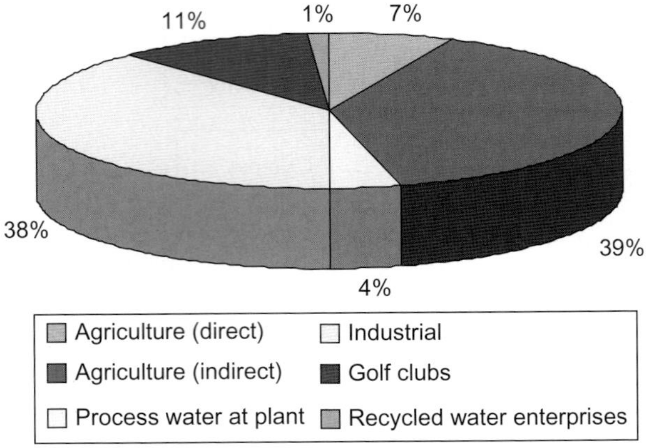

Figure 6 Relative volumes of recycled water used in Newcastle, New South Wales for different purposes.

Queensland is more or less similar, with 27,400 ML/year reuse on a total of 340,000 ML/year of effluent (8%). Other parts of Australia show the same picture, with large plants often close to and discharging into the ocean (e.g., Adelaide) and many small-scale inland recycling applications, mainly for irrigation purposes.

8. WATER RECYCLING IN OTHER PARTS OF THE WORLD

Water recycling is an obvious choice in water-stressed regions where it is the only option to grow or even to survive. Many of these regions can be often found in less developed regions in Africa, Asia, and Latin America.

In Africa, the Goreangab reclamation plant in Windhoek, Namibia, is the typical example. Namibia has ca. 1.8 million inhabitants and has an average annual rainfall of 285 mm. There are large differences in rainfall. In the south, bordering South Africa, the Oranje River is an important water source. In the north, the Kunene River from Angola can be used; this part of Namibia has plenty of water as it is within the catchment area of the Zambezi and Okavango rivers (with even direct access to the Zambezi through the Caprivi Strip). In the west, the coastal plain quickly transforms to the Namib Desert, where rainfall is virtually zero. The Namib Desert has been hyperarid for 17 million years; local people, the Topnaar, survive by making use of temporary rivers and vegetation adapted to extreme conditions such as a plant called *!nara*. These conditions are barely sufficient for a very small number of people; growing pressure from tourism in this extraordinary region poses a severe threat. In the east of Namibia, the

Kalahari Desert is also dry. Windhoek is located in between these deserts, too far away from the rivers in north and south (750 and 900 km, respectively). It has ca. 250 000 inhabitants and relied originally on groundwater. When the city grew, other sources had to be found by constructing dams catching water from ephemeral rivers. The Avis Dam was constructed in 1933 (2.4 Mm^3), the Goreangab Dam in 1958 (3.6 Mm^3). Potable water was produced in a conventional treatment plant. From 1969 on, the effluent from the Gammams Wastewater Treatment Plant was also treated in this plant. This was the start of the first, and to date still the only water recycling plant for direct potable reuse [23]. The initial capacity was 4300 m^3/day; after a number of upgrades this became 7500 m^3/day. The effluent used for reclamation originates from domestic and business wastewater; industrial effluents are treated in a separate plant. Treatment occurred by coagulation/flocculation with $FeCl_3$, dissolved air flotation, rapid sand filtration, adsorption on granular activated carbon, chlorination, and pH correction.

In 2002, a new plant was started, with a capacity of 21,000 m^3/day. This plant was based on thirty years of experience, but the previously applied scheme was extended with ozonation and ultrafiltration Fig. 7). The policy is to blend the wastewater with other water sources, so that the final drinking water has a maximum of 35% reclaimed water. Monitoring of the quality of raw water, intermediate water, and drinking water is obviously of extreme importance; communication and public perception is another main point of attention [23]. Safety measures can be taken, ranging from diversion of raw water to addition of powdered activated carbon.

The Goreangab Water Reclamation Plant was (and still is) a pioneer in water recycling. Elsewhere in Africa, water recycling is not at all developed, with a few exceptions: Zambia and South Africa reused up to 16% of their wastewater in the late 1980s [12], and this was even up to 75% in Tunisia's capital Tunis. Bixio et al. [10] identified a total of 20 recycling sites in sub-Saharan Africa, which is a remarkably low number.

In Latin America, the overall picture is similar to Africa. The number of sites identified by Bixio et al. [10] is 50, and this low number says all. According to the Pan-American Health Organization, less than 14% of the wastewater in Latin America receives any sort of treatment before discharge to rivers or the sea. The situation is roughly the same in all Latin American countries, with dramatic numbers such as 1% sewage treatment in Suriname, 5% in Venezuela, and 13% in Mexico [12]. It is understood that in these circumstances, water reuse is not feasible, at least not in the sense that recycled water should be a safe and reliable water source.

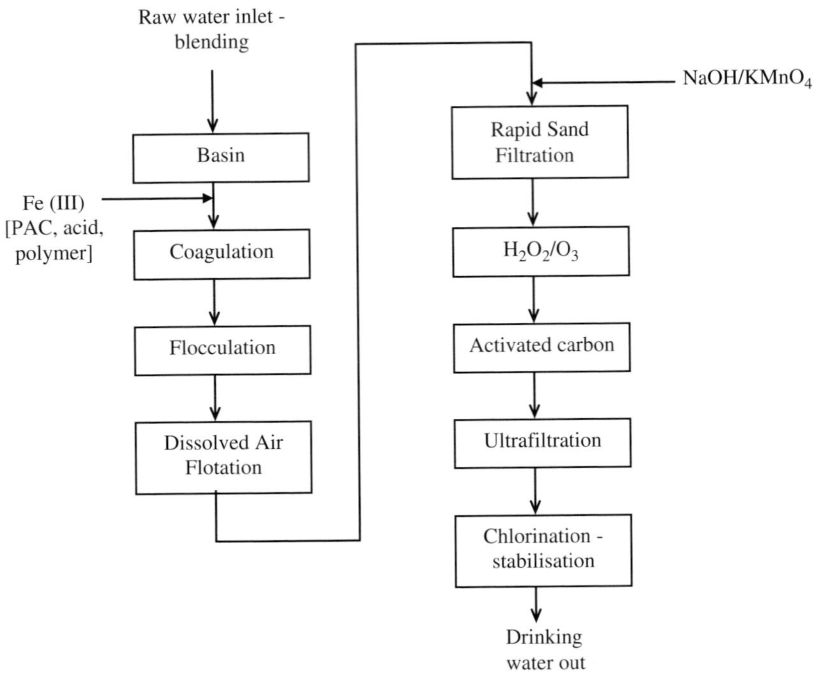

Figure 7 Water reclamation scheme used in the Goreangab Water Reclamation Plant, Windhoek, Namibia.

Irrigation with raw wastewater, however, is common practice; more than 500,000 ha of agricultural land in Latin America is irrigated with untreated wastewater.

9. CONCLUSIONS AND FURTHER CHALLENGES

From the number and nature of water recycling projects throughout the world, it can be concluded that the principle of treating wastewater as a valuable source becomes increasingly accepted. The larger demand of water due to population growth and increased comfort is translated into better protection of sources and eventually, zero discharge or nearly zero discharge systems. A remarkable observation is that in none of these projects adverse health effects were obtained, even on a long timescale. This shows that the technologies for producing clean water from wastewater are available, but not everyone is convinced (as proven in San Diego, California, and in Toowoomba, Australia). A good direction for the future to stimulate water recycling is therefore not in technology but rather in education programs and information campaigns. The success of the NEWater project in

Singapore is a good example of how it should be done: by using the appropriate technologies, but also providing the right information. The project in Koksijde, Belgium, is another example of public acceptance for water reuse.

Another conclusion to be made is the difference between developed and developing countries. Water recycling requires (expensive) technology if the effluent is to be safe and reliable. Developing countries lack infrastructure and financial means to do this, so that water recycling, where applied, is unsafe due to remaining microbiological and chemical contaminations. When used for irrigation, pollutants are spread over a large area and may affect the crops. This is common practice in parts of Asia and Africa. Ethiopia is a typical example: waste and wastewater is discharged into rivers without any treatment; pollutants accumulate and cause problems downstream where the water is to be used as the only source. Drinking water production interferes with this approach, in some cases even through direct contact between river water and drinking water (shown in Fig. 5: the pipeline for drinking water supply is below the level in the river during the wet season, contaminating drinking water with diluted wastewater).

Another typical example is Hanoi, Vietnam, where there is hardly any wastewater treatment. The continuous massive discharge of unprocessed wastewater in the Nhue-Day River (from domestic, industrial and agriculture activities from the city of Hanoi, Ha Tay and peri-urban area) has resulted into surface water that is heavily polluted. This water is used as a source for drinking water for more than 60,000 people who live near the river bank. Again, this is a case of wastewater reuse without any treatment. Solving this problem is probably the most important challenge for water recycling in the next decades.

REFERENCES

[1] A.D. Levine, T. Asano, Water reclamation, recycling and reuse in industry, in: P. Lens, L. Hulshoff Pol, P. Wilderer, T. Asano (Eds.), Water Recycling and Resource Recovery in Industry, IWA Publishing, London, UK, 2002.

[2] C.P. Chu, S.R. Jiao, H.M. Lin, C.H. Yang, Y.J. Chung, Recycling the wastewater of the industrial park in Northern Taiwan using UF-RO system: in-situ pilot testing and cost analysis, J. Water Supply Res. Technol. – Aqua 56(8) (2007) 533–540.

[3] G.B. Keremane, J. McKay, Successful wastewater reuse scheme and sustainable development: a case study in Adelaide, Water Environ. J. 21(2) (2007) 83–91.

[4] J.F. Scamehorn, E.E. Tucker, H. Gecol, N. Kornesvarakul, L.K. Serventi, K.H. Raney, S.F. Capps, Industrial home laundry closed loop cleaning process, J. Surfactants Detergents 10(4) (2007) 287–299.

[5] K.Y. Wang, S. Liu, H.Y. Ju, Recycling of reclaimed tannery wastewater for use in pickling and chrome tanning, J. Soc. Leather Technol. Chem. 91(3) (2007) 128–131.
[6] K.G. Wilkinson, R.B. Brooks, C. Balmer, D. Halliwell, L. Palmowski, J.G. Issa, S. Jeyaseelan, B. Meehan, K. Baskaran, A survey of waste management practices in Victorian dairy factories, Austr. J. Dairy Technol. 62(3) (2007) 154–158.
[7] EPA (Environmental Protection Agency and U.S. Agency for International Development), Office of Water, Office of Research and Development, Manual of Guidelines for Water Reuse, 2004, EPA/625/R-92/004.
[8] P. Ratnaweer, Water reuse, in: J.R. Pfafflin, E.N. Ziegler (Eds.), Encyclopedia of Environmental Science and Engineering, Vol. 2, 5th ed., CRC Taylor & Francis, Boca Raton, FL, 2006.
[9] J. Marsalek, K. Schaefer, K. Excall, L. Brannen, B. Aidun, Water reuse and recycling. Canadian Council of Ministers of the Environment, Winnipeg, Mannitoba, CCME Linking Water Science to Policy Workshop Series. Report No. 3, 2002.
[10] D. Bixio, B. De heyder, H. Cikurel, M. Muston, V. Miska, D. Joksimovic, A.I. Schäfer, A. Ravazzini, A. Aharoni, D. Savic, C. Thoeye, Municipal wastewater reclamation: where do we stand? An overview of treatment technology and management practice, Water Sci. Technol.: Water Supply 5(1) (2005) 77–85.
[11] M. Ogoshi, Y. Suzuki, T. Asano, Water reuse in Japan, Water Sci. Technol. 43(10) (2001) 17–23.
[12] J.C. Radcliffe, Water Recycling in Australia, Australian Academy of Technological Sciences and Engineering, Parkville, Vic., Australia, 2004.
[13] DHS., California Health Laws Related to Recycled Water, State of California Department of Health Services, Sacramento, CA, 2001.
[14] R. Williams, Reclaimed water reuse in residential and commercial situations, Report to the Churchill Fellowship Trust, Canberra, 1996.
[15] A.N. Puah, NEWater: Singapore's experience in water reclamation, in: ECI Conference on Water Treatment and Reuse II, Tomar, Portugal, February 11–16, 2007, 34 pp.
[16] V. Lazarova, B. Levine, J. Sack, G. Cirelli, P. Jeffrey, H. Muntau, M. Salgot, F. Brissaud, Role of water reuse for enhancing water management in Europe and Mediterranean countries, Water Sci. Technol. 43(10) (2001) 25–33.
[17] A. Aharoni, H. Cikurel, Mekorot's research activity in technological improvements for the production of unrestricted irrigation quality effluents, in: S.J. Khan, M.H. Muston, A.I. Schaefer (Eds.), Conference on Integrated Concepts in Water Recycling, Wollongong, Australia, February 2005, pp. 19–31.
[18] S. Barbaballo, G.L. Cirelli, S. Indelicato, Wastewater reuse in Italy, Water Sci. Technol. 43(10) (2001) 43–50.
[19] B. Janosova, J. Miklankova, P. Hlavinek, T. Wintgens, Drivers for wastewater reuse: regional analysis in the Czech Republic, in: S.J. Khan, M.H. Muston, A.I. Schaefer (Eds.), Conference on Integrated Concepts in Water Recycling, Wollongong, Australia, February 2005, pp. 331–342.
[20] J.C. Radcliffe, The future directions for water recycling in Australia, in: S.J. Khan, M.H. Muston, A.I. Schaefer (Eds.), Conference on Integrated Concepts in Water Recycling, Wollongong, Australia, February 2005, pp. 575–586.
[21] M.J. Boake, Recycled water – case study: Gerringong Gerroa, in: S.J. Khan, M.H. Muston, A.I. Schaefer (Eds.), Conference on Integrated Concepts in Water Recycling, Wollongong, Australia, February 2005, pp. 93–100.
[22] W. Hird, Recycled water – case study: BlueScope steel, Pt Kembla steelworks, in: S.J. Khan, M.H. Muston, A.I. Schaefer (Eds.), Conference on Integrated Concepts in Water Recycling, Wollongong, Australia, February 2005, pp. 274–280.
[23] P.L. Du Pisani, Direct reclamation of potable water at Windhoek's Goreangab reclamation plant, Desalination 188(1–3) (2006) 79–88.

PART Two

Desalination Technologies

CHAPTER 4

Desalination: Reverse Osmosis and Membrane Distillation

Tilak Gullinkala, Brett Digman, Colleen Gorey, Richard Hausman *and* **Isabel C. Escobar***

Chemical and Environmental Engineering Department, The University of Toledo, Toledo, OH 43606-3390, USA

Contents

1. REVERSE OSMOSIS

1.1 Introduction

Reverse osmosis (RO) is a membrane filtration process for removing solvent from a solution by leaving a concentrated solution behind. When a semipermeable membrane separates a dilute solution from a concentrated solution, the solvent crosses from the dilute to the concentrated side of the

*Corresponding author.
E-mail address: Isabel.Escobar@utoledo.edu

Sustainability Science and Engineering, Volume 2 ISSN 1871-2711, DOI 10.1016/S1871-2711(09)00204-9

membrane in an attempt to equalize concentrations. The flow of solvent can be prevented by applying an opposing hydrostatic pressure to the concentrated solution.

The magnitude of the pressure required to completely impede the flow of solvent is defined as the "osmotic pressure." If the applied hydrostatic pressure exceeds the osmotic, flow of solvent will be reversed, that is, solvent will flow from the concentrated to the dilute solution. This phenomenon is referred to as "reverse osmosis." Fig. 1 illustrates the concepts of osmosis, osmotic pressure, and RO schematically.

Osmotic pressure (π_i) is a thermodynamic property of a solution and is related to the mole fraction of the solvent, X_{Bi}, as

$$\pi_i = -\left(\frac{RT}{V_B}\right) \ln X_{Bi} \tag{1}$$

where R is the universal gas constant [J/(K mol)], T the temperature (K), and V_B the volume of the solvent (m^3).

For dilute solutions, the osmotic pressure is found to obey the ideal gas law making Eq. (1) simplify to the van't Hoff equation as

$$\pi_i = C_{Ai}RT \tag{2}$$

where C_{Ai} is the concentration of solute (mol/m^3).

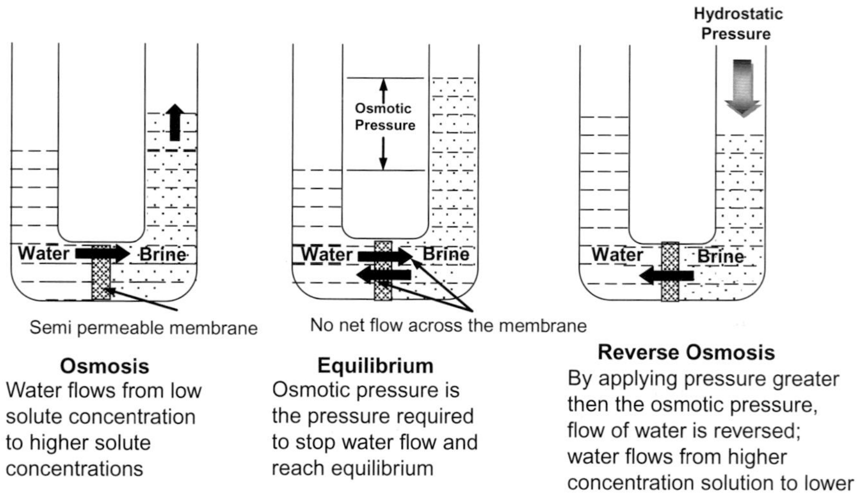

Figure 1 Concept of osmosis, osmotic pressure, and reverse osmosis.

Therefore, the osmotic pressure difference across a membrane, $\Delta\pi$, is related to the concentration difference, $C_{A2} - C_{A3}$. The osmotic pressure depends neither on the solute type nor on its molecular size but only on molar concentration as shown in the formula.

In order to use RO as a water purification process, the feedwater is pressurized on one side of a semipermeable membrane. The pressure must be high enough to exceed the osmotic pressure to cause reverse osmotic flow of water. If the membrane is highly permeable to water, but essentially impermeable to dissolved solutes, pure water crosses the membrane and is known as product water. As product water crosses the membrane, the concentration of dissolved impurities increases near the membrane surface, leading to a condition known as concentration polarization. Concentration polarization is associated with the accumulation of solute particles on the membrane surface, resulting in a higher solute concentration at the membrane surface as compared to the bulk solution, which in turn increases the osmotic pressure of the system. A point is reached at which the applied pressure is no longer able to overcome the osmotic pressure and no further flow of product water occurs. Moreover, if the applied pressure is increased in an attempt to gain more product water, the membrane becomes fouled by precipitated salts and other undissolved material from the water. Therefore, there is a limit to the fraction of feedwater, which can be recovered as pure water. RO units are operated in a configuration where only a portion of the feedwater passes through the membrane with the remainder being directed to the drain (cross-flow configuration).

The water flowing to the drain contains concentrated solutes and other insoluble materials, such as bacteria, endotoxin, and particles, and is referred to as the reject stream. The product water to feedwater ratio can range from 10% to 50% for purification of water depending on the characteristics of the incoming water as well as other conditions, such as the membrane material and fouling potential.

1.2 Membrane synthesis

RO membranes are generally prepared by the classical phase inversion method, in which the polymer solution is formed in two phases from which the selective layer and porous support are fabricated by selective solvent evaporation. Integrally skinned phase inversion membranes are prepared in this method. Interfacial composite membranes (IFC) are prepared in general by generating the selective layer by chemical reaction on the already prepared porous support [1]. The advantage of the IFC membrane is that the chemistry of the critical selective layer can be chosen independently from the underlying porous support as thin-film chemistry and morphology determines the membranes transport properties.

Syntheses of IFC membranes involve impregnating the porous polymer support, preferably polysulfone, with aqueous solutions containing a multifunctional amine [1,2]. The impregnated membrane is then contacted with a hexane solution containing a multifunctional acid chloride. As the two solutions are immiscible, polycondensation reaction occurs at the interface, leading to the formation of a thin selective layer at the surface of porous support. Various parameters such as diffusion rates of amines into organic solvent, polymer film permeability, interfacial tension, concentration of the reactants, and polymer film growth rate play a crucial role in the synthesis of IFC membranes. Acid produced during the condensation reaction is neutralized by using suitable bases, such as sodium hydroxide or excess diamines in the reaction. Porous polymeric supports are prepared by phase inversion method in the IFC membranes.

Integrally skinned phase inversion membranes are casted from various organic polymers such as cellulose acetate [3], nylon 4, and polyvinyl alcohol. Once the desired polymer material is prepared through melt phase polymerization, membranes are casted by phase inversion process. The polymers are dissolved in suitable polar organic solvents to form appropriate weight percent solutions. The membranes are prepared by casting these solutions on dry Pyrex glass plates to a uniform thickness with the aid of a specially designed glass rod. The plates are then introduced into an oven at a desired temperature for a selected period to facilitate the partial evaporation of the solvent. These two parameters, solvent evaporation temperature and solvent evaporation period, play a vital role in determining the flux and salt rejection properties of the resulting RO membranes [4]. The casted film side facing the air has higher solvent evaporation than the side facing the glass, which results in smaller pore size on the former side. The glass plates are then immersed in deionized water overnight to complete the exchange of residual solvent with water. The appearance of the membrane at this stage is an indication of the residual solvent present in the membrane. A transparent membrane indicates low solvent content in the membrane. This can be achieved by optimizing solvent evaporation temperature and period. The solvent-free membranes are optimized after annealing in deionized water at preselected temperatures.

Apart from the above mentioned parameters, the chosen polymer and its structure play a very important role in determining the properties of RO membranes. Uniformity and symmetry in the polymer chain structure provide tighter and uniform pore structures, resulting in improved salt rejections [4,5]. Lack of uniformity in the molecular structure can result in reduced rejection due to openings in the polymer structure [4].

Pore size distribution (PSD) of a membrane predominantly determines its filtration characteristics. The thin selective layer is characterized by the presence of two distinct kinds of pore sizes on the membrane surface [3,6]. These are tight "polymer network pores" or wider "polymer aggregate pores," which facilitate the transfer of matter across the membrane. In most cases, the membrane is designed to only allow water to pass through. The water goes into solution in the polymer of which the membrane is manufactured, and crosses it mostly by diffusion. However, convection may aid filtration when membranes with wider pores are used. Water transport across membranes requires that a high pressure be exerted on the feed side of the membrane, usually 5–20 MPa (50–200 bar).

1.3 Driving forces for transport

The main driving forces for transport are pressure, concentration, electrical potential, and temperature, each of which primarily influences the flux of solvent, solute, electrical current, and thermal energy, respectively. In addition to the primary effects, each of the driving forces has a cross-influence on the other fluxes. For instance, the pressure driving force can cause a flux of current, called the streaming current.

In RO systems, the only driving forces of interest are pressure and concentration, which lead to flux of solvent and solute, respectively (Table 1). The cross-influence of solute concentration driving force on solvent flux is represented by the osmotic pressure term in the solvent flux equation. The cross-influence of pressure driving force on solute flux is often small, for high separation membranes, and is therefore neglected.

The solvent flux equation, written here for both volume flux, J_V, and molar flux of solvent, N_B, indicates that flux is directly proportional to the effective pressure driving force:

$$N_B = J_V\, C = A\,(\Delta P - \Delta \pi) \tag{3}$$

Table 1 Driving forces of interest in reverse osmosis

Driving force		Pressure gradient	Concentration gradient
Flux	Solvent flow	Solvent permeability	Osmosis
	Solute flow	Ultrafiltration	Diffusion

where C is the molar density (kmol/m^3), A the pure water permeability coefficient [kmol/(m s kPa)], ΔP the pressure difference across the membrane (kPa), and $\Delta\pi$ the osmotic pressure difference across the membrane (kPa).

1.4 Types of reverse osmosis membranes

A RO membrane must be freely permeable to water, highly impermeable to solutes, and able to withstand high operating pressures. It should ideally be tolerant of wide ranges of pH and temperature, and should be resistant to attack by chemicals like free chlorine and by bacteria. Ideally, it should also be resistant to scaling and fouling by contaminants in the feedwater. There are three major types of RO membranes: cellulosic, fully aromatic polyamide, and thin-film composite. A comparison of characteristics of these three membrane types is given in Table 2.

Table 2 Comparison of reverse osmosis membranes

Comparison of reverse osmosis membranes			
Feature	**Cellulosic**	**Aromatic polyamide**	**Thin film composite**[a]
Rejection of organic	L	M	H
Rejection of low-molecular-weight organics	M	H	H
Water flux	M	L	H
pH tolerance	4–8	4–11	2–11
Temperature stability	Maximum 35 °C	Maximum 35 °C	Maximum 45 °C
Oxidant tolerance(e.g., free chlorine	H	L	L
Compaction tendency	H	H	L
Biodegradability	H	L	L
Cost	L	M	H

L = Low; M = Medium; H = High.

[a] Thin film composite type having polyamide surface layer.

1.4.1 Cellulosic membranes

The concept of RO was first demonstrated in the late 1950s with cellulose acetate membranes. These membranes are asymmetric, composed of a thin dense surface layer (0.2–0.5 μm) and a thick porous substructure. Solute rejection is accomplished by the thin dense layer and the porous substructure provides structural strength. Cellulose acetate membranes can be cast in sheets or as hollow fibers.

Cellulose acetate membranes are inexpensive and easy to manufacture but suffer from several limitations. Their asymmetric structure makes them susceptible to compaction under high operating pressures, especially at elevated temperatures. Compaction occurs when the thin dense layer of the membrane thickens by merging with the thicker porous substructure, leading to a reduction in product flux. Cellulose acetate membranes are susceptible to hydrolysis and can only be used over a limited pH range (low pH 3–5 and high pH 6–8, depending on the manufacturers). They also undergo degradation at temperatures above 35 °C. They are vulnerable to attack by bacteria. Cellulose acetate membranes have high water permeability but reject low-molecular-weight contaminants poorly. Cellulose triacetate membranes have been developed with improved salt rejection characteristics and reduced susceptibility to pH, high temperature, and microbial attack. However, cellulose triacetate membranes have lower water permeability than cellulose acetate membranes. Blends of cellulose triacetate and cellulose acetate have been developed to take advantage of the desirable characteristics of both membranes.

1.4.2 Aromatic polyamide membranes

Aromatic polyamide membranes were first developed by DuPont in a hollow fiber configuration. Like the cellulosic membranes, these membranes also have an asymmetric structure with a thin (0.1–1.0 μm) dense skin and a porous substructure. Polyamide membranes have better resistance to hydrolysis and biological attack than do cellulosic membranes. They can be operated over a pH range of 4–11, but extended use at the extremes of this range can cause irreversible membrane degradation. They can withstand higher temperatures than cellulosic membranes. However, like cellulosic membranes, they are subject to compaction at high pressures and temperatures. They have better salt rejection characteristics than cellulosic membranes as well as better rejection of water-soluble organics. A major drawback of polyamide membranes is that they are subject to degradation by oxidants, such as free chlorine.

1.4.3 Thin-film composites

As the name indicates, these membranes are made by forming a thin, dense, solute-rejecting surface film on top of a porous substructure. The materials of construction and the manufacturing processes for these two layers can be different and optimized for the best combination of high water flux and low solute permeability. The water flux and solute rejection characteristics are predominantly determined by the thin surface layer, whose thickness ranges from 0.01 to 0.1 μm.

Several types of thin-film composite (TFC) membranes have been developed, including aromatic polyamide, alkyl-aryl polyurea/polyamide, and polyfurane cyanurate. The supporting porous sublayer is usually made of polysulfone. Polyamide thin-film composites, like polyamide asymmetric membranes, are highly susceptible to degradation by oxidants, such as free chlorine. Consumers must be consistent in their maintenance of the TFC systems, particularly the carbon prefiltration element, which is present to remove free chlorine (and other oxidative organics) and prevent damage and premature destruction of the TFC membrane. Although the stability of these membranes to free chlorine has been improved by modifications of the polymer formulation and the processing technique, exposure to oxidants must be minimized.

1.5 Common causes of fouling

1.5.1 Concentration polarization

Concentration polarization is the term used to describe the accumulation of rejected solute at the surface of a membrane so that the solute concentration at the membrane wall is much higher than that of the bulk feed solution. As water passes through the membrane, the convective flow of solute to the membrane surface is much larger than the diffusion of the solute back to the bulk feed solution; as a result, the concentration of the solute at the membrane wall increases. This is shown schematically in Fig. 2.

Possible negative effects of concentration polarization include: (1) decreases in water flux due to increased osmotic pressure at the membrane wall; (2) increases in solute flux through the membrane because of increased concentration gradient across the membrane; (3) precipitation of the solute if the surface concentration exceeds its solubility limit, leading to scaling or particle fouling of the membrane and reduced water flux; (4) changes in membrane separation properties; and (5) enhanced fouling by particulate or colloidal materials in the feed, which block the membrane surface and reduce water flux. The extent of concentration polarization can be reduced

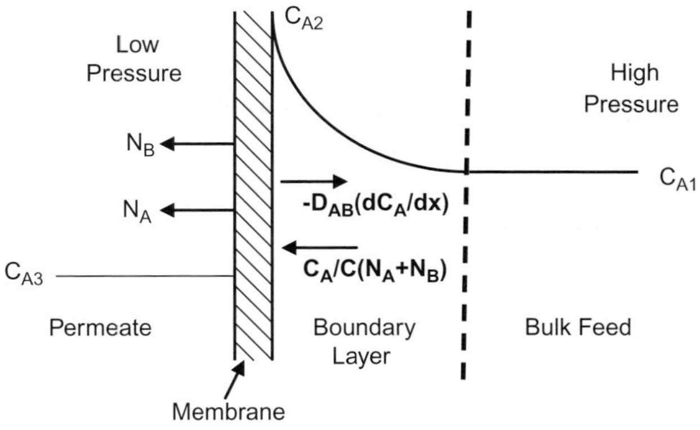

Figure 2 Concentration polarization.

by promoting good mixing of the bulk feed solution with the solution near the membrane wall. Mixing can be enhanced through membrane module optimization of turbulence promoters, spacer placement, hollow fiber diameter, and so on or by simply increasing axial velocity to promote turbulent flow.

At steady state, the flux of solute to the membrane, $(C_A/C)(N_A+N_B)$, the flux of solute through the membrane, N_A, and the solute back diffusion, $D_{AB}(dC_A/dx)$, are balanced:

$$N_A = \frac{C_A}{C}(N_A + N_B) - D_{AB}\frac{dC_A}{dx} \quad (4)$$

where C_A is the concentration of A at any x position (kmol/m^3), D_{AB} the diffusivity of solute in solvent (m^2/s), N_A, N_B the flux of solute and solvent through the membrane [kmol/(m^2 s)], and x the coordinate direction perpendicular to the membrane (m).

Eq. (4) is in the form of Fick's first law. Solving this equation with appropriate boundary conditions gives:

$$C_{A2} = C_{A3} + (C_{A1} - C_{A3})\exp\left(\frac{N_T}{kC}\right) \quad (5)$$

where N_T is the molar flux through the membrane [kmol/(m^2 s)], C_{A1}, C_{A2}, C_{A3} the feed, boundary layer, and permeate concentrations (kmol/m^3), and k the mass transfer coefficient (m/s).

1.5.2 Gel formation

Although pretreatment processes such as coagulation/flocculation and low-pressure membrane filtrations such as ultrafiltration (UF) and microfiltration (MF) have been used to remove larger particles and colloids, sometimes finer suspended particles and small colloidal matter still plague RO application [7,8]. Cake layer formation in RO process is influenced by various hydrodynamic and physiochemical parameters such as transmembrane pressure (TMP), cross-flow velocity (CFV), particle size, pH, and ionic strength [9]. According to that study, a significant permeate flux decline was observed as TMP increased and CFV decreased, which was attributed to the higher accumulative mass of particles on the membrane surface. The rate of flux decline decreased significantly with an increase in the ionic strength as well as particle size, while the flux decline rate was unaffected with the change in pH of the solution. The effect of hydrodynamic parameters on membrane performance deteriorated with an increase in the size of the solute particles. Various other studies on RO membrane filtration also determined that large particles do not contribute significantly to the membrane fouling and that fouling is usually controlled by small colloid particles [10,11]. Studies also showed that RO fouling was significantly caused by secondary water effluents than large suspended particles over size of 5.0 μm, leading to the conclusion that particles smaller that 0.45 μm, including true colloids and dissolved solids, contribute the most to the RO fouling [10].

Development of a colloidal cake layer on the membrane surface also significantly reduces the rejection of various kinds of solute particles such as salts and inert organics by restricting back diffusion of these solute particles from the membrane surface to bulk of the solution [12]. In this case, filtration of larger molecular weight (>100 g/gmol) inert organic solutes was controlled by steric exclusion, and colloid fouling had little influence on their rejection. This study also showed the inability of the RO membrane to prolonged rejection of hormones as adsorbed hormones diffuse through the membrane matrix to the permeate side, resulting in gradual loss of rejection. This hormone breakthrough is accelerated significantly in the presence of colloidal fouling.

Compressible gel foulants. Membrane fouling is often caused by the formation of gels at the film surface. The extent of compressible gel formation depends on the membrane/feed combination under consideration, as well as the operating regime of the membrane processing system, for example, transient versus steady-state operation, recovery, and so on.

Compressible gel formation often occurs when molecules with very low diffusion coefficient are present. The most common species of this include: humic substances, bioslimes, phenols, pesticides (and other industrial compounds), and macromolecules (proteins, carbohydrates, cheesy whey, greases, oils, surfactants, and tannins).

These compounds block the membrane pores as they leave the bulk solution. In the case of proteins and other macromolecules, diffusion rates are extremely low: once these molecules enter the boundary layer of the membrane, they tend to stay there. Charged species, such as surfactants, have additional fouling potential because they possess some hydrophobic properties. Since most membranes are partially charged, an oppositely charged surfactant is attracted to the membrane surface, changing the barrier layer so that the water flux of the membrane is greatly reduced. As for the mechanism of fouling by humic substances, there is no definitive work that relates the concentration of humic substances in the feed solution and the rate of fouling.

Incompressible gel foulants. Incompressible gel formation is related closely to precipitation at the membrane surface. Typically the most common compounds that form gels are $CaSO_4$, $CaCO_3$, $Mg(OH)_2$, $SrSO_4$, $Fe(OH)_3$, and $BaSO_4$. Most other metal hydroxides have some tendency for gel formation also. Amorphous silica is one of the worst foulants of any type because it is very difficult to remove from the membrane once gel formation occurs. The maximum nonfouling concentration of silica to be fed to a RO module is 10 parts per million (ppm) for the crystalline form, and 120 ppm for the amorphous state. Other sparingly soluble minerals have varying propensities for fouling membranes. Certain operating conditions, such as high recovery and pH, can greatly aggravate the fouling process.

1.5.3 Precipitation

Precipitation is similar to noncompressible gel formation in the sense that many of the same foulants are involved. In this case, however, fouling occurs through formation of scales that decrease membrane permeability. It is generally agreed that the causal mechanism is exceeding the solubility of the foulants in water. Solution becomes supersaturated with the dissolved salts at the membrane surface as the filtration proceeds and results in the precipitation of salts. The time required for the feedwater to reach supersaturation at the membrane surface is known as induction time and these times are short for filtration systems with high recovery. It follows that precipitation is worst in a high-recovery membrane system. Among the species commonly encountered in precipitants are Ca^{2+}, Mg^{2+}, CO_3^{2-}, SO_4^{2-}, silicas, and most forms of iron. Scaling is usually prevented by acidifying the feedwater to prevent the precipitation of carbonates and by the use of antiscalants to prevent the precipitation of sulfates of Ca^{2+}, Mg^{2+}, and Sr^{2+} [13]. Precipitation of iron on membranes is generally associated with Fe_2O_3 or $Fe(OH)_3$ formed by condensation reactions of hydrolyzable Fe^{3+} ions present in the feed solution. Reactions appear to occur in both bulk and concentration polarization regimes. Precipitation of this kind is generally not a problem when concentrations are approximately 4 mg/L and recovery is below 80%. When the recovery limit becomes 80–90% despite the presence of antiscalants, the supersaturation ratio becomes too large (e.g., up to 400% of calcium sulfate). Cases have also been reported in which antiscalants themselves have contributed to the fouling of the membranes. In some cases, reversing the flow before reaching the induction time of the system replaces the supersaturated brine at the exit with unsaturated feed and thus "zeroes the induction clock" [14].

1.5.4 Plugging

Plugging in membrane processes is generally caused by finely dispersed or suspended solids. TFC membranes basically contain two distinct distributions of pore sizes: tight "polymer network pores" and wider "polymer aggregate pores" [3,6]. The PSD curves and the effective number of pores in the membrane surface indicate plugging of the tight network pores and even their disappearance during fouling of RO membranes [6]. Fouling also results in the shift of the aggregate pores toward larger values, resulting in noticeable reduction in salt rejection. Iron, organic and inorganic colloids, and humic substances are among the major foulants that plug membranes. Humic acids are troublesome because they can react with chlorine and chloride to form haloforms, that is, $CHCl_3$. Some colloids, including silica, tend to be more troublesome than humic substances because they are very difficult to remove from the feed. Fouling can be worsened by allowing particles that are larger than one-fifth the size of the water channel into the RO module. The cut-off size for colloids is generally 40 Å, although the larger supracolloids can form in the size range of 0.45–2 μm. The behavior of the supracolloids is controlled by composition, while the more common foulants tend to be more heavily influenced by small size and high surface area. Most colloids tend to be hydrophobic.

1.5.5 Biological fouling

Biological fouling has two general mechanisms. One is blocking, much like that due to particulates and precipitants. The difference lies in the type of matter that blocks the membrane pores. In biofouling, the principal foulants are comprised of cells, cell wall debris, viruses, and the humic acids cited in the section on plugging. These form a flux-inhibiting layer at either the surface of a membrane, or within its pores. Representative thicknesses of biofouling layers are approximately 10–20 μm. A second mechanism of biofouling is direct microbial attack of the membrane, causing decomposition. Study of biological fouling, with radioactive glucose as a substrate, indicates the formation and growth of fouling microorganisms is possible under normal operating conditions. Biofouling is of especially great concern in the electronics industry, where there is critical need for ultrapure water. Although membranes are usually impermeable to bacteria, contamination of other parts of the RO process is undesirable.

1.5.6 Membrane degradation

The most common cause of membrane degradation is chemical interaction with organic and inorganic species. Chlorine is often used in pretreatment prior to RO when biofouling is of concern. Use in membrane process applications can change the polymeric structure of the cellulose acetate membrane. A high-chlorine membrane is generally more brittle than a low-chlorine membrane. Low ammonia concentrations lead to formation of free chlorine, which can quickly damage RO membranes, especially those made from aromatic polyamides. Free chlorine destroys polyamide membranes by dissolving the semipermeable layer, creating depressions and holes in the polymer.

1.6 Predicting fouling

Because of the complexity of the fouling phenomenon, predictive correlations of flux decline with time are important in process design. Most of the studies made in this area were conducted with the assumption that mathematical modeling is nearly impossible due to the nature of the fouling process. Only two models, the silt density index (SDI) and permanganate demand (PD) have become widely accepted.

Silt is composed of suspended particulates of all types that accumulate on the membrane surface. Sources of silt are usually organic colloids, iron corrosion products, precipitated iron hydroxide, algae, and fine particulate matter (Table 3). The SDI has been very popular for the prediction of fouling caused by colloidal and particulate matter in most of the RO and nanofiltration (NF) membrane processes. It is based on the time required to filter a volume of feed solution through a membrane filter at a fixed hydrostatic pressure or the rate at which a membrane becomes plugged at

Table 3 Silt density index values for selected common feedwaters

Feedwater type	Maximum SDI	Minimum SDI
Well waters	<3	<2
Surface waters	175	10

the feed pressure. In the accepted method, the filtration apparatus consists of a 0.45 μm cellulose acetate membrane filter and a feed pressure of 30 psig. Two variables are measured in the SDI procedure: the first is the length of time required to filter 500 mL of feed through a membrane filter. The filtration is continued for an additional 5–15 min, depending on feedwater quality. The time to filter a second 500 mL of the same solution is recorded.

The two times recorded are translated onto a plugging factor (PF):

$$PF = 100\left(1 - \frac{t_1}{t_2}\right) \tag{6}$$

or

$$SDI = \frac{100[1 - (t_1/t_2)]}{T} \tag{7}$$

where t_1 is the time required to filter the first 500 mL of feed solution, t_2 the time required to filter the second 500 mL of the same solution, and T the time of continuous filtration. The value of T is reduced, if $[1 - (t_1/t_2)]$ is greater than 0.75.

SDI could be a good parameter to evaluate the fouling potential of water with low turbidity and suspended solids. However, it does not identify every kind of fouling potential as in the case of dissolved iron [15]. There were also reports of RO membranes getting fouled by using seawater with an SDI of less than 1 that underwent chemical pretreatment [8].

The PD test was devised after the fouling dependence was studied on the concentration of aromatic hydroxyl species in membrane feed was studied. Permanganate lowers fouling potential through many reaction pathways. Among the reaction possibilities are electron abstraction, hydroxide ion removal, oxygen donation to organic compounds, and formation of manganous ion in acid solutions. Studies have shown that permanganate oxidizes many substances. The advantage of the PD test is versatility. Permanganate concentrations are readily detectable through spectrophotometric analysis, at $\lambda = 522$ nm, and the amount of manganate is measured with similar ease, at $\lambda = 426$ nm. Both ions obey Beer's law up to 100 mg/L.

The first step in the PD test is the introduction of a known amount of MnO_4^- into a sample of the feedwater or a dilution thereof. This mixture is allowed to react in a boiling water bath (100 °C) for 2 h. The mixture is cooled in water for 10 min, followed by glass fiber filtration to remove manganate from the sample. The absorbance of the mixture is measured. The procedure is repeated for the permeate stream.

1.7 Fouling control

1.7.1 Pretreatment

To control the RO membrane fouling, all the organic, colloidal, and biological matter needs to be removed from the feedwater to the RO system. Hence, a proper pretreatment process capable of producing a substantial reduction in the fouling potential of the membrane is very important to the functioning of a RO filtration process. Conventionally disinfection, fast mixing, coagulation, flocculation, sedimentation, and deep bed filter application are used together as a pretreatment approach after which SDI measurements are used as a criterion to evaluate the efficiency of the pretreatment. However, minute changes occurring in conventional treatment can have adverse effect on the RO filtration process. Factors such as chemical overdose, improper chemical use in pretreatment will result in irreversible fouling, resulting in the increase of TMP, power consumption and increased cleaning operations. Due to these limitations, many seawater reverse osmosis (SWRO) plants are using membrane filtrations such as MF and UF as pretreatment techniques. Application of UF as a pretreatment technique results in the usage of fewer chemicals, less floor space, and higher water recovery than conventional methods. It also eliminates the need for cartridge filter/sludge disposal with similar energy requirements as conventional pretreatment processes [16].

1.7.2 Membrane cleaning

An important technique for membrane regeneration is chemical cleaning of the fouled membrane. Membrane fouling can be prevented by feedwater pretreatment but it cannot be completely avoided. Fouling at the membrane surface is often controlled by employing periodic cleaning processes. The objective of these cleaning processes is to restore the original pore size distribution. This cannot always be achieved if cleaning cycles are delayed or in cases where fouling is severe. However, periodic cleaning of the desalination membranes is critical to the performance of wastewater treatment plants. Cost associated with the membrane cleaning constitutes 5–20% of the operating cost of the RO process [17]. Efficiency of cleaning generally depends on the type of cleaning agent and its concentration. It also depends on the understanding of specific interactions occurring between cleaning agents and membrane foulants. The basis for choosing a cleaning agent is the type of the foulant deposited on the membrane surface. Various kinds of agents such as acid, alkaline, surfactants such as sodium dodecyl sulfate (SDS), and commercial detergents are used for RO membrane cleaning depending on the foulant nature. In the case of fouling caused by calcium phosphate and calcium silicate, acids were the weakest cleaning agents. However, alkali had a moderate effect and the combination of chelating agents and surfactants with alkali (EDTA+SDS+NaOH) provided the best cleaning efficiency [18]. Similar results were reported in the cleaning of organic-fouled RO membranes. Fouling caused by organic feed solutions containing alginate and natural organic matter (NOM) can be effectively cleaned by either EDTA or SDS by optimizing chemical (concentration and pH) and physical (time, CFV, and temperature) conditions during cleaning [19]. Salt cleaning can be used for the cleaning of RO membranes fouled by gel forming hydrophilic organics [20]. It is assumed that cleaning agents diffuse into the deposited fouling cake layer on the membrane surface [18]. Diffusion rate depends on different factors including turbulence, shear factor, and concentration of the cleaning agent. Various studies indicated the occurrence of a reaction between cleaning agent and foulants present in the cake layer leads to the removal of fouling matter from the membrane surface [18,20]. These reactions could be hydrolysis, dissolution, or dispersion depending on the nature of foulants.

1.8 Applications

RO membranes reject dissolved inorganic solutes, larger organic solutes (molecular weight greater than 200), a portion of microbiological contaminants such as endotoxin, viruses and bacteria, and particles. Because of this broad spectrum of solute rejection, RO is an important process in a wide variety of water treatment processes.

1.8.1 Removal of inorganic contaminants

The removal of inorganic contaminants by RO membranes has been studied in great detail by many researchers using a variety of membrane types. Complex interactions occur in feedwaters containing mixtures of ionic species. Ionic contaminants are more readily rejected than neutral species. For most membrane types, polyvalent ions are rejected to a greater extent than monovalent ions. If the polyvalent ion is strongly hydrated, rejection is even higher. Because electrical neutrality must be preserved, ions diffuse across the membrane as a cation–anion pair. As a consequence, rejection of a particular ion depends on the rejection of its counterion.

Variations in pH influence the water flux and rejection characteristics of RO membranes exposed to a mixture of monovalent and polyvalent solutes. The effect of pH varies with membrane composition and ionic species. For example, fluoride rejection increases from 45% to 90% as pH increases from 5.5 to 7.2, whereas nitrate rejection decreases slightly as pH increases from 5.2 to 7.0. It is thought that the high pH causes chloramines to dissociate into ammonium and hypochlorite ions. The ammonium ions, which are poorly removed by activated carbon, interact with the polyamide membranes, causing their rejection characteristics to deteriorate. The decrease in rejection can generally be reversed by lowering the pH of the water supply.

1.8.2 Removal of organic contaminants

While RO membranes have a wide spectrum of removal of organic contaminants, the nature and extent of rejection will depend upon the nature of the organic solute. RO is effective in rejecting organic solutes with molecular weights greater than 200–300, such as fulvic acids, lignins, humic acids, and detergents. Low-molecular-weight, nonpolar, water-soluble solutes (for example, methanol, ethanol, and ethylene glycol) are poorly rejected.

Undissociated organic acids and amines are poorly rejected while their salts are readily rejected. For example, phenol is poorly rejected by RO membranes but when converted to its salt, rejections as high as 95–99% are observed. Also, rejection of acetic acid is only of the order of 50% but that of sodium acetate is as high as 90–95%.

The variable (and in some cases poor) removal characteristics of RO membranes dictate the use of auxiliary carbon filtration components either before or after (or both) the membrane. As in steam distillation, which has similar problems with organic materials, both RO and distillation require some type of organic removal mechanism such as replaceable carbon filters. The placement of carbon filters in RO systems depends on the type of membrane in use: for cellulose acetate or cellulose triacetate membranes, the carbon element is usually placed after the membrane and captive air tank, and just before the dispensing faucet. For thin-film membranes, a carbon filter is usually placed before and after the membrane. The carbon filter placed in front of the membrane is necessary since various types of organic materials and chlorine are detrimental to the structure of the thin-film membrane. Extra caution must be taken to regularly replace the carbon pre filter so as to ensure reasonable performance and lifetime for the TFC membrane.

1.8.3 Removal of microbiological contaminants

RO manufacturers claim to reduce levels of bacterial and viral contamination in the feedwater by factors of 10^3–10^5. However, in reality RO should not be relied upon to produce sterile water, much less water with reduced bacterial levels. Using the biological process, such as mitosis, bacteria and viruses may rapidly penetrate the RO membrane through defects and imperfections in the membrane as well as through tiny leaks in seals of the membrane module. In order to prevent colonization of the product waterside with bacteria and proliferation of these bacteria, regular disinfection procedures are necessary.

1.8.4 Selection criterion for reverse osmosis membrane

The ideal membrane for RO consists of a thin impermeable film. The transport properties of the material allow water to pass through with very little hindrance, while presenting a virtually impermeable barrier to salts. A membrane has to be extremely thin so that it has a large surface area for a maximum flow to occur across it, while simultaneously it is strong in order to withstand the feed driving pressure.

In choosing a RO membrane, a practical RO membrane for water applications should be permeable to water in preference to all other components of the feed stream. Also, the rate of permeation of water per unit surface area, or water flux, should be high enough to produce reasonable product volumes per unit time. Furthermore, the membrane must be durable physically, chemically, and biologically, and it must have a sufficiently extended life. Lifetimes for commercial RO membranes are in the order of 1–5 years, and waters with high total dissolved solids (TDS) reduce the life of membranes.

1.9 Transport modeling

Various mathematical models have been proposed to describe RO theory. These models can be divided into three groups: irreversible thermodynamic models, nonporous membrane models, and porous membrane models [21]. The nonporous or homogeneous models assume that transport across the membrane takes place through the interstitial spaces of the polymer chains by diffusion mechanism. Whereas in porous models it is assumed that apart from previously mentioned mode, the transportation also occurs via convection through pores.

2. MEMBRANE DISTILLATION

2.1 Introduction

Some membrane processes such as RO, UF, MF, and electrodialysis are used commercially while others such as thermo-osmosis, pervaporation, and membrane distillation (MD) are still in the research and developmental stages. Membrane distillation is an emerging technology for separations that are normally accomplished via conventional distillation or RO [22]. As applied to desalination, MD involves the transport of water vapor from a saline solution through the pores of a hydrophobic membrane.

Membranes used in MD are commonly hydrophobic polymers with pore sizes on the order of micrometers. The large contact angle of water with the hydrophobic membrane prevents liquid water from penetrating the pores, and water vapor is transported across the membrane in response to a change in partial pressure due to a thermal gradient. Permeate flux can be as high as 120 kg/(m^2 h) [for direct contact MD (DCMD)], comparable to RO membranes, and salt rejection is typically $>99\%$.

The efficiency of an MD process depends highly on membrane and module design, and thermal management. Heat and mass transport across the membrane must be optimized to obtain maximum permeate flux with minimal energy loss, and heat recovery from the permeate stream is essential for optimal operation [23]. Although energy consumption is quite high, the process is typically run at relatively low temperature ($\sim$70 °C) and thus can make use of waste heat or other relatively low-grade heat sources.

Potential advantages of MD are the ability to use low-grade and inexpensive heat sources, smaller plant footprint, and lower capital costs than conventional distillation processes. Membrane fouling is a problem, but is thought to be less severe than conventional RO [24]. Membrane degradation (loss of hydrophobicity) is also known to occur, but composite hydrophilic/hydrophobic membranes may overcome this problem [25].

2.2 Membrane distillation configurations

Among MD processes, variations exist as to the method by which the vapor is recovered once it has migrated through the membrane. MD systems can be configured in a number of ways, depending on the nature of the cold side of the membrane. But DCMD in which the membrane is in direct contact with the feed on one side and permeate on the other, and air gap

MD (AGMD) in which an air gap is interposed between the membrane and a cold condensation surface are perhaps most appropriate for desalination applications. Other configurations, such as vacuum MD (VMD) and sweeping gas MD (SGMD) methods, are typically used for stripping of volatile organics or dissolved gases. These alternatives are shown in Fig. 3.

2.2.1 Direct contact membrane distillation

DCMD, the oldest and most widely used process, has liquid phases in direct contact with both sides of the membrane. The vapor diffusion path is limited to the thickness of the membrane, thereby reducing mass and heat transfer resistances. Condensation within the pores is avoided by selecting appropriate temperature differences across the membrane.

2.2.2 Vacuum membrane distillation

In VMD, the vapor is withdrawn by applying a vacuum on the permeate side. The permeate-side pressure is lower than the saturation pressure of the

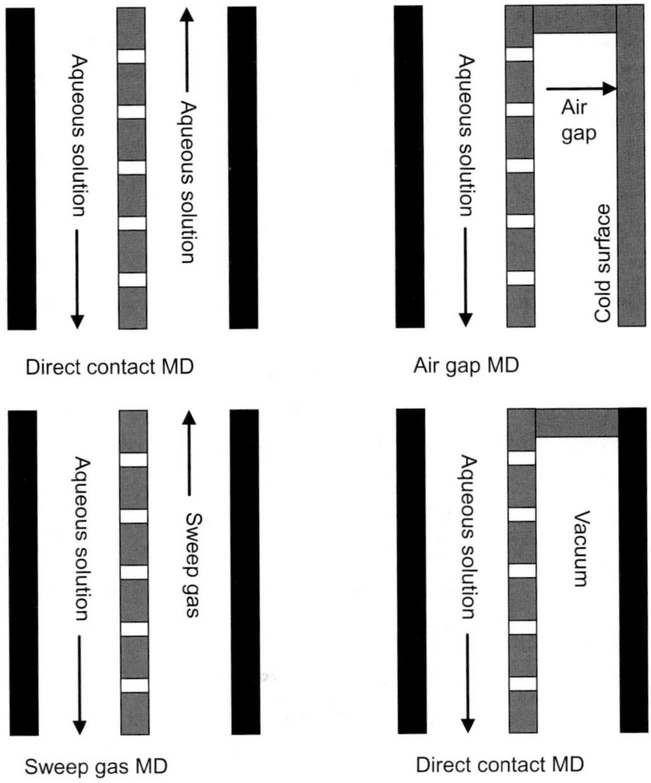

Figure 3 Schematic of various MD configurations. Adapted from Ref [26].

evaporating species and the condensation of the permeate takes place outside the module.

2.2.3 Air gap membrane distillation

AGMD has an additional air gap interposed between the membrane and the condensation surface. This gives rise to higher heat and mass transfer resistances. Although heat loss by conduction is reduced, the penalty is flux reduction. The use of an air gap configuration allows larger temperature differences to be applied across the membrane, which can compensate in part for the greater transfer resistances.

2.2.4 Sweeping gas membrane distillation

In SGMD, the permeating vapor is removed in an inert gas stream, which passes on the permeate side of the membrane. Condensation is done externally and involves large volumes of the sweep and vapor stream.

2.3 Mass transfer in membrane distillation processes

In MD process, a microporous hydrophobic membrane is in contact with an aqueous heated solution on one side (feed or retentate). The hydrophobic nature of the membrane prevents a mass transfer in liquid phase and creates a vapor–liquid interface at the pore entrance. Here, volatile compounds evaporate, diffuse, and/or convect across the membrane pores, and are condensed and/or removed on the opposite side (permeate or distillate) of the system [26].

2.3.1 General

A detailed description of MD principle is given by Banat (1994) [27] and a summary of that is provided here. Conventionally, MD is a thermally driven process in which a microporous hydrophobic membrane separates a warm solution from a cooler chamber, which contains either a liquid or a gas. As the process is nonisothermal, vapor molecules migrate through the membrane pores from the high to the low vapor pressure side, that is, from the warmer to the cooler compartment. It is also possible to lower the vapor pressure isothermally by using concentrated solutions or applying vacuum in the downstream side. The separation mechanism of MD is based on vapor–liquid equilibrium. The principle of MD is illustrated in Fig. 4.

The transport mechanism of MD involves four steps [27]:

1. Movement of the volatile components from the bulk of the feed stream to the membrane surface
2. Evaporation of the volatiles in the warm feed at the membrane surface

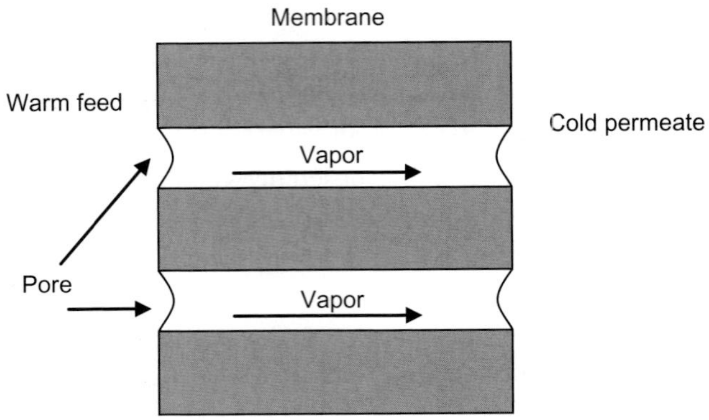

Figure 4 Membrane distillation concept. Adapted from Ref [27].

3. Migration of the vapor through the nonwetted pores
4. Condensation of the vapor at the cold permeate side either in a liquid or in a condenser

The nature of the driving force, in synergy with the hydrorepellent character of the membrane, allows the complete rejection of nonvolatile solutes such as macromolecules, colloidal species, ions, and so on. Typical feed temperatures vary in the range of 30–60 °C, thus permitting the efficient recycle of low-grade or waste heat streams, as well as the use of alternative energy sources (solar, wind, or geothermal). If compared to RO process, MD does not suffer from limitations arising from concentration polarization phenomenon and can be preferentially employed whenever elevated permeate recovery factors or high retentate concentrations are requested.

The main requirement of this process is that the membrane must not be wetted and only vapor is present in the pores. When used for desalination, saltwater is the hot feed solution. Pure water vapor passes through the membrane pores while the salts and other nonvolatiles remain on the warm side of the membrane. When volatile components are to be removed from water, the separation depends on their relative volatilities. As in ordinary distillation, the relative volatilities of compounds at the operating conditions determine their presence in the recondensed phase. When used for ethanol, acetone, or benzene removal, these compounds and some water vapor migrate through the membrane pores.

Vapor pressure gradient. A detailed discussion of MD transport can be found elsewhere [26], and only a summary is provided here. Heat and mass transport through membranes occur only if the overall system is not in thermodynamic equilibrium. In membrane processes, two homogeneous subsystems (with defined chemical potentials of μ_i' and μ_i'') are separated by a membrane. For small changes of the number of moles in the two phases (caused by the mass transfer across the membrane), the variation of the Gibbs free energy (G) is:

$$dG = \sum_i (\mu_i' - \mu_i'')\, dn_i' \quad (8)$$

Relation (8) expresses a general concept: the driving force for the mass transport of a component from one phase to the other is given by the difference in the chemical potential of the two phases caused by changes in temperature, pressure, and activity. In Eq. (8), n_i' is the mole of i-th component transferred and is related to transmembrane flux J_i by:

$$\frac{dn_i'}{dt} = AJ_i \quad (9)$$

where t indicates the time and A the membrane area. The hydrostatic pressure gradient across the membrane is negligible in MD, and the driving force of process is the partial pressure difference across the membrane, established by a temperature difference between the two contacting solutions, or by vacuum, air gap, or sweep gas in the permeate side. In the frequent case of nonideal mixtures, the vapor–liquid equilibrium is mathematically described in terms of partial pressure (p_i), vapor pressure of pure i (p_i°), and activity coefficient ζ_i, according to the thermodynamic relationship:

$$p_i = Py_i = p_i^{\circ} a_i = p_i^{\circ} \zeta_i x_i \quad (10)$$

In Eq. (10), P is the total pressure, a_i the activity, and x_i and y_i are the liquid and vapor mole fractions, respectively. The vapor pressure p° of a pure substance varies with temperature according to the Clausius–Clapeyron equation:

$$\frac{dp^{\circ}}{dT} = \frac{p^{\circ}\lambda}{RT^2} \quad (11)$$

where λ is the latent heat of vaporization ($\lambda = 9.7$ cal/mol for water at 100 °C) [26], R the gas constant, and T the absolute temperature. At the pore entrance, the curvature of the vapor–liquid interface is generally assumed to have a negligible effect on the equilibrium; however, possible influences on the vapor pressure value can be estimated by the

Kelvin equation:

$$p^{\mathrm{o}}_{\mathrm{convex}} = p^{\mathrm{o}} \exp\left[\frac{2\gamma_{\mathrm{L}}}{rcRT}\right] \tag{12}$$

where r is the curvature radius, γ_{L} the liquid surface tension, and c the liquid molar density. Activity coefficients ζ_i can be deduced by a large variety of equations aiming to evaluate the excess Gibbs function of mixtures; the most popular of them are reported in elsewhere [26,28].

2.4 Membranes and modules for membrane distillation

2.4.1 Membranes for membrane distillation

The major requirement of MD membranes is that they not be wetted by the process liquids. To avoid liquid invasion of the pores, highly hydrophobic membranes with an appropriate pore size are used. The liquid surface tension also affects wetting. Organic solutes present in an aqueous solution reduce the surface tension to the point where spontaneous membrane wetting may occur. At this point, the surface tension is called the critical surface tension at which MD is no longer possible. Franken et al. found that the maximum allowable concentration of organic material in water cannot be calculated but has to be determined experimentally [29].

The second major consideration in membrane selection for this process is pore size and porosity. High porosities are of special interest since the area available for evaporation is directly related to flux. However, high porosities are usually associated with large pore sizes, which are undesirable as they increase the risk of membrane wetting.

In MD, the membrane is not involved in the transport phenomena on the basis of its selective properties. Volatile compounds are transferred across the membrane according to vapor–liquid equilibrium principia, whereas the microporous polymeric material acts as physical barrier between two phases and sustains the interfaces where heat and matter are simultaneously exchanged. Since the hydrophobic character of the membrane represents a crucial requirement, membranes have to be made from polymers with a low value of surface energy. Polymers such as polypropylene (PP), polytetrafluoroethylene (PTFE), and polyvinylidene fluoride (PVDF) are commonly employed in the preparation of membranes for MD applications [30].

2.4.2 Modules for membrane distillation

Shell-and-tube modules. The tubular configuration resembles the shell-and-tube heat exchanger except that membranes replace the tubes through which a radial mass flux takes place. This is the most popular configuration for commercial UF, NF, and MF units. A tubular membrane module consists of membrane tubes placed into porous stainless steel fiber glass–reinforced plastic pipes. The diameter of tubular membranes typically varies between 1.0 and 2.5 cm, with a packing density, which is the ratio between the membrane area and the given packing volume of approximately 300 m^2/m^3. In MD operations, such kinds of modules are used for high-viscosity fluids; they also allow the achievement of high feed flow rates that reduce fouling tendency and polarization phenomena. In a capillary membrane module, a large number of membrane capillaries (inner diameter of 0.2–3 mm) are arranged in parallel as a bundle in a shell tube; packing density is in the order of 600–1200 m^2/m^3 [26]. The biggest disadvantage of shell-and-tube modules is that damaged membranes cannot be replaced as easily as in flat sheet apparatuses. Therefore, the module use is limited by the membrane life.

Hollow fiber membranes (diameter <0.5 mm) provide high surface area per unit volume, making the flux density greater than in other configurations. However, the softness of the membrane and the small fiber diameter make it susceptible to fouling and damage. The outer diameter typically ranges between 50 and 100 mm, and several thousands of fibers are installed in the vessel. This configuration has the highest packing density ($\sim$3000 m^2/m^3). The basic features of thermal MD modules [31] include that housing and membranes must be resistant to temperature and chemicals, and that capillaries have to be adequately potted free of cracks and with a good adhesion. It must be ensured there is uniform flow through capillaries avoiding dead corners or channel formation.

In MD, it is necessary to achieve high heat transfer coefficients in the tube and on the shell side. Liquid channeling makes this difficult to achieve on the shell side especially when large bundles of fibers are involved. Various modifications have been proposed to the standard shell-and-tube configuration in order to promote mixing so as to reduce fouling and promote turbulence at the membrane surface. Schnider et al. suggested that membrane twisting or braiding promotes mixing [31]. Modifications of module hydrodynamics to reduce fouling are also an active research area [32].

Flat sheet modules. Flat sheet membranes are used in cross-flow and stirred cells where the membrane needs to be easily removed for replacement and treatment. The packing density is considered low for these modules. Therefore, flat membranes are usually incorporated into plate-and-frame or spiral-wound modules. In plate-and-frame modules, the membranes, the porous support plates, and the spacers are stacked between two endplates and placed in an appropriate housing. In this configuration, the packing density is about 100–400 m^2/m^3, depending on the number of membrane used. Various cassettes stacked together, each consisting of injection molded plastic frames containing two membranes, intermediate feed channel for warm saltwater, and condensing walls have been used by Andersson et al. (1987) for desalination purposes [33].

In spiral-wound modules, the feed flow channel spacer, the membrane, and the porous support are enveloped and rolled around a perforated central collection tube. The feed solution moves in axial direction through the feed channel across the membrane surface. The permeate flows radially toward the central pipe. The packing density of this setup is about 300–1000 m^2/m^3, depending on the channel height. The use of spiral-wound MD modules at industrial level has been proposed for desalination [34,35].

2.5 Applications of membrane distillation

The possible applications of MD are limited by the wettability of the membrane, which is a function of the feed surface tension. Therefore, aqueous solutions containing inorganic solutes or low concentrations of volatile organic compounds can be treated while solutions with surface-active components cannot. In MD, the desired product can be either the permeate or the concentrate solution. Tests for the following applications have been reported in the literature:

1. Production of pure water from brackish or seawater [36,37]. The fact that salt is nonvolatile means that the permeate will be pure water.
2. Concentration of juice, grape juice, milk, sugar, and gelatin solutions [38,39].
3. Blood concentration [40].
4. Extraction of dilute ethanol from aqueous solutions [41] or from fermentation broths [42].

2.5.1 Membrane fouling

The phenomenon of flux decay in MD has been often observed in long-term operation and transmembrane flux declines as a consequence of fouling [43,44]. Membrane biofouling due to growth of microorganisms present in raw water often causes pore clogging, as well as an increase in pressure drop along the module. Chemical disinfection associated with UV treatment has been proposed to control biofouling; however, bacteria embedded on a polymeric surface show a significant resistance to biocides. The presence of fungi in the membrane pores was observed only on the feed side [26]. Flux is also reduced by scaling, occurring whenever the concentration of dissolved salts and minerals overcomes the solubility limit. Solid precipitation on the membrane surface can lead to both pore clogging and pore wetting. The presence of particulate and colloids in the processed liquid can also induce fouling because these particles are preferentially trapped at the membrane–liquid interface by interfacial tension forces [39]. In these cases, a prefiltering of the feed solution is generally sufficient to limit the flux decay effect [45]. Membrane fouling is a severe problem particularly in foods concentration; again, a preliminary UF treatment for heavy fouling feeds can be useful in order to remove larger particles that could increases the viscosity of the stream through MD units [38].

2.5.2 Cleaning

Cleaning procedures for hydrophobic membranes impose specific problems, particularly in processes involving fats and proteins that can adhere and foul the membrane, or alcohols and surfactants that can cause leakage. The most effective cleaner for membranes with a surface tension greater than 23 mN/m was 1% NaOH. The most effective cleaner for membranes with a surface tension less than 23 mN/m was P3 Ultrasil 56; water vapor flux was maintained and there was not salt leakage during repeated fouling/ cleaning runs [26].

2.6 Advantages of membrane distillation

The main advantages of membrane distillation over conventional distillation processes are: lower operating temperatures, compact modules, mist elimination, and the possibility of overcoming corrosion problems by using plastic equipment. This process can use available energy sources such as solar energy or waste energy in industrial processes. However, the process still has shortcomings such as membrane wetting and high membrane cost. Development of new membranes, membrane coatings, and increased competition among manufacturers should reduce the latter drawback.

REFERENCES

[1] J.E. Cadotte, R.J. Piersen, Thin-film composite membranes, in: A.F. Turbak (Ed.), Origin, Development and Recent Advances, ACS Symposium Series 153, Washington, DC, 1981, pp. 305–326.
[2] J.E. Tomaschke, Interfacially synthesized reverse osmosis membrane containing an amine salt and processes for preparing the same, US Patent 4,872,984, October 10, 1989.
[3] K. Chan, L. Tinghui, T. Matsuura, S. Sourirajan, Determination of interaction forces and average pore size and pore size distribution and their effects on fouling of ultrafiltration membranes, Ind. Eng. Chem. Prod. Res. Dev. 23 (1984) 124–133.

[4] N.A. Mohamed, Novel wholly aromatic polyamide-hydrazides: 6. Dependence of membrane reverse osmosis performance on processing parameters and polymer structural variations, Polymer 38 (1997) 4705–4713.
[5] M. Hirose, Y. Minamizaki, Y. Kamiyama, The relationship between polymer molecular structure of RO membrane skin layers and their RO performances, J. Membr. Sci. 123 (1997) 151–156.
[6] K. Kosutic, B. Kunst, RO and NF membrane fouling and cleaning and pore size distribution variations, Desalination 150 (2002) 113–120.
[7] E.M.V. Hoek, M. Elimelech, Cake-enhanced concentration polarization: a new fouling mechanism for salt-rejecting membranes, Environ. Sci. Technol. 37(24) (2003) 5581–5588.
[8] H. Winters, Twenty years experience in sea water reverse osmosis and how chemicals in pretreatment affect fouling of membranes, Desalination 110 (1997) 93–96.
[9] C. Park, Y.H. Lee, S. Lee, S. Hong, Effect of cake layer on colloidal fouling in reverse osmosis membranes, Desalination 220 (2008) 335–344.
[10] M. Sugahara, T. Kitao, Y. Terashima, S. Iwai, Transport properties of thermally conditioned sludge liquors in reverse osmosis, Int. Chem. Eng. 19 (1979) 322–328.
[11] B.A. Winfield, A study of the factors affecting the rate of fouling of reverse osmosis membranes treating secondary sewage effluents, Water Res. 13 (1979) 565–569.
[12] H.Y. Ng, M. Elimelech, Influence of colloidal fouling on rejection of trace organic contaminants by reverse osmosis, J. Membr. Sci. 244 (2004) 215–226.
[13] M. Wilf, J. Ricklis, RO desalting of brackish water oversaturated with CaSo4, Desalination 47 (1983) 209–219.
[14] J. Gilron, M. Waisman, N. Daltrophe, N. Pomerantz, M. Milman, I. Ladizhansky, E. Korin, Prevention of precipitation fouling in NF/RO by reverse flow operation, Desalination 199 (2006) 29–30.
[15] A. Mosset, V. Bonnelye, M. Petry, M.A. Sanz, The sensitivity of SDI analysis: from RO feed water to raw water, Desalination 222 (2008) 17–23.
[16] M. Galloway, J. Mahony, Ultrafiltration for seawater reverse osmosis pretreatment, Membr. Technol. 2004 (2004) 5–8.
[17] A.G. Fane, Proceedings of Symposium on Characterization of Polymers with Surfaces, Lappeenranta, Finland, 1997, p. 51.
[18] S.S. Madaeni, T. Mohamamdi, M.K. Moghadam, Chemical cleaning of reverse osmosis membranes, Desalination 134 (2001) 77–82.
[19] W.S. Ang, S. Lee, M. Elimelech, Chemical and physical aspects of cleaning of organic-fouled reverse osmosis membranes, J. Membr. Sci. 272 (2006) 198–210.
[20] S. Lee, M. Elimelech, Salt cleaning of organic-fouled reverse osmosis membranes, Water Res. 41 (2007) 1134–1142.
[21] M.E. Williams, A Review of Reverse Osmosis Theory, EET Corporation and William Engineering Services Company, Inc., Harriman, TN, 2003.
[22] A.M. Alklaibi, N. Lior, Membrane-distillation desalination: status and potential, Desalination 171 (2004) 111–131.
[23] H. Kurokawa, T. Sawa, Heat recovery characteristics of membrane distillation, Heat Transfer Jpn. Res. 25 (1996) 135–150.
[24] S. Srisuricharan, R. Jiraratananon, A.G. Fane, Humic acid fouling in the membrane distillation process, Desalination 174 (2005) 63–72.
[25] P. Peng, A.G. Fane, X. Li, Desalination by membrane distillation adopting a hydrophilic membrane, Desalination 173 (2005) 45–54.
[26] E. Curcio, E. Drioli, Membrane distillation and related operations – a review, Sep. Purif. Rev. 34 (2005) 35–86.
[27] F.A. Banat, Membrane Distillation for Desalination and Removal of Volatile Organic Compounds from Water, Doctoral Dissertation, McGill University, Montreal, Canada, 1994.

[28] R.H. Perry, Perry's Chemical Engineering Handbook, 6th ed., McGraw-Hill, Singapore, 1984.
[29] A. Franken, I. Nolten, M. Mulders, D. Bargeman, Wetting criteria for the applicability of membrane distillation, J. Membr. Sci. 33 (1987) 315–328.
[30] M. Mulder, Basic Principles of Membrane Technology, Kluwer Academic Publishers, Dordrecht, 1996.
[31] K. Schnider, W. Holz, R. Wollbeck, S. Ripperger, Membranes and modules for transmembrane distillation, J. Membr. Sci. 39 (1988) 25–42.
[32] M.E. Brewster, K.Y. Chung, G. Belfort, Dean vortices with wall flux in a curved channel membrane system. 1. A new approach to membrane module design, J. Membr. Sci. 81 (1993) 127–137.
[33] S.I. Andersson, N. Kjellander, B. Rodesjo, Design and field tests of a membrane distillation system for seawater desalination process, Desalination 61 (1987) 237–243.
[34] D.W. Gore, Gore-Tex membrane distillation, in: Proceedings of 10th Annual Conference of Water Supply Improvement Association, Honolulu, Hawaii, July 25–29, 1982.
[35] J. Koschikowski, M. Wieghaus, M. Rommel, Solar thermal-driven desalination plants based on membrane distillation, Desalination 156 (2003) 295–304.
[36] S.I. Andersson, N. Kjellander, B. Rodesjo, Design and field tests of a new membrane distillation desalination process, Desalination 56 (1985) 345–354.
[37] S. Kubota, K. Ohta, I. Hayaro, M. Hirai, K. Kikuchi, M. Murayama, Experiments on seawater desalination by membrane distillation, Desalination 69 (1988) 19–26.
[38] V. Calabro, B. Jiao, E. Drioli, Theoretical and experimental study on membrane distillation in the concentration of orange juice, Ind. Eng. Res. 33 (1994) 1803–1808.
[39] S. Kimura, S. Nakao, Transport phenomena in membrane distillation, J. Membr. Sci. 33 (1987) 285–298.
[40] K. Sakai, T. Koyano, T. Muroi, Effects of temperature and concentration polarization on water vapor permeability for blood in membrane distillation, Chem. Eng. J. 38 (1988) B33–B39.
[41] C. Gostoli, G.C. Sarti, Separation of liquid mixtures by membrane distillation, J. Membr. Sci. 41 (1989) 211–224.
[42] H. Udriot, S. Ampuero, I. Marrison, U. Stockar, Extractive fermentation of ethanol using membrane distillation, Biotechnol. Lett. II (1984) 509–514.
[43] B.A. Okoniewski, Remove VOCs from wastewater by air stripping, Chem. Eng. Prog. 88 (1992) 89–93.
[44] B.R. Bali, M.D. Edward, Air stripping of VOCs from ground water: process design considerations, J. Environ. Prog. 11 (1992) 39–48.
[45] E. Drioli, Y. Wu, V. Calabro, Membrane distillation in the treatment of aqueous solutions, J. Membr. Sci. 33 (1987) 277–284.

CHAPTER 5

Salinity Gradient Energy

Kitty Nijmeijer[1,*] *and* **Sybrand Metz**[2]
[1]Membrane Technology Group, Institute of Mechanics, Processes and Control Twente (IMPACT), University of Twente, P.O. Box 217, 7500 AE Enschede, The Netherlands
[2]Wetsus, Centre of Excellence for Sustainable Water Technology, P.O. Box 1113, 8900 CC Leeuwarden, The Netherlands

Contents

1. INTRODUCTION

Since the Kyoto protocol and the report of the Intergovernmental Panel on Climate Change (IPCC) [1] on carbon dioxide capture and storage, there is an emerging need to reduce the emission of CO_2 to the atmosphere. In principle, three possible routes can be envisioned focusing on (1) the reduction of the energy consumption, (2) the efficient use of energy sources (if desired combined with capture and storage of CO_2), and (3) the use of alternative energy sources with reduced or no CO_2 emission. In addition to

*Corresponding author.
E-mail address: d.c.nijmeijer@utwente.nl

Sustainability Science and Engineering, Volume 2 ISSN 1871-2711, DOI 10.1016/S1871-2711(09)00205-0

that, the limited amount of fossil fuels forces the developments in the direction of alternative energy sources.

Salinity gradient energy has a huge potential as alternative and sustainable energy source. It uses the Gibbs energy of mixing of two salt solutions with different concentrations to generate electrical energy. It is a nonpolluting (no emissions of CO_2, SO_2, or NO_x), sustainable technology to generate energy by mixing water streams with different salinity. Salinity gradient power is available worldwide, everywhere where salt solutions of different salinity mix, for example, where fresh river water flows into the sea, or where industrial brine is discharged. The estimated global energy potential from estuaries alone is estimated to be 2.6 TW [2], which is approximately 20% of the worldwide energy demand [3] and more than the global electricity consumption (2.0 TW).

Pressure-retarded osmosis (PRO) and reverse electrodialysis (RED) are the most frequently studied processes to extract the potential energy available from the mixing of freshwater and saltwater, although some other membrane-based processes are proposed as well. In PRO, two solutions of different salinities are brought into contact by a semipermeable membrane that only allows the transport of the solvent (water) and retains the solute (dissolved salts). In RED, a number of anion and cation exchange membranes (CEM) are stacked together in an alternating pattern between an anode and a cathode and allow the selective transport of salt ions only.

Although the potential of salinity gradient power was already recognized in the 1950s [4], until now, commercialization and industrial use are still limited; however, several initiatives are currently employed for pilot plant construction and upscaling of both technologies (see later in this chapter).

This chapter describes the process of salinity gradient energy and its potential. It first gives a thermodynamic overview of the theoretical amount of energy available from the mixing of a diluted and a concentrated salt solution, which in principle is independent of the used technology (PRO or RED). After that, the chapter continues with a section especially dedicated to PRO and a section only focusing on RED. Both sections describe the principle and theory of the specific technology and are followed by a detailed description of the literature and membranes used for PRO or RED. It also mentions the challenges for membrane development in this respect. After that, both sections address process design considerations. The last part of both sections is dedicated to the upscaling and commercialization of both processes. The chapter finally ends with some concluding remarks.

2. THEORETICAL POTENTIAL OF SALINITY GRADIENT ENERGY

The driving force for transport of a component in salinity gradient power is a gradient in Gibbs energy or a potential difference between the two salt solutions. The Gibbs energy of a system reflects that part of the energy of the system that is available for work. The total amount of energy available from mixing 1 m^3 of a concentrated and 1 m^3 of a diluted salt solution can be determined from the chemical potential difference of the system after mixing, subtracted by the chemical potential of the system before mixing (Fig. 1):

$$\Delta G_{\text{mix}} = G_{\text{b}} - (G_{\text{c}} + G_{\text{d}}) \tag{1}$$

where ΔG_{mix} is the change in Gibbs energy (J/mol) and G_{b}, G_{c}, and G_{d} are the Gibbs energies of the brackish, the concentrated, and the diluted solution, respectively (J/mol). The Gibbs energy of an ideal solution is equal to

$$G = \sum \mu_i n_i \tag{2}$$

where G is the Gibbs energy of the system (J/mol), μ_i the chemical potential of component i in the solution (J/mol), and n_i the number of moles of component i in the solution.

The chemical potential of a component i (μ_i) in an ideal solution can be written as (e.g., [5])

$$\mu_i = \mu_i^0 + \bar{V}_i \Delta p + RT \ln x_i + |z_i| F \Delta\varphi \tag{3}$$

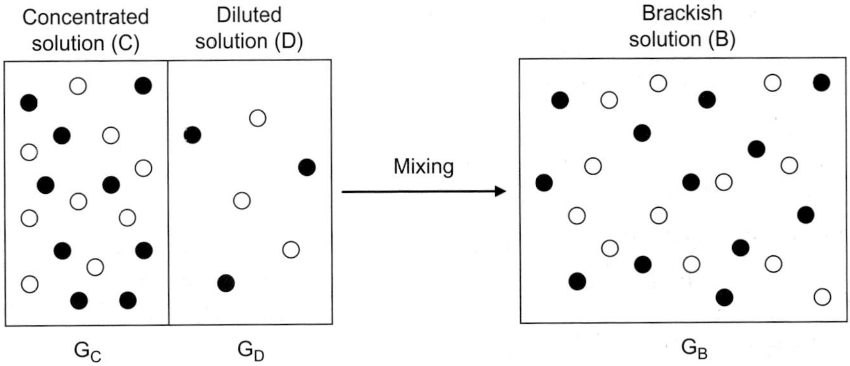

Figure 1 The mixing of a concentrated and a diluted solution to a brackish solution.

where μ_i^0 is the molar free energy under standard conditions (J/mol), Δp the pressure change compared to atmospheric conditions (Pa), $\bar{V}_i$ the molar or specific volume of component i (m^3/mol), R the universal gas constant [8.314 J/(mol K)], T the absolute temperature (K), x_i the mol fraction of component i, z the valence of an ion (eq/mol), F the Faraday constant (96,485 C/eq), and $\Delta\varphi$ the electrical potential difference (V). Since there is no pressure change or charge transport, when the concentrated and the diluted solution are mixed, Eq. (3) reduces to

$$\mu_i = \mu_i^0 + RT \ln x_i \tag{4}$$

When Eq. (4) is substituted in Eqs. (2) and (1), the standard chemical potential (μ_i^0) is eliminated and the final equation describes the Gibbs energy of mixing of a concentrated and a diluted salt solution:

$$\begin{aligned}\Delta G_{\mathrm{mix}} &= \sum_i [G_{i,\mathrm{b}} - (G_{i,\mathrm{c}} + G_{i,\mathrm{d}})] \\ &= \sum_i [\{(n_{i,\mathrm{c}} + n_{i,\mathrm{d}})RT \ln x_{i,\mathrm{b}}\} \\ &\quad - (n_{i,\mathrm{c}}RT \ln x_{i,\mathrm{c}} + n_{i,\mathrm{d}}RT \ln x_{i,\mathrm{d}})]\end{aligned} \tag{5}$$

And when n is replaced by cV, this changes into

$$\begin{aligned}\Delta G_{\mathrm{mix}} = \sum_i [&c_{i,\mathrm{c}} V_{\mathrm{c}} RT \ \ln(x_{i,\mathrm{c}}) + c_{i,\mathrm{d}} V_{\mathrm{d}} RT \ \ln(x_{i,\mathrm{d}}) \\ &- c_{i,\mathrm{b}} V_{\mathrm{b}} RT \ \ln(x_{i,\mathrm{b}})]\end{aligned} \tag{6}$$

Because the mixing of two solutions is a spontaneous process, the Gibbs energy of mixing is negative: energy is released when two solutions are mixed. With Eq. (6), the theoretical available amount of energy available from the mixing of two salt solutions can be calculated and thus the theoretical potential of salinity gradient energy can be evaluated. This theoretically available amount of energy for an extensive range of sodium chloride concentrations is presented in Fig. 2 [3]. (*Note*: Because the figure shows the theoretical amount of energy *available* from the mixing of a diluted and a concentrated solution, the energy has a positive sign.)

Fig. 2 shows an extensive range of salt concentrations and the theoretically available amount of energy that can be obtained from the mixing of the two solutions. Values as high as ~17 MJ can be obtained, depending on the concentration difference between the two solutions. Of course, this amount of energy strongly depends on the difference in

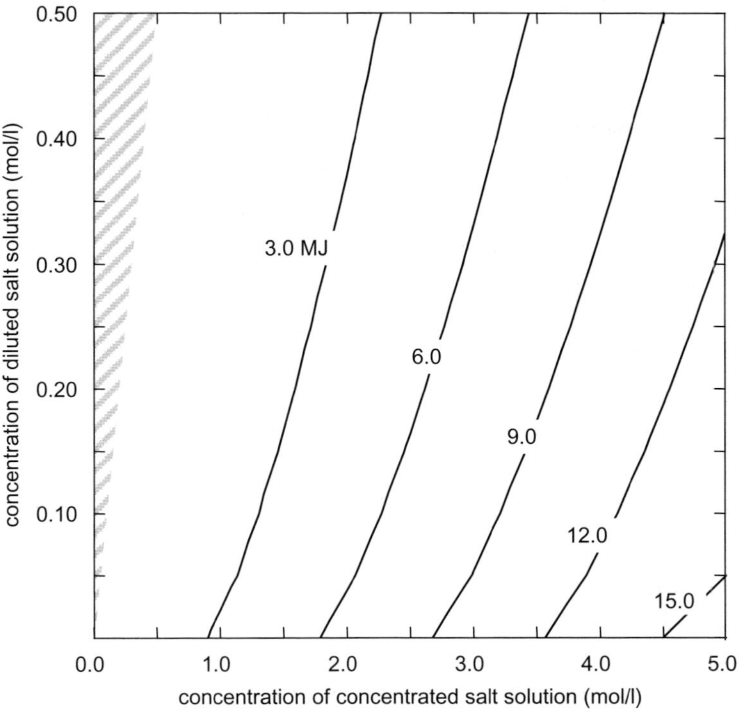

Figure 2 Theoretical available amount of energy (MJ) from mixing 1 m^3 of a diluted and 1 m^3 of a concentrated sodium chloride solution ($T = 293$ K). The shaded area is not taken into account because in this area the salt concentration of the concentrated solution is lower than that of the diluted solution [3].

concentration (or chemical potential) between the concentrated and the diluted salt solution. The higher this difference, the more energy can be extracted from the system. For example, the theoretically available amount of energy from mixing 1 m^3 seawater (comparable to 0.5 mol/L NaCl) and 1 m^3 river water (comparable to 0.01 mol/L NaCl) both at a temperature of 293 K is 1.7 MJ, whereas the theoretically available amount of energy from mixing 1 m^3 brine (5 mol/L NaCl) and 1 m^3 river water (0.01 mol/L NaCl) at 293 K is more than 16.9 MJ. When mixed with a large surplus of seawater, 2.5 MJ is theoretically available from 1 m^3 of river water (Table 1) [6]. Table 1 shows the amount of Gibbs energy theoretically available from the mixing of different volumes of a diluted and a concentrated salt solution [6]. This table clearly shows that when the amount of saltwater limits the process, the use of an excess of river water can be very beneficial

Table 1 Gibbs energy theoretically available from mixing different volumes of NaCl solutions at 298 K [6]

V_d (m^3)	V_c (m^3)	ΔG_{mix} (MJ)
∞	1	∞
10	1	6.1
2	1	2.8
1	1	1.76
1.26	0.74	1.87
1	2	2.06
1	10	2.43
1	?	2.55

V_d is the volume of the diluted solution (0.01 M NaCl), V_c the volume of the concentrated solution (0.5 M NaCl), and ΔG_{mix} the change in Gibbs energy.

(compare an available amount of energy of 6.1 MJ at $V_d = 10\,m^3$ and $V_c = 1\,m^3$ to an available amount of energy of only 1.76 MJ when both V_d and V_c are $1\,m^3$).

Although the above-presented equations provide a good first approximation for the theoretical amount of energy obtainable from salinity gradient energy, the calculations assume that the feed solutions consist of pure sodium chloride and behave ideal (no distinction between concentrations and activities). In practice, however, sea and river water are much more complex solutions and do not behave ideal, which makes the calculations much more complex. The numbers presented here represent the theoretical, maximum amount of energy available from the mixing of fresh and saltwater. Of course, in practice, it will not be possible to harvest this total theoretically available amount of energy, due to for example, mass transfer limitations, pressure drop, nonideal behavior, and so on. In addition, depending on the location and situation, there can be also several other limitations to use the total resources available, which are related to, for example, environmental impact, shipping, recreation, and tourism. But even if only part of the available energy can be recovered, the potential of salinity gradient energy remains huge.

3. PRESSURE-RETARDED OSMOSIS

3.1 Introduction

In PRO, the free energy of mixing from two solutions with different salinity is converted into energy by water transport through a semipermeable membrane from the diluted solution to the concentrated solution. In literature, PRO is defined as [7] "The process of osmosis through a semipermeable membrane at a hydrostatic pressure difference between 0 and the osmotic pressure difference of the separated solutions, which generates a water flux against the hydrostatic pressure difference." This transport of water causes an increase of the pressure of the concentrated solution, which can be converted into electrical energy. Much of the pioneering work is published by Loeb [7–12] and Metha [9,13–15] and coworkers. They introduced the concept and published the first experimental results. Loeb et al. not only focused on the mixing of sea and river water, but also explored the possibility of applying PRO for the mixing of high saline solutions like Dead Sea water with seawater. Lee et al. [16] developed a theoretical model, which describes the PRO performance of a membrane based on osmosis and reverse osmosis (RO) measurements. They concluded that "membranes with significantly improved performance will be needed if PRO is to become an economically feasible method for power generation using seawater–freshwater as the salinity gradient resource. However, the economics of a brine/freshwater system appear competitive with conventional power generation technologies."

Due to ineffective membranes, which are the key component of PRO, not much effort took place to establish this technology. Since 1997, Statkraft, a Norwegian energy company, is engaged in the development of PRO [17]. They expect that the cost of osmotic power production will be in line with the cost of offshore wind generation and below wave and tidal power generation in 2010–2015. Statkraft is targeting for a membrane with high water flux and a low salt permeability. The performance of such a membrane should be close to 5 W of power generated per square meter of membrane area (W/m^2). Statkraft together with GKSS improved the performance of an asymmetric cellulose membrane from 0.6 to 1.3 W/m^2 [18]. Over 50 support materials have been tested for the development of a thin film composite membrane, which resulted in a power increase from 0.1 to 3.5 W/m^2 [18]. It is believed that the performance of these membranes could be improved even further. McCutcheon and Elimelech demonstrated the importance of a hydrophilic support for osmotically driven processes [19]. A hydrophobic support layer significantly hinders the water flux not only through increased internal concentration polarization phenomena but also through disrupting the pathway by poor wetting of the structure. Recently, Thorsen and Holt [20] presented improved experimental results for PRO. They determined the PRO performance of a commercial cellulose acetate membrane without fabric reinforcement and obtained a power density of 1.6 W/m^2 for a solution containing 23.5 g/L NaCl. They predict for this membrane a power density of 2.1 W/m^2 for a 28.0 g/L NaCl solution. Even a much higher power density was obtained for a thin film composite membrane developed by GKSS [18], 2.7 W/m^2 for a 30.6 g/L NaCl [20]. This is a large increment in the performance of PRO.

Water transport from a less concentrated solution toward a more concentrated solution also occurs in forward osmosis (FO). There are many similarities in the desired membrane properties of FO and PRO and both aim for a highly selective membrane with a high water flux. In FO, an artificial salt solution (also called draw solution) is used to create a driving force for water transport through a semipermeable membrane. FO is used to recover the water from a saline or polluted water source. An advantage of this process is that no pressure is applied in FO equipment. The draw solution is either consumed (glucose/fructose draw solution), discarded, or regenerated and separated from the product water. Most of the recent work on FO is published by Elimelech and coworkers [21–26]. FO can also be used with the effluent of a freshwater sewage treatment plant and seawater (draw solution) [27]. In this case, the seawater is diluted with the water of the sewage treatment plant and is easier to treat due to its lower osmotic pressure.

3.2 Principle

In PRO, solutions of different salt concentrations are brought into contact through a membrane that allows the transport of water and retains the passage of salts. The chemical potential difference between both solutions creates a driving force [Eq. (5)]; water diffuses from the less concentrated solution through the membrane toward the concentrated salt solution, equalizing the chemical potential difference. If the concentrated solution is pressurized, then the transport of water would be lowered until the pressure reaches the osmotic pressure between both solutions. If the saltwater compartment would be further pressurized, RO would occur: transport of water from the concentrated salt solution toward the freshwater side. This process is schematically depicted in Fig. 3.

In PRO, the pressure on the concentrated salt solution is partly retarding the water flow through the membrane. This higher pressure allows the generation of electricity by a turbine.

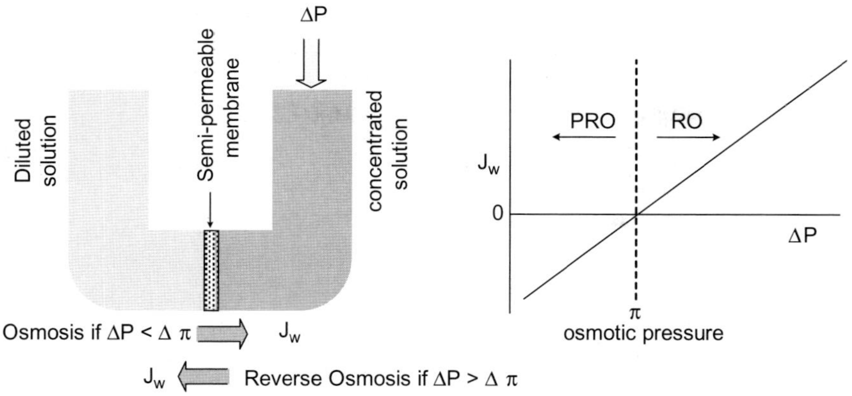

Figure 3 Transport of water (J_w) through a semipermeable membrane. If the hydrostatic pressure (*P*) on the concentrated solution is larger than the osmotic pressure (Π) transport of water from the concentrated solution to the diluted solution occurs (reverse osmosis). If the hydrostatic pressure is lower than the osmotic pressure, water is transported toward the concentrated solution. The increase in pressure at the concentrated solution can be converted into energy.

The osmotic pressure of a solution can be calculated by the van't Hoff equation:

$$\Pi = c_j RT \tag{7}$$

where c_j is the concentration of the solute (mol/m^3), R the universal gas constant [J/(mol K)], and T the absolute temperature (K). When the solute dissociates, the osmotic pressure increases proportionally. The osmotic pressure of a 35 g/L NaCl solution (comparable to seawater) is equal to 29.7 bar.

3.2.1 Concentration polarization

Fig. 4 shows the transport of water and salt through a semipermeable membrane in PRO. The membrane consists of a thin selective top layer supported by a porous support. The selective layer faces the high-pressure side in order to prevent release of the selective top layer from its support due to the pressure differences.

Because the membrane is not 100% selective, some salt will also be transported from the saltwater side to the freshwater side. In RO the transport of salt and water are in the same direction.

In PRO water is transported from the low-pressure freshwater side to the high-pressure saltwater side due to the osmotic pressure difference.

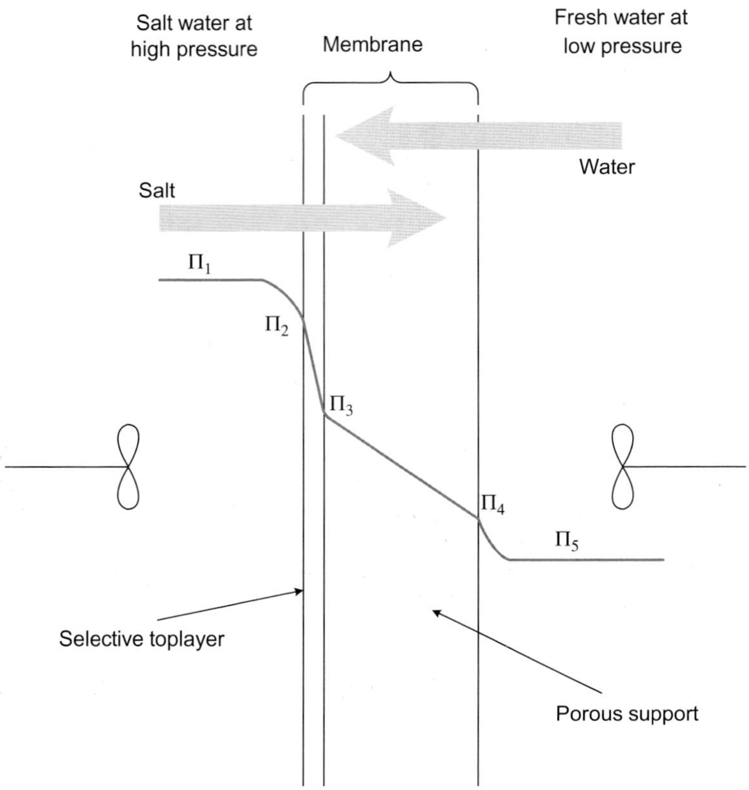

Figure 4 Schematic representations of the osmotic profiles of a PRO membrane. Π_1 is the osmotic pressure of the bulk of the concentrated salt solution, Π_2 the osmotic pressure at the dense top layer of the membrane, Π_3 the osmotic pressure inside the membrane between the dense top-layer and the porous support, Π_4 the osmotic pressure at the surface of the membrane in the diluted solution, Π_5 the osmotic pressure in the bulk of the diluted solution.

However, salt is transported in the opposite direction due to its concentration difference. The salt transport is limited by several resistances: external concentration polarization due to stagnant layers caused by reduced mixing on the membrane surface at the saltwater side and the freshwater side; internal concentration polarization due to resistance against salt transport in the thin selective top layer of the membrane and in the porous support layer.

Intensified mixing due to high cross-flow rates at the membrane surface can lower external concentration polarization. Internal concentration polarization arises from the resistance against mass transfer that salt experiences from the dense top layer and the stagnant boundary layer in the porous support. This porous support creates a stagnant zone through which salt can only be transported by diffusion. These resistances against salt transport lower the effective osmotic pressure (Π_2–Π_3) over the selective top layer of the membrane. A good PRO membrane has a thin high selective top layer with a high resistance for salt transport and a very open (preferably thin) support layer. Loeb et al. showed that the support has a large contribution to the overall transport resistance and that the removal of the nonwoven/woven support from the membrane caused a higher osmotic water flux through the membrane [11]. A capillary membrane (with a thin porous layer) might be very beneficial for PRO applications [16].

When external concentration polarization is neglected, then the water transport can be described as follows:

$$J_{\mathrm{w}} = A(\Delta\Pi_{\mathrm{eff}} - \Delta P) = A(\Pi_2 - \Pi_3 - \Delta P) \tag{8}$$

where J_{w} is the water flux through the membrane [$\mathrm{m}^3/(\mathrm{m}^2\,\mathrm{day})$], A is a specific membrane transport parameter [$\mathrm{m}^3/(\mathrm{m}^2\,\mathrm{day\,bar})$], Π the osmotic pressure (bar), and ΔP the pressure difference between the fresh and saltwater solution (bar).

The osmotic pressure Π_3 is not known but can be calculated from the salt leakage through the membrane. This can be described as follows:

$$-J_{\mathrm{s}} = B(C_2 - C_3) \tag{9}$$

where J_{s} is the salt flux through the membrane [$\mathrm{mol}/(\mathrm{m}^2\,\mathrm{day})$], B the salt permeability constant (m/day), and C the concentration ($\mathrm{mol}/\mathrm{m}^3$).

This salt flux is negative since its transport is in the opposite direction of the water flow. In the porous support, the diffusion of salt is counteracted by the flow of water. The salt flux through the support can be written as follows:

$$-J_{\mathrm{s}} = D_{\mathrm{s}}\varepsilon\frac{\mathrm{d}C(x)}{\mathrm{d}x} - J_{\mathrm{w}}C(x) \tag{10}$$

where D_{s} is the diffusion coefficient of the salt in the membrane substrate (m^2/s), ε the porosity of the membrane substrate (–), and x the thickness of the porous support.

Lee et al. [16] solved this problem resulting in

$$J_{\mathrm{w}} = A\left(\pi_2\frac{1 - (C_4/C_2)\exp(J_{\mathrm{w}}K)}{1 + (B/J_{\mathrm{w}})(\exp(J_{\mathrm{w}}K) - 1)} - \Delta P\right) \tag{11}$$

For the special case of $C_4 = 0$ with only water on the freshwater side of the membrane reduces Eq. (11) to

$$J_w = A\left(\frac{\pi_2}{1 + (B/J_w)(\exp(J_w K) - 1)} - \Delta P\right) \tag{12}$$

Both equations can be solved numerically. In these equations A and B can be obtained from RO experiment, concentrations are known and J_w is measured during osmosis experiments allowing for the determination of K. K (s/m) refers to the solute diffusion in the porous support structure and is given as

$$K = \frac{t\tau}{D_s \varepsilon} \tag{13}$$

where t is the thickness of the membrane (m), τ the tortuosity of the pores in the support (–), D_s the solute diffusion coefficient (m^2/s), and ε the porosity of the membrane (–).

Eq. (11) can be further simplified by assuming that $\pi_2/\pi_4 = C_2/C_4$ [6] resulting in

$$K = \frac{1}{J_w}\left(\ln \frac{B + A\pi_2 - J_w}{B + A\pi_4}\right) \tag{14}$$

This equation is valid when the concentrated salt solution is facing the active dense layer. If the concentrated salt solution is facing the porous support, which is sometimes applied in FO [24], then the following equation is valid:

$$K = \frac{1}{J_w}\left(\ln \frac{B + A\pi_4}{B + J_w + A\pi_2}\right) \tag{15}$$

3.3 Membranes for pressure-retarded osmosis

PRO is the most studied membrane technology exploiting a salinity gradient. However, the amount of experimental data is scarce and difficult to compare with each other. Metha and Loeb and recently Thorsen and Holt [20] are the only authors who published experimental determined power densities for PRO at real conditions. Some PRO values are determined from osmosis experiments without applying a hydrostatic pressure. Such a pressure can have a significant effect on the water flux, since a very open support structure allows a high water flux, but is also susceptible to compaction. Table 2 shows the available experimental results obtained from literature.

The reported power densities in Table 2 are derived from the reported flux and feed pressure. All the experimental results were obtained by Loeb and Metha from 1976 to 1982 and mainly for concentrated salt streams [8,9,14,15] and from Thorsen and Holt (2009) [20]. Loeb and Metha determined only one value (0.21 W/m^2) obtained for a feed concentration of 30 g/L NaCl, which represents seawater. Furthermore, it should be noted that the feed pressure is not chosen optimal [half of the osmotic pressure as indicated in Eq. (18)] necessary for a maximal power density. However, Thorsen and Holt [20] systematically varied the feed pressure and obtained for a NaCl concentration representing seawater the highest reported power densities 1.6 W/m^2 for a cellulose acetate membrane and 2.7 W/m^2 for a thin film composite membrane. Higher power densities (1.76–5 W/m^2) can be obtained for concentrated brine streams. However, the performance of the fibers deteriorates when exposed to high salt concentrations, probably caused by a change in the porous substructure. Based on Table 2 no clear conclusion can be drawn for the optimal membrane properties for PRO, mostly because the experimental conditions are difficult to compare with each other because of different process condition: feed pressure, salt concentration, and flow rates (external concentration polarization).

Proper PRO experiments are difficult to perform: feed pressure chosen should be optimal, feed flow rate should be high in order to minimize concentration polarization, and the amount of permeated water should be determined accurately. Therefore, Lee et al. [16] generated a theoretical model in order to predict the PRO performance from osmosis experiment [parameters A and B in Eqs. (8) and (9)] and from direct osmosis (K derived from the osmotic flow) as input parameters for their model. The results of direct osmosis experiments might be too optimistic, since these experiments do not take compaction phenomena into account [16]. The results of Lee et al. [16] and Loeb et al. [11] are shown in Table 3.

All the membrane parameters from Table 3 are derived from experiments with NaCl solutions, except for the Toray CA-3000 values, which were determined with $MgCl_2$. These latter values might be too optimistic since the parameters B and K depend on the type of salt. The asymmetric membranes show a higher performance compared to the composite membranes. These composite membranes have a lower projected power density due to their denser support layer, resulting in a high K value [16]. The retention and the resistance of the support layer play a crucial role in

Table 2 Measured PRO performance for various membranes at different feed concentrations and pressures for the mixing with freshwater (0 g/L NaCl)

Membrane	Type	Concentration feed		Water flux $[m^3/(m^2\,d)]$	P_{feed} (bar)	Power (W/m^2)	Reference
		NaCl (g/L)	Π_{feed} (bar)				
Du Pont permasep B-10	Asymmetric polyamide fiber	96	81	0.056	41	2.62	[14]
		108	91	0.042	51	2.46	
		143	122	0.070	51	4.10[a]	
		191	162	0.084	51	4.90[a]	
		191	162	0.081	51	4.77[a]	
		96	81	0.038	41	1.78[a]	
		96	81	0.070	41	3.26[b]	
		96	81	0.081	20	1.90[b]	
		143	122	0.045	61	3.17[b]	

FRL thin film composite	Composite polysulfone with furan skin fiber	72	61	0.032	3	0.11	[9]
		24	20	0.006	3	0.02	
		119	101	0.070	19	1.56	
UOP CA/SW	Asymmetric cellulose acetate flat sheet spiral wound	47	40	0.035	21	0.85	[15]
		94	80	0.090	21	2.14	
		139	118	0.081	24	2.26	
		51	44	0.037	21	0.89	
Du Pont permasep B-10	Asymmetric polyamide fiber	239	203	0.100	30	3.52	[8]
		119	101	0.050	30	1.76	
		30	25	0.012	15	0.21	
Osmonics SS10	Asymmetric cellulose acetate	23.5	16	0.390	8	1.60	[20]
Thin Film Composite	Composite	30.6	21.5	0.229	12	2.70	[20]

[a] Change in performance when $\Pi_f - P_f > 50$ bar.
[b] Different module.

Table 3 Calculated power densities (W/m^2) for various membranes derived from osmosis experiments (*A* and *B*) and osmosis experiments (K)

Membrane	Type	Membrane parameters						Projected	Reference
		A [m^3/(m^2 day bar)]	B (m/day)	K (day/m)	$\Delta P_o/\Delta\pi$	$\Delta P_o/2$ (bar)	J_w [m^3/(m^2 day)]	W_{max} (W/m^2)	
CA-80	Asymmetric	0.0088	0.173	0.75	0.88	12.60	0.109	1.59	[16]
CA-70	Asymmetric	0.0289	7.517	0.44	0.23	3.31	0.092	0.35	[16]
BM-05	Asymmetric	0.0035	0.020	21.99	0.70	9.90	0.028	0.32	[16]
PBIL	Asymmetric	0.0057	0.028	7.99	0.82	11.66	0.060	0.81	[16]
PA-300	Composite	0.0096	0.015	65.97	0.51	7.23	0.020	0.17	[16]
NS-101	Composite	0.0105	0.038	335.65	0.07	1.03	0.003	0.00	[16]
BM-1-C	Composite	0.0072	0.053	46.30	0.29	4.14	0.015	0.07	[16]
Toray CA-3000[a]	Asymmetric	0.0324	0.018	104.00	0.34	4.88	0.014	0.08[a]	[11]
Toray CA-3000[a]	Asymmetric without support fabric	0.0324	0.018	17.00	0.76	10.84	0.115	1.45[a]	[11]

[a] Values determined with $MgCl_2$ too optimistic for NaCl.

the performance of a PRO membrane. As can be seen from Table 3 membranes with a low water permeability (*A*) but with a high selectivity toward salts (low *B*) and an open support structure (low *K*) can exhibit a high power density (CA-80 membrane). A membrane with a high selective top layer for salts and an open porous support allow a higher optimal pressure stemming from a larger osmotic pressure difference over the selective layer of the membrane.

Support layers have a tremendous effect on the performance of a PRO membrane. The support layer should be as thin and open as possible without a support fabric. Asymmetric fibers are very attractive for PRO since they possess a thin porous support layer and no support fabric. However, these open structures should also be able to withstand the hydrostatic pressure during PRO operation and should not compact. Compaction of the open CA-80 fiber is observed by Lee et al. [16] and is not taken into account in their model. Real PRO experiments would most likely show a lower power density as compared to the values reported in Table 3.

Summarizing the optimal PRO membrane should have the following characteristics:

- A high water permeability [high *A*, Eq. (8)]
- Low salt permeability [low *B*, Eq. (9)]
- Low resistance in the porous support, very open or no support fabric [11] [low *K*, Eq. (13)]
- Hydrophilic porous support [19]
- Resistant against compaction
- Minimal external concentration polarization (high flow rates)

These parameters might be conflicting with each other and an optimal membrane is optimized with respect to these variables.

3.4 Process design

The basic process of an osmotic power plant is shown in Fig. 5.

A pressure exchanger is used in order to maintain a high pressure at the feed side of the membrane. The pressure of the brackish water leaving the system is used to pressurize the incoming seawater. The flows of the brackish water leaving the system and the seawater entering the system should be equal. The amount of water permeating through the membrane is used to generate electricity via a turbine. It should be noted that the pressure exchange should work very efficiently at low pressures (14.8 bar half the osmotic pressure) in order not to lose too much energy. Statkraft a Norwegian energy company found a very elegant solution for this problem by placing the osmotic power plant below sea level at such a depth that the

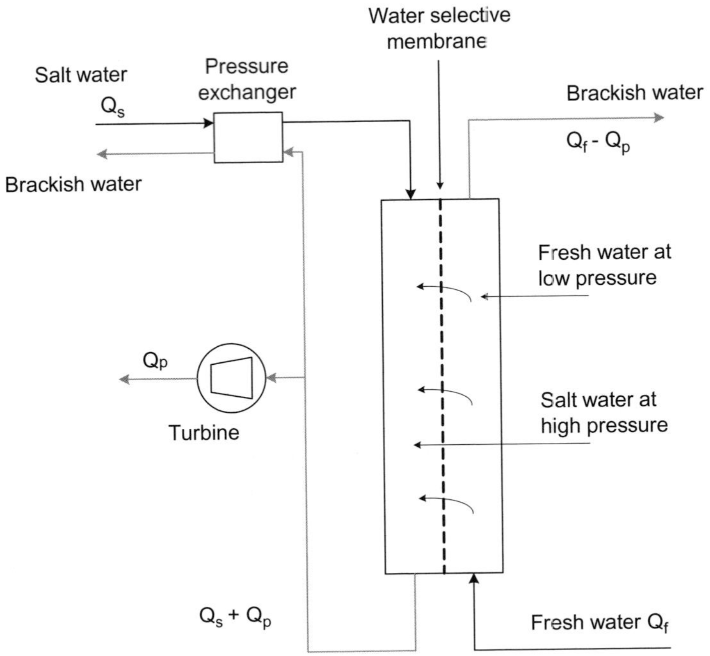

Figure 5 Basic principle of PRO water transport from freshwater toward a pressurized saltwater solution. *Q* is the flow of water (m^3/s), the subscripts f, s, and p stand for freshwater, seawater, and permeated water, respectively.

hydrostatic pressure equals the optimal operation pressure [Eq. (18)] for PRO [17,18].

The water selective membrane consists of a dense selective top layer (which is permeable for water and not for salts) and a porous support backing this thin layer. The selective top layer is facing the pressurized seawater.

The flux of water occurs due to an osmotic pressure difference between the freshwater and the saltwater and is retarded by the higher pressure of the saltwater. This can be described by the following relationship:

$$J_{H_2O} = A(\Delta\Pi - \Delta p) \tag{16}$$

where J_{H_2O} is the water flux in $m^3/(m^2\,s)$; A a specific membrane constant, $\Delta\Pi$ the osmotic pressure, and Δp the pressure difference between both solutions.

The amount of energy produced per square meter of membrane (E) is obtained by multiplying the water flux with the hydrostatic pressure difference:

$$E = J_{H_2O}\Delta P \tag{17}$$

The maximal power density is obtained when $dE/dP = 0$, resulting in

$$\Delta P_{max} = \frac{1}{2}\Delta\Pi \tag{18}$$

which is the optimal pressure of the concentrated salt solution at the feed side of the membrane giving the highest power output. For PRO on river water and seawater this would mean an optimal pressure at the seawater side of 14.8 bar.

The maximal obtainable amount of power can be derived by substitution of ΔP_{max} in Eq. (17) resulting in

$$E_{max} = \frac{1}{4}A\Delta\Pi^2 \tag{19}$$

This equation clearly shows the effect of the osmotic pressure and membrane properties (A) on the energy production of PRO.

3.5 Pilot testing and upscaling[1]

Statkraft, an energy utility owned by the Norwegian government, is today the largest generator of renewable energy in Europe. With generation capacity within hydropower, wind power, gas power, and soon also solar power, the company has a large portfolio of environmental energy solutions. But it is clear to the company that to maintain a leading position within renewable energy it is necessary to focus on innovation with a clear ambition to deliver the energy solutions of the future. With over 100 years of tradition in hydropower, working with pressurized water and sustainable project development, it was natural that Statkraft turned the focus toward PRO already in 1997.

When Statkraft started working on PRO, the first efforts were to understand the realistic potential of this concept provided the technology would be made available. Calculations and surveys of the availability of the resources – freshwater and seawater – were executed, and the result showed that a significant amount of clean, renewable energy could be produced by

[1]The information given in this section is provided by and property of the company Statkraft AS, Norway and used with permission. The authors would like to acknowledge Statkraft AS for the contribution.

osmotic power. In addition, there are specific characteristics of this technology that give it its unique character not only among the new sources of renewable energy that are currently under development, such as tidal and wave power, but also in regard to more established technologies such as solar and wind. Since the generation of power is based on the availability of freshwater and seawater, resources that usually will be available all year round, osmotic power has the characteristics of a base load source of renewable energy. This is very different from the other technologies that are dependent on the present weather conditions, hence require back up supplies from other sources.

Another interesting characteristic is that after making a survey of the rivers running into the ocean worldwide, one found that these sites usually also have either settlements or industry, and mostly both. This means that the consumer of the electricity produced by osmotic power will be just next door to the power plant. When reflecting on the situation that most new sources of renewable energy, such as wind, wave, and so on, usually have huge challenges and significant investments related to the connection of the power generation device to the grid, this adds another advantage for the generation of osmotic power as a contribution to the total energy mix.

Based on the previously stated advantages of this new technology, Statkraft made a detailed study of the state of the technology necessary to exploit these possibilities of PRO. Although there is a lot of resemblance with components used in other processes, it became clear that the membranes are one of the crucial components, where significant improvements both in efficiency and in cost were necessary. The membranes produced at that time were not in a position to produce power at a competitive level. Hence extensive efforts to design a membrane suitable for PRO were made, and this was done together with partners with long experience in membrane development both in the United States and in Europe. As described earlier, this is not an easy task, but today the best results produced by Statkraft are in the range of $3\,W/m^2$. This result shows the significant progress made in membrane development and it made Statkraft to decide to expand their efforts to the maximum towards a full-scale osmotic power system.

In the fall of 2007, Statkraft decided, due to the promising improvement in the critical components, such as membrane and pressure recovery devices, the time had come for a full-scale proof of the concept for a complete PRO system. A plant with a sufficiently large amount of membrane area is currently built to transfer the salinity gradient into work and also further into electricity. At the same time, the interface for, and integration of, all the components in the system can be studied together in operation, not only as individual parts of a system.

After a little more than a year of development and construction, the world's first prototype plant has been put into operation in spring 2009 in the southeast of Norway (Fig. 6). The location is within the facility of a pulp factory in operation, which simplifies the approval process and at the same time gives good access to the existing infrastructure. In addition, the location has good access to seawater from the ocean and freshwater from a nearby lake.

The prototype plant is designed as a typical plant placed at sea level. Freshwater is taken from a river close to its outlet. Seawater is fed into the plant by underground pipes, and the brackish water is led to the natural brackish water zone.

The main objectives of the prototype PRO plant are twofold. First, it confirms that the designed system can produce power on a reliable 24-h/day production. Second, the plant will be used for further testing of the technology achieved from parallel research activities to substantially increase the efficiency. These activities will mainly be focused on membrane modules, pressure exchanger equipment, and power generation (turbine and generator). In addition, there will be a focus on further development of

Figure 6 Prototype PRO plant at the east coast of Norway.

control systems, water pretreatment equipment, as well as infrastructure with regard to water inlets and outlets.

The plant is equipped with 2000 m^2 of specially designed PRO membranes. A miniature hydropower turbine and devices for recovery of hydraulic pressure are installed. Although the design capacity is in the range of 10 kW, the expectations for the capacity in the first phase are somewhat less. The membranes have room for improvement, and there are high expectations for optimizations for the full system as such.

Since this is the first plant built for PRO operation, several precautions have been taken to make sure that possible pollution in the water does not destroy the membranes (Fig. 7). For the seawater regular pressure screens are used, and for the freshwater from the lake the pretreatment is similar to that being used for drinking water. The ambition is that the freshwater can be treated similar to the seawater. This will however be based on the operational experiences.

After the start-up, operation, and further testing the experience gained will be based on both operational changes as well as changes to the system and replacement of parts. This is in order to increase the efficiency and optimize the power generation. In a longer perspective, this would be used as a basis to develop a power plant with an installed capacity between 1 and 2 MW, bringing the technology one step further toward commercialization.

The prototype plant put into operation during 2009 is also intended as a meeting place for parties from both government and industry with ambitions in osmotic power. With the increasing focus on the environmental

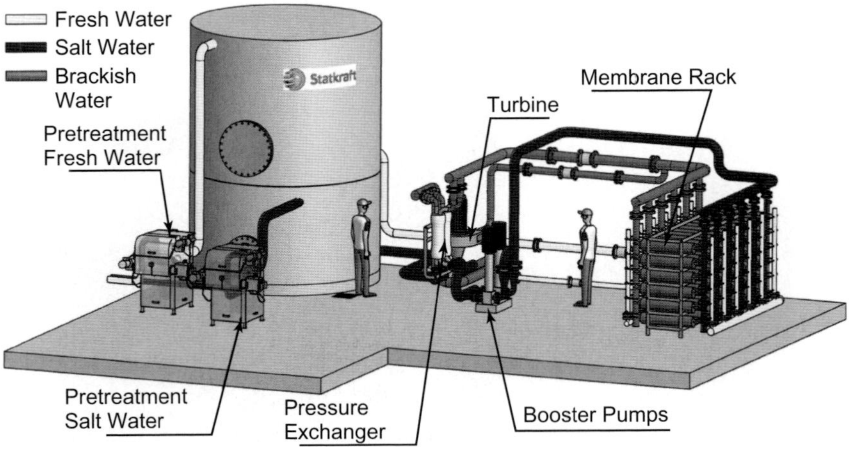

Figure 7 Prototype PRO plant illustration.

challenges and the need for more renewable energy, this can give a significant contribution to increase the momentum in development of new clean technologies.

Statkraft has specified that in order to be competitive to other new, renewable sources of energy, a power output of 5 W/m^2 for flat-sheet membranes is required, whereas due to the higher packing densities obtainable, a target in the range of 3 W/m^2 should be sufficient for hollow fiber membranes. This is based on the water flux trough the membrane, in relation to the salt retention that creates the driving force. The estimated costs of producing one MW based on a number of detailed investment analyses are that osmotic power will be able to produce electricity at a cost level of Euro 50–100 MW^{-1}, which is in a similar range as other renewable technologies such as wind power, wave and tidal power, and power based on biomass.

These calculations are based on existing hydropower knowledge, general RO desalination engineering information, and with a membrane target as a prerequisite. The capital costs of installed capacity are high compared to other renewable energy sources. However, each MW installed is very productive, with an average operation time above 8000 h a year. This should generate approximately twice the energy supplied (GWh) per installed MW per year compared to a wind mill.

To achieve competitiveness, given the large volumes of membranes, the membrane pricing is important. For an average 25 MW plant, it is calculated that 5 million m^2 of membrane area is required, meaning that the industry would see a demand of PRO membranes exceeding the current RO membrane market.

There are still significant improvements and verifications of the technology required before osmotic power can be represented among the currently commercial renewable energy technologies. But it is not only the technology itself that need to be put into place to exploit this huge potential; in the following sections some of the major topics to be assessed will be discussed, and it is known from the history of developing both wind power and solar power that these topics are not trivial. For wind and solar power, the technology was long past the proof of concept, but it took still several decades before these were able to gain a significant market share.

A new technology such as osmotic power can only be developed to a certain level by researchers and especially dedicated companies such as Statkraft. But to exploit the full potential of such a technology, one will be dependent on external factors as well, such as that several organizations have sufficient demand for this specific power technology. When several companies and governments around the world commit themselves to utilize the technology, whether it is solar, wind, or osmotic power, this provides strong signals to the supplier industry and the competition for developing and supplying the best solution will go up to full speed.

4. REVERSE ELECTRODIALYSIS

4.1 Introduction

In RED, the energy of mixing two solutions with different salinity is extracted through the transport of ions (this in contrast to PRO, where the transport of water accounts for the generation of power). Pattle was the first researcher who proved the principle of RED [4]. With his pioneering work, he was the first one to be able to generate power from the mixing of fresh and saltwater through the selective transport of ions. In the 1970s, Weinstein and Leitz [28] investigated the effect of the composition of the salt solutions on the power output. The main conclusion of their work was that large-scale application of RED could become feasible, but only if major improvements regarding the manufacturing of ion exchange membranes and careful optimization of the operating conditions are possible. In the early 1980s, Lacey [29] prepared a comprehensive review on RED and concluded that to make RED economically viable minimization of the internal stack resistance and maximization of the net power output from the cell are a prerequisite for success. The main conclusion of Lacey's work is that membranes for RED should have a low electrical resistance and a high selectivity combined with a long service life time, acceptable strength, dimensional stability, and low costs. In the early 1980s, Audinos [30] compared two different types of electrodialysis membranes for their applicability in RED (one pair of homogeneous and one pair of heterogeneous membranes) and investigated the effect of the type of salt solution (NaCl vs. $ZnSO_4$). The maximum power output obtained was 400 MW/m^2. In the mid 1980s, Jagur-Grodzinski [31] investigated the effect of hydrodynamics, that is, different salt solution streams and membrane spacer modifications, as a method to increase the power output. Although promising, the number of papers on RED in the 1990s and in the beginning of the 21st century was very limited. However, since a few years RED has been recognized again as a potentially attractive technology for the production of sustainable energy and as such it has regained the interest of many researchers [32–39], industrial partners, and the public. In this part, we first discuss the principle and the fundamentals of RED. It continues with a closer look at the membranes used for RED.

After that we focus on the different elements in RED, which are subsequently the membranes and the feed compartments including spacers. Although electrodes and electrode reactions are also major elements in a RED stack, the available literature and research on this topic is very limited, and therefore this topic will not be addressed here. This part is followed by a paragraph that focuses on process and stack design. This chapter finally ends with a description of the state-of-the-art and current status of RED and also gives a glimpse on pilot testing and upscaling.

4.2 Principle

In RED, a concentrated salt solution and a less concentrated salt solution are brought into contact through an alternating series of anion exchange membranes (AEM) and CEM (Fig. 8).

The concentrated and the diluted salt solution are separated by an alternating series of AEMs and CEMs. The AEM contain fixed positive charges and only allow the selective transport of anions toward the anode, whereas the CEM contain fixed negative charges and only allow the selective passage of cations towards the cathode. Both the concentrated

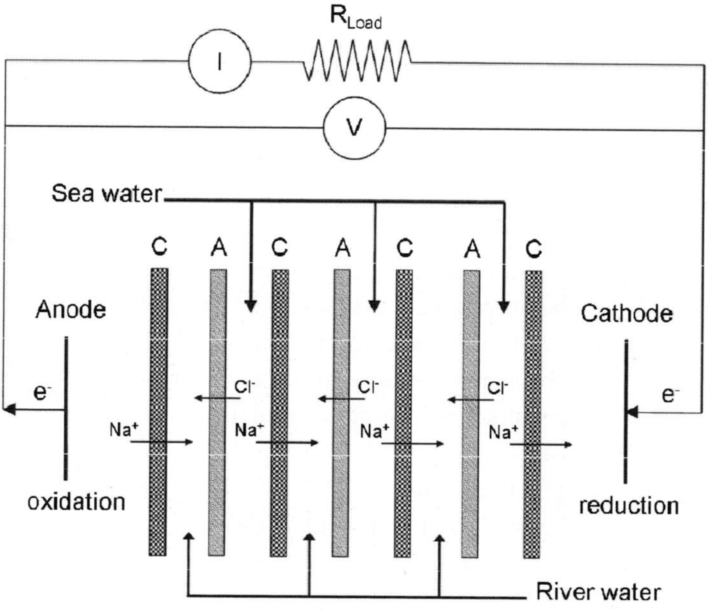

Figure 8 Principle of RED. *A* is an anion exchange membrane, *C* a cation exchange membrane, *V* the potential difference over the applied external load (V), *I* the electrical current (A) and R_{Load} the resistance of the external load (Ω). A redox couple is used at the electrodes to mitigate the transfer of electrons from anode to cathode [34].

and the diluted feed compartment contain a spacer to control the hydrodynamics. The electrons released at the anode are subsequently transported through an external circuit containing an external load, to the cathode. In the internal circuit in the stack, charge is carried by ions, while in the external circuit, electrons carry the charge. The ionic current is converted into electrical current by redox reactions that occur at the electrodes at the outer side of the stack. The redox couple is used to mitigate the transfer of electrons. A typical redox couple currently often used for RED is a solution of $K_4Fe(CN)_6$ and $K_3Fe(CN)_6$ (potassium iron(II) hexacyanoferrate and potassium iron(III) hexacyanoferrate) in a bulk solution of NaCl. At the cathode, the iron(III) complex is reduced and the iron(II) complex is reoxidized at the anode:

$$Fe(CN)_6^{3-} + e^- \leftrightarrow Fe(CN)_6^{4-} \quad E_0 = 0.36\ \mathrm{V}$$

The solution is recirculated between both electrode compartments to maintain the original iron(III)/iron(II) ratio.

The chemical potential difference between the two salt solutions with different concentrations is the driving force for this process and generates a voltage difference over each pair of membranes. The theoretical value of this potential difference over the membrane for an aqueous monovalent electrolyte (e.g., NaCl) can be calculated using the Nerst equation:

$$\Delta V_{\mathrm{theo}} = \frac{RT}{zF} \ln\left(\frac{a_c}{a_d}\right) \tag{20}$$

where ΔV_{theo} is the theoretical membrane potential for a 100% selective membrane (V), R the universal gas constant [8.314 J/(mol K)], T the absolute temperature (K), z the electrochemical valence, F the Faraday constant (96,485 C/mol), a_c the activity of the concentrated salt solution (mol/L), and a_d the activity of the diluted salt solution (mol/L). For freshwater (0.017 M NaCl, $\gamma_{\pm} = 0.878$) and seawater (0.5 M NaCl, $\gamma_{\pm} = 0.686$), the theoretical voltage difference per membrane is 80.3 mV. The overall, total potential of the system is the sum of the potential differences over each pair of membranes (e.g., 100 membrane pairs provide a voltage difference of $100 \times 80.3 = 8030$ mV or 8 V).

The power density obtainable from RED (defined as the power generated per unit of total membrane area) is equal to the product of half the current and the potential difference over an external load (comparable to PRO, where the power is equal to the product of the pressure and the flux):

$$P^{\text{RED}} = \frac{I}{2}\Delta V = \frac{1}{2r}(\Delta\phi - \Delta V)\Delta V \tag{21}$$

where P^{RED} is the power density obtainable in RED (W/m^2), ΔV the potential difference over an external load (V), r the area resistance ($\Omega\,m^2$), and $\Delta\phi$ the electrochemical potential difference between the two solutions (V). The maximum power density obtainable from RED can be calculated when Eq. (21) is differentiated with respect to the potential difference over the external load. At the maximum power output, $dP/d\Delta V$ is 0 and, as a result, the maximum power output can be obtained when ΔV is equal to $\Delta\phi/2$. In this situation, when substituting this value of ΔV in Eq. (21), the maximum power density obtainable is equal to

$$P^{\text{RED}} = \frac{1}{2r}\frac{\Delta\phi^2}{4} \tag{22}$$

4.3 Membranes for RED

In 2007, Turek [32] studied the effect of the solution velocity on cell power output and process economy and observed that the main bottleneck for successful market introduction of RED is the membrane price. Nevertheless, most of the earlier work was dedicated to stack design and the effect of solution flow and composition, but not to ion exchange membrane characterization and performance testing. Ion exchange membranes are membranes with fixed anionic or cationic exchange groups that are able to transport cations or anions. The presence of these charged groups gives these membranes their specific properties and amount, type, and distribution of the ion exchange groups determine the overall membrane properties. Based on the type of fixed charge groups, ion exchange membranes can be classified as strong acidic and strong basic, or weak acidic and weak basic membranes. In strong acidic CEMs, sulfon groups serve as the fixed charged group in the membrane. Weak acidic membranes contain carboxylic acid as the fixed charged group. Quaternary and tertiary amines, respectively, provide the fixed positive charged groups in strong and weak basic AEMs (Fig. 9).

Two different types of ion exchange membranes can be distinguished, a classification that is based on the structure of the membrane: homogeneous and heterogeneous membranes. In homogenous ion exchange membranes, the fixed charge groups are evenly distributed over the entire membrane matrix. Homogenous membranes can be manufactured by polymerization

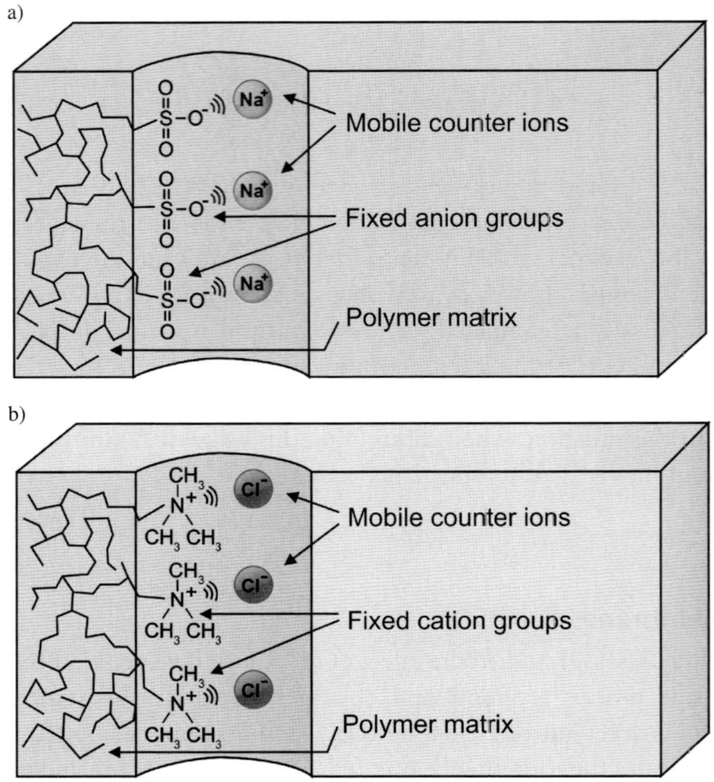

Figure 9 Typical example of (a) a cation exchange membrane (CEM) with SO_3^- groups as the cation exchange group and (b) an anion exchange membrane (AEM) with $N(CH_3)_3^+$ as the typical anion exchange group.

and polycondensation of functional monomers (e.g., fenylosulfonic acid with formaldehyde) or functionalization by, for example, postsulfonation [40–43]. Heterogeneous membranes have distinct macroscopic charged domains of ion exchange resins in a basically uncharged polymer membrane matrix. These membranes are usually produced by melting and pressing a dry ion exchange resin with a granulated polymer (e.g., polyvinylchloride) [44] or by dispersing the ion exchange resin in a polymer solution [45]. The distinct difference in structure between homogenous and heterogeneous ion exchange membranes also influences the properties of the specific membrane, as will be shown later.

Ion exchange membranes are the key elements in RED and the electrical resistance of the membrane and its permselectivity (the ability of the membrane to discriminate between cations and anions) are the most important membrane properties for RED because these properties directly influence the overall RED performance and power output. Because these properties are directly determined by the number of fixed charges inside the ion exchange membrane, the ion exchange capacity (IEC), the swelling degree (SD), and the fixed charge density of a membrane also play a crucial role.

The IEC [expressed in milliequivalent of fixed groups per gram of dry membrane (meq/g membrane)] is the number of fixed charges inside the ion exchange membrane per unit weight of dry polymer. The fixed charge density, expressed in milliequivalent of fixed groups per volume of water in the membrane (meq/L), is determined by this IEC and the SD of the membrane. The fixed charge density is lower in the swollen state than in the dry state because the distance between the charged groups is increased upon swelling of the membrane, while the number of charged groups remains unchanged. The concentration and the type of these fixed charged groups determine the electrical resistance and the permselectivity of the membrane, and these properties are directly related to the maximum power output obtainable in RED.

When an ion exchange membrane is in contact with an electrolyte (salt solution), ions with the same charge as the fixed charges in the membrane (co-ions) are excluded and cannot pass through the membrane, while the oppositely charged ions (counterions) can freely move through the membrane. This effect is known as Donnan exclusion [46]. Ion exchange membranes are never 100% selective and the permselectivity of an ion exchange membrane quantifies the ability of that membrane to discriminate between co-ions and the oppositely charged counterions.

Although the charge density has a strong influence on both the permselectivity and the membrane resistance, a straightforward relationship between the permselectivity and the membrane resistance does not exist as can be seen in Fig. 10 [32] (values for both AEMs and CEMs and homogeneous and heterogeneous membranes are shown).

In general, the resistance of heterogeneous ion exchange membranes is significantly higher than that of the homogenous types. This phenomenon can be related to the structure of the heterogeneous membranes: heterogeneous ion exchange membranes have distinct macroscopic domains of ion exchange resins in an uncharged polymer matrix. Consequently, the resistance of these heterogeneous membranes is higher. In general, less selective membranes have a lower membrane resistance than more selective ones, although this is only a general trend and several exceptions exist. In general, the permselectivity of CEMs is higher than the corresponding

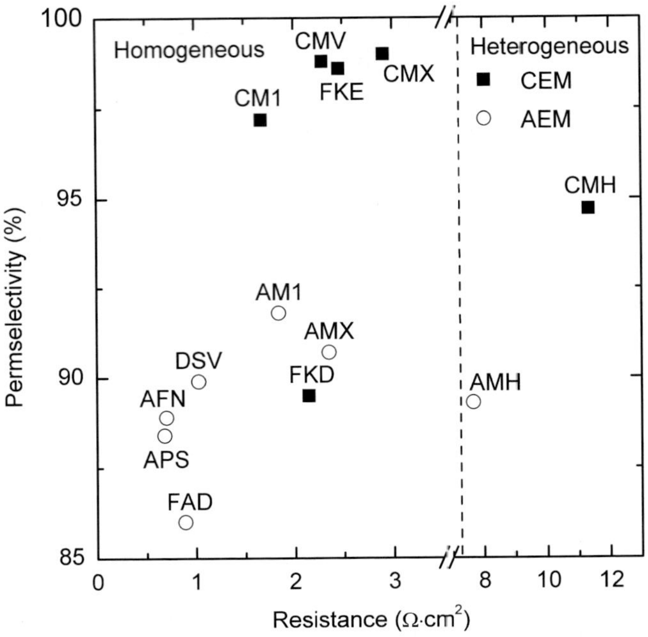

Figure 10 Membrane permselectivity as a function of the membrane resistance (at 25 °C). CEM is a cation exchange membrane (■) and AEM is an anion exchange membrane (□) [34].

values for AEMs. This is mainly due to the higher SD of AEMs, which reduces the effective fixed charge density and thus reduces the permselectivity.

Audinos [30], who was one of the first who systematically investigated the effect of two types of anion and CEM pairs on the power output in RED, already mentioned explicitly the importance of membranes specially developed for RED. Nevertheless, mainly due to limitations in availability of such membranes, most scientists use the above-presented standard electrodialysis membranes to study the performance of a RED system [30,31,38,47]. The manufacturer data available for these membranes do not offer sufficient information on the membrane properties relevant for RED and do not always allow mutual comparison of the different commercially available membranes, because of the different conditions often used for membrane characterization. Długołęçki et al. [34] made a comprehensive overview of membrane benchmarking for RED. They experimentally determined a range of membrane properties of commercially available membranes relevant for RED under equivalent conditions to enable a fair comparison of the results and a proper evaluation of the different membranes for application in RED. Table 4 shows the experimentally determined values of these properties [34]. For comparison, the data of the manufacturers are also presented, although they are not always determined under equal conditions [48–51].

Table 4 clearly shows that the membrane characteristics vary over a wide range and strongly depend on the type of membrane and the differences in molecular structure and composition of the membranes. In general, the data provided by the manufacturers are in reasonable good agreement with the experimentally determined values, with some exceptions.

The IEC presented in Table 4 represent the number of strong acidic ($-SO_3^-$) groups in CEMs and strong basic ($-NR_3^+$) groups in the AEMs. Although the experimentally determined IEC is generally in good agreement with the data supplied by the manufacturers, strong deviations are visible for the APS membranes of Selemion and the FAD membranes from Fumasep. Both AEMs consist of a mixture of weak and strong ion exchange groups, but the experimental method used to determine the IEC only allows the detection of strong basic groups, whereas weak basic groups are not recognized. This results in significantly lower experimental values for the IEC, compared to the manufacturer's data. In general, SD values are similar to the data of the manufacturers, although the experimentally determined SD of the Selemion APS membrane is extremely high, which is probably due to the rough membrane surface of the APS membrane, which affects the wiping off of water from the membrane surface before measuring the weight. The thickness of the membrane strongly depends on the type of the membrane: Homogenous membranes are generally thinner than heterogeneous membranes, which is due to the structure of the membrane and its preparation method [40–43,52,53].

Based on these experimentally determined data, Długołęçki et al. [34] applied a theoretical model to evaluate these specific membrane properties in relation to the expected performance of these membranes under RED conditions [28,34]. This model relates the membrane resistance (R_{aem} and R_{cem}) and its permselectivity (α_{av}) directly to the maximum power output in RED [W_{max} (W)]. Membrane resistance and membrane permselectivity are the two most important parameters in this respect because they indirectly also include the membrane thickness and structure, its IEC and SD, and thus the fixed charge density:

$$W_{max} = NA\frac{\left[\alpha_{av}RT/F\ \ln(a_c/a_d)\right]^2}{R_{aem} + R_{cem} + (d_c/\kappa_c) + (d_d/\kappa_d)} \tag{23}$$

Table 4 Experimentally determined membrane characteristics of several commercially available ion exchange membranes (bold) [34]

Membrane	IEC (meq/g dry)		Permselectivity[a] (%)		Resistance[b] ($\Omega.cm^2$)		SD (%)		Thickness (μm)		Properties
Cation exchange membranes											
Fumasep®											
FKE	**1.36**	>1.0	**98.6**	>98	**2.46**	<3.0	**12**	15	**34**	50–70	Electrolysis, high selectivity
FKD	**1.14**	>1.0	**89.5**	>95	**2.14**	<3.0	**29**	25–30	**113**	90–100	Diffusion dialysis for NaOH
Neosepta®											
CM-1	**2.30**	2.0–2.5	**97.2**	>96[c]	**1.67**	1.2–2.0	**20**	35–40	**133**	120–170	Low electric resistance
CMX	**1.62**	1.5–1.8	**99.0**	>96[c]	**2.91**	1.8 3.8	**18**	25 30	**164**	140 200	High mechanical strength
Ralex® (Heterogeneous)											
CMH-PES	**2.34**	2.2	**94.7**	>92	**11.33**	<10	**31**	<55	**764**	<700	Electrodialysis, Electrodeionization
Selemion®											
CMV	**2.01**	N/A	**98.8**	>92	**2.29**	3.0[d]	**20**	N/A	**101**	130.0	Electrodialysis

Anion exchange membranes											
Fumasep®											
FAD	**0.13**	>1.5	**86.0**	>91	**0.89**	<0.8	**34**	25	**74**	80–100	Diffusion dialysis for acid
Neosepta®											
AM-1	**1.77**	1.8–2.2	**91.8**	>96[c]	**1.84**	1.3–2.0	**19**	25–35	**126**	130–160	Low electric resistance
AFN	**3.02**	2.0–3.5	**88.9**	>96[c]	**0.70**	0.4–1.5	**43**	40–55	**163**	150–200	Resistant against organic fouling
AMX	**1.25**	1.4–1.7	**90.7**	>96[c]	**2.35**	2.5–3.5	**16**	25–30	**134**	160–180	High mechanical strength
Ralex® (Heterogeneous)											
AMH-PES	**1.97**	1.8	**89.3**	>90	**7.66**	<8	**56**	<65	**714**	<850	Electrodialysis, Electro deionization
Selemion®											
DSV	**1.89**	N/A	**89.9**	N/A	**1.03**	1.0[d]	**28**	N/A	**121**	100.0	Diffusion dialysis, low resistance
APS	**0.29**	N/A	**88.4**	N/A	**0.68**	0.5[d]	**147**	N/A	**138**	150.0	Diffusion dialysis, oxidant proof

Note: For comparison the data given by the membrane manufacturers are also presented [48–51].
[a] Membrane potential measured across the membrane between 0.5 and 0.1 M solutions.
[b] Measured in 0.5 M NaCl solution at 25°C.
[c] Measured by electrophoresis, 2 mA/cm^2.
[d] Determined by 1 kHz AC measurement in the 0.5 N NaCl solution at 25°C.

where N is the number of membrane pairs (one cell pair consist of one anion and one CEM), α_{av} the average membrane pair permselectivity (–), R the universal gas constant [8.314 J/(mol K)], T the absolute temperature (K), F the Faraday constant (96,485 C/mol), a_c the concentrated solution activity (mol/L), a_d the diluted solution activity (mol/L), R_{aem} the AEM resistance ($\Omega\,m^2$), R_{cem} the CEM resistance ($\Omega\,m^2$), A the effective membrane area (m^2), d_c the thickness of the concentrated compartment (m), d_d the thickness of the diluted compartment (m), κ_c the concentrated compartment conductivity (S/m), and κ_d the diluted compartment conductivity (S/m).

In order to compare commercially available membranes with each other, it is more convenient to convert the power output into power density, which is the power output normalized for the membrane area (W/m^2):

$$P_{max} = \frac{W_{max}}{AN_m} \tag{24}$$

where P_{max} is the maximum power density (W/m^2), W_{max} maximum power output (W), A the effective membrane area (m^2), and N_m the number of membranes (–).

As Eqs. (23) and (24) predict the theoretical power output of the total system under RED conditions in relation to the individual membrane characteristics, it can be used as a tool to evaluate and compare the different anion and cation exchange membranes with respect to their performance in RED. Długołęçki et al. [34] evaluated the relative importance of membrane resistance and permselectivity on the power density in a RED stack. Fig. 11 shows the power density as a function of the membrane resistance and permselectivity for two different spacer thicknesses (a) 600 and (b) 150 μm.

When thicker spacers (>600 μm, Fig. 11a) are used in the system, the power density of the system is hardly dependent on the membrane resistance or permselectivity. In this case, the resistance of the dilute compartment dominates the overall process resistance and maximum power densities of only 2 W/m^2 can be obtained. When the distance between the membranes is decreased (Fig. 11b), the effect of the membrane properties and thus the difference in power density of the different membranes becomes more pronounced. With increasing permselectivity and especially decreasing membrane resistance, the power density significantly increases and values as high as 7 W/m^2 can be obtained with properly designed stacks. Nevertheless, the process requires a minimum in spacer thickness

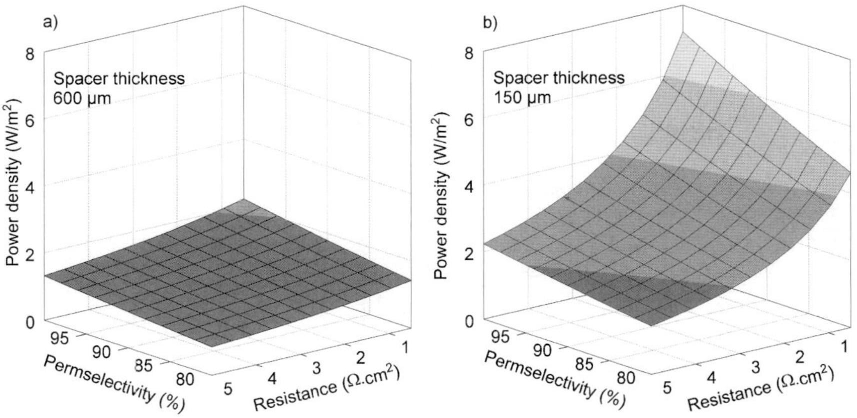

Figure 11 Relationship between the power density, the membrane permselectivity, and the membrane cell pair resistance for membrane pair with (a) 600 μm and (b) 150 μm thick spacers. Model calculations are based on seawater (0.5 M NaCl) as concentrated salt solution and river water (0.05 M NaCl) as diluted stream ($T = 25\,°C$) [34].

because at too thin spacer thicknesses, the energy consumption for solution pumping increases tremendously due to the high pressure drop over the compartments.

Because Eqs. (23) and (24) can also be used to predict the performance of only a cation or only an AEM in RED, Długołęçki et al. [34] used their experimental data presented in Table 4 as input values for the model calculations to predict the maximum power density obtainable with each specific membrane (Fig. 12a and b). In this case, the average membrane pair selectivity (α_{av}) is replaced by the individual membrane selectivity of the cation or anion exchange membrane, respectively, whereas in the case of a CEM the corresponding resistance of the AEM is neglected, and vice versa when only an AEM is used. The thickness of the concentrated and diluted compartment is divided by a factor 2. Długołęçki et al. assumed that seawater has a NaCl concentration of 0.5 M ($\gamma_{\pm} = 0.686$ and $\kappa_c = 4.648$ S/m, $T = 25\,°C$) and river water has a concentration of 0.05 M NaCl ($\gamma_{\pm} = 0.820$ and $\kappa_c = 0.551$ S/m, $T = 25\,°C$).

It is obvious that the power density strongly depends on the spacer thickness (as presented before) and also on the type of membrane. The resistance of the heterogeneous membranes investigated is too high to be useful in RED. Even in a perfectly designed RED stack (extremely thin spacers), it is not possible to obtain power densities higher than 1.5 W/m^2.

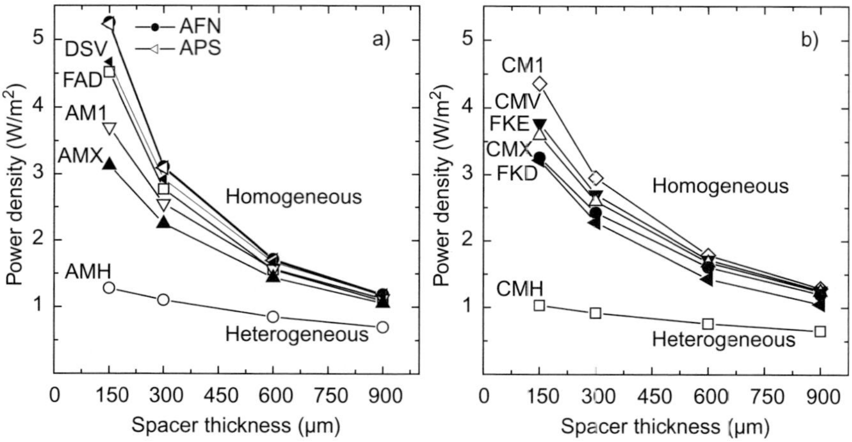

Figure 12 Prediction of the maximum obtainable power density based on experimental membrane characterization for (a) anion exchange membranes and (b) cation exchange membranes [34].

Homogeneous membranes are more suitable for RED. Based on these results, the best benchmarked AEMs are Neosepta AFN from Tokuyama Co. (Japan) and Selemion APS from Asahi Glass Co. Ltd. (Japan), with a predicted power density of more than $5\,W/m^2$ (at a spacer thickness of $150\,\mu m$). The Neosepta CM-1 CEM from Tokuyama Co. (Japan) shows the best performance as CEM for RED and reaches a theoretical power density of more than $4\,W/m^2$.

Although this model is a very useful tool to make a rough estimation of the performance of the different membranes under RED conditions, it is a theoretical model that includes several assumptions [34]: (i) concentration polarization phenomena near the membrane surface are negligible due to the small current densities obtained through the membranes and (ii) the resistance of the electrodes is assumed to be negligible compared to the membrane resistance. This assumption is allowed when the resistance of the membranes is large compared to the resistance of the electrodes, which can be obtained when a large number of membrane cell-pairs is used (as will be required anyway to generate sufficient power at low costs), and (iii) the feed solution does not change in concentration along the channels. This assumption has a strong relationship with the feed channel design. Although assumptions (i) and (iii) are valid assumptions for a first initial comparison under laboratory conditions, they will become an important issue in the real application where real river and seawater are used.

4.4 Process and stack design

The final performance of the stack depends on a variety of parameters: (1) membrane properties (conductivity, selectivity, osmotic behavior), (2) cell properties (compartment thickness, spacer type), (3) stack design parameters (way of feed, electrodes), (4) operating conditions (flow rate, electrical load), and (5) water quality (salt content, impurities, temperature, composition). These different parameters often conflict with each other and all together they determine the final power output. Veerman et al. [6] systematically investigated the performance of a real RED stack with respect to power density and energy efficiency, especially focusing on the effect of the current density, the membrane and spacer resistance, and the feed flow rate. They used a custom-made RED stack with an adaptable number of cells, with a maximum of 50 cells (total effective membrane area of 1 m^2). Each cell consisted of an anion and a CEM with an effective membrane area of 100 cm^2 per membrane. Commercially available membranes from Fumasep (Germany) were used: FAD as AEM and FKD as CEM. These membranes have a thickness of 0.082 mm. Polyamide woven sheets with a thickness of 200 μm were used as spacer. As electrode system, the authors used a solution of 1 M NaCl with 0.05 M $K_4Fe(CN)_6$ and 0.05 M $K_3Fe(CN)_6$. Sea and river water were represented by NaCl solutions of, respectively, 30 and 1 g/L. The 50-cell stack generated a power output of 0.93 W, which is the highest power output reported for RED using sea and river water. Fig. 13 shows the power output of the stack as a function of the current density for different numbers of cells (N) [6].

The obtained power increases almost linearly with the number of cells, which indicates that the losses due to limiting currents are limited [36]. The maximum power obtainable in this stack is 0.93 W/m^2, which is the highest power reported in literature.

Not only the power output is an important parameter, the energy efficiency also plays a significant role. It represents the fraction of the total available energy available from the mixing of river and seawater that is really used to generate power. In the case of the stack experiments of Veerman et al. [6], the highest power density reported could be obtained. However, the energy efficiency at that point is no more than 50% [6]. So optimization with respect to obtained power only would result generally in low energy efficiencies and loss of potentially available energy. Post et al. [35] show that, in principle, no fundamental limitations restrict the energy efficient use of the resources and values as high as 80% can be obtained.

In the real application, the power density obtainable in a RED stack is often reduced due to parasitic currents, or also called current leakage in the

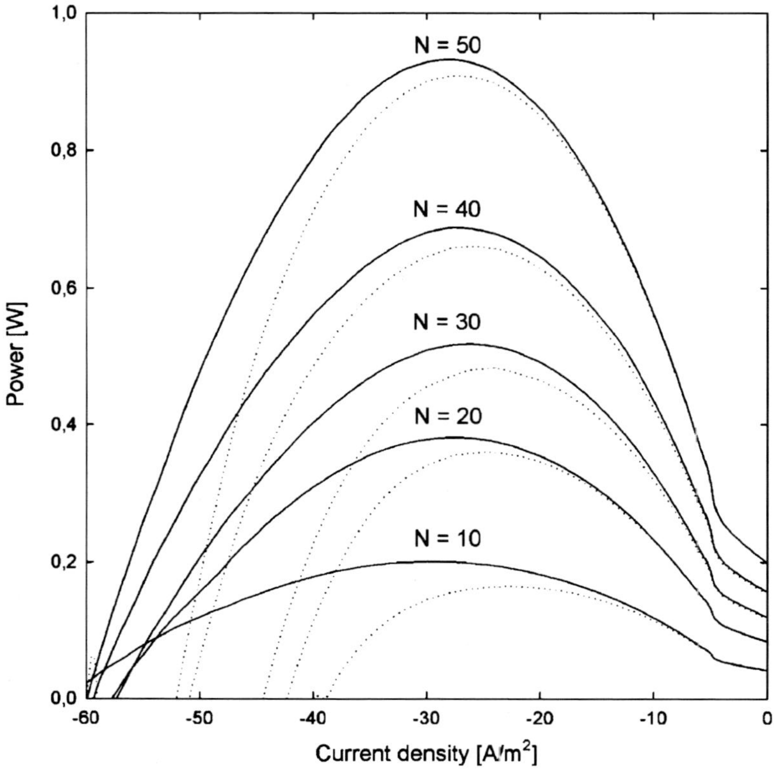

Figure 13 Experimentally determined power obtainable in RED when using sea and river water, as a function of the current density for different numbers of cells. The solid line represents the power generated by the stack and the dashed line is the output at the working electrodes [6].

stack. There are two sources of these losses [36]: (1) ion exchange membranes are never 100% selective, which apart from generating the transport of counterions, also generates a transport of co-ions, which reduces the power output. This issue is related to membrane design and optimization. (2) Ionic shortcut currents occur due to the transport of ions in feed and drain channels and this effect is more severe at higher salt concentrations. These ionic shortcut losses are strongly related to stack design.

In principle, three different ionic shortcut currents can be distinguished in the stack [36]:

1. Ionic shortcut currents in the electrode solution (the electrode solution connects the anode and the cathode compartment). These losses can be easily reduced by increasing the length of the tubing that connects the electrodes.

2. Ionic shortcut currents between the river water compartments. Generally, this shortcut current can be neglected because the salt concentration in the river water compartment is too low to cause significant leakages.
3. Shortcut currents between the seawater compartments.

Veerman et al. [36] investigated the possibilities to reduce the shortcut currents between the seawater compartments. Model calculations show that the effect of these losses can be significantly reduced through proper stack design. Especially important in this respect are the number of cells (*N*), the channel resistance in relation to the cell resistance (R/r), and the lateral spacer resistance in relation to the cell resistance (ρ/r), where the latter two are the critical design parameters that need to be optimized [36]. In medium-size stacks, the number of cells and the ratio R/r and ρ/r need to be as high as possible [36]. Possibilities to do so include (i) increasing the channel resistance (*R*) by narrowing the channels; (ii) increasing the lateral spacer resistance (ρ) by using thinner spacers (in the seawater compartment); and (iii) decreasing the cell resistance *r* by using membranes with low resistances and thin spacers (in the river water compartment). Possibilities to increase *R* are limited: with narrowing the channels, the hydrodynamical resistance in the channels also increases. An increase in the spacer resistance only induces a very small change in power output and the only way to increase the power output is to decrease the cell resistance, as it increases the efficiency and the power output of the system [36]. In very large stacks, (R/r) (the channel resistance in relation to the cell resistance) should be maximized to obtain the largest power output [36]. A decrease in *r* can be induced by minimization of the membrane thickness and the thickness of both the sea and the river water compartments, but this results in an equal decrease in the channel resistance, and consequently has no effect. But, at given membrane thickness and river water compartment dimensions, a decrease of the thickness of the seawater compartment induces a lower *r* and a higher R/r ratio and consequently a higher efficiency and power output [36].

4.5 Pilot testing and scale-up[2]

Wetsus – Center for Sustainable Water Technology in the Netherlands– started with the "Blue Energy" project in 2005 with a focus on RED. At that time, only a few scientific papers were published [4,28,31] about

[2]The information given in this paragraph section is provided by and property of the company REDstack B.V., The the Netherlands, and is used with permission. The authors would like to acknowledge REDstack B.V. for the contribution.

results of the RED system in a period of 50 years. Over the past few years, the performance of RED on laboratory scale has improved considerably. However, thus far, RED experiments have typically been performed on a small scale, varying from current-passing areas of just a few square centimeters [32] to hundreds of square centimeters [36] and from four cell-pairs [33] to fifty cell-pairs [38]. State-of-the-art is a stack with an active membrane area of $25 \times 75\,cm^2$ and 50 cell-pairs with a power output of about 16 W (Fig. 14; drawing prepared by REDstack B.V., the Netherlands, and belongs to the company; used with permission).

To achieve practical implementation, RED still needs to be scaled up by several orders of magnitude. This upscaling and practical implementation is beyond the academic expertise and needs to be done in close cooperation with industry. For this reason, REDstack B.V. was founded by Magneto Special Anodes B.V., the Netherlands and Landustrie/Hubert, two industrial companies participating within the Blue Energy research of Wetsus. The challenges still faced by REDstack B.V., concerning the economics, technological feasibility, and the developing path of RED, are the development of low-cost membranes, the pretreatment in relation to stack design and operation, and the upscaling.

Figure 14 Reverse electrodialysis stack for research purposes with a total membrane area of $25 \times 75\,cm^2$ and 50 cell-pairs as manufactured by REDstack B.V., the Netherlands.

Although the technical requirements are already met by currently available membranes, the cost prices are out of range to make RED affordable. According to Turek and Bandura [32], it is hard to believe that the price of low-resistance ion exchange membranes may be reduced 100 times, which seems to be the desired cost level [32]. Nevertheless, for several reasons, REDstack B.V. is more optimistic that membrane prices for (reverse) electrodialysis can be reduced tremendously [31]. This is because of the fact that electrodialysis membranes have never had a considerable market share. Even then, on the global market, heterogeneous ion exchange membranes can be found with very low cost prices (<5 US$/$m^2$). Of course, low-resistance ion exchange membranes command higher prices of 100 US$/$m^2$ or more [32], but these prices can also be expected to fall, as manufacturing techniques improve, and the range of applications expands. Market research for related membrane applications show unit prices of installed membranes falling by an order of magnitude in 10 years, and this made Sutherland [54] to predict that the 1 US$/$m^2$ of installed membrane is not far off. Second, it should be noticed that – apart from different technical requirements – the current membrane market would never be able to match the demand of required membrane area for power production. This implicates that besides the expertise in manufacturing of membranes the expertise of bulk production is also needed. While at the start of the membrane development, REDstack B.V. was dedicated to the technical requirements (as described previously) and cost prices of base materials, nowadays REDstack B.V. focuses on the scalability of the production process with focus on labor-extensive reel-to-reel production lines operating at high speeds.

Although addressed in scientific papers, challenges often not considered are the pretreatment of river water and seawater [34] and the hydrodynamic aspects of RED [55]. The required water quality parameters are still unknown. It is not likely to look at experiences in desalination stacks because the usually applied pretreatment steps [56] would be too capital-intensive to be viable for RED. Nevertheless, RED would require an extensive pretreatment as the distance between the membranes is less than in conventional flat-sheet membrane systems. It requires a more robust system design using the developed CFD model for flat-sheet membrane configurations [55]. Besides the cost aspect, also the footprint, energy consumption, and use of chemicals should be taken into account regarding the feasibility of RED.

The promising results raised the interest of different industrial and power supply companies and water authorities to invest in pilot tests. At this

Figure 15 Artist impression of a salinity-gradient power plant at the IJsselmeer (by Rijkswaterstaat); inset top right: reverse electrodialysis pilot in Harlingen; inset bottom right: pretreatment pilot (REDstack B.V., The Netherlands).

stage of the project, focus is on consortium building, with customers entering into technical development agreements with suppliers, joint designs, and test programs. Parties agreed on the following development path for scale-up of the system (Fig. 15):

- Industrial pilot (kW-scale) on saline flows in a salt factory (financially supported by SenterNovem, Innowator project; 2008–2010).
- Feasibility study and definition of requirements for a communal power plant of 200 MW at the Afsluitdijk, The Netherlands (private funding, 2008).
- Communal pilot (10–40 kW) on seawater and river water (2009–2010) at the Afsluitdijk, The Netherlands.
- Communal demonstration plant (1 MW) on seawater and river water (2010–2012) at the Afsluitdijk, The Netherlands.

5. CONCLUDING REMARKS

This chapter clearly shows the huge potential of salinity gradient energy and the significant progress that has been made during recent years on both

PRO and RED. This has led to an increased power density for both technologies. Nevertheless, significant efforts are still required to make salinity gradient energy competitive with other new, renewable energy sources. For PRO, the reduction of external and internal concentration polarization is the main challenge, whereas for RED the main challenge is the reduction of the internal stack and membrane resistance. In addition, system design and pilot and demonstration plant testing using real feed waters are important issues to investigate. And, although not thoroughly investigated yet, but definitely important to consider is the positioning of such a system in the local environment without harming the ecological system, shipping traffic, and recreational activities.

REFERENCES

[1] O. Davidson, B. Metz, Special report on carbon dioxide capture and storage. Retrieved December 23, 2008, from Intergovernmental Panel on Climate Change Web site: http://www.ipcc.ch/pdf/special-reports/srccs/srccs_wholereport.pdf, 2005.

[2] G.L. Wick, W.R. Schmitt, Prospects for renewable energy from the sea, Marine Technol. Soc. J. 11(5 and 6) (1977) 16–21.

[3] J.W. Post, J. Veerman, H.V.M. Hamelers, G.J.W. Euverink, S.J. Metz, K. Nijmeijer, C.J.N. Buisman, Salinity-gradient power: evaluation of pressure-retarded osmosis and reverse electrodialysis, J. Membr. Sci. 288 (2007) 218–230.

[4] R.E. Pattle, Production of electric power by mixing fresh and salt water in the hydroelectric pile, Nature 174(4431) (1954) 660–661.

[5] H. Strathmann, Ion-exchange Membrane Separation Processes, 1st ed., Elsevier, Amsterdam, 2004.

[6] J. Veerman, M. Saakes, S.J. Metz, G.J. Harmsen, Reverse electrodialysis: performance of a stack with 50 cells o the mixing of sea, river water, J. Membr. Sci. 327 (2008) 136–144.

[7] S. Loeb, Production of energy from concentrated brines by pressure-retarded osmosis. 1. Preliminary technical and economic correlations, J. Membr. Sci. 1 (1976) 49–63.

[8] S. Loeb, F. Vanhessen, D. Shahaf, Production of energy from concentrated brines by pressure-retarded osmosis. 2. Experimental results and projected energy costs, J. Membr. Sci. 1 (1976) 249–269.

[9] S. Loeb, G.D. Mehta, 2-Coefficient water transport-equation for pressure retarded osmosis, J. Membr. Sci. 4 (1979) 351–362.

[10] S. Loeb, Energy production at the Dead Sea by pressure-retarded osmosis: challenge or chimera? Desalination 120 (1998) 247–262.

[11] S. Loeb, L. Titelman, E. Korngold, J. Freiman, Effect of porous support fabric on osmosis through a Loeb-Sourirajan type asymmetric membrane, J. Membr. Sci. 129 (1997) 243–249.

[12] S. Loeb, Large-scale power production by pressure-retarded osmosis, using river water and sea water passing through spiral modules, Desalination 143 (2002) 115–122.

[13] G.D. Mehta, S. Loeb, Internal polarization in the porous substructure of a semipermeable membrane under pressure-retarded osmosis, J. Membr. Sci. 4 (1978) 261–265.

[14] G.D. Mehta, S. Loeb, Performance of permasep b-9 and b-10 membranes in various osmotic regions and at high osmotic pressures, J. Membr. Sci. 4 (1979) 335–349.

[15] G.D. Mehta, Further results on the performance of present-day osmotic membranes in various osmotic regions, J. Membr. Sci. 10 (1982) 3–19.
[16] K.L. Lee, R.W. Baker, H.K. Lonsdale, Membranes for power-generation by pressure-retarded osmosis, J. Membr. Sci. 8 (1981) 141–171.
[17] S.E. Skilhagen, J.E. Dugstad, R.J. Aaberg, Osmotic power – power production based on the osmotic pressure difference between waters with varying salt gradients, Desalination 220 (2008) 476–482.
[18] K. Gerstandt, K.-V. Peinemann, S.E. Skilhagen, T. Thorsen, T. Holt, Membrane processes in energy supply for an osmotic power plant, Desalination 224 (2008) 64–70.
[19] J.R. McCutcheon, M. Elimelech, Influence of membrane support layer hydrophobicity on water flux in osmotically driven membrane processes, J. Membr. Sci. 318 (2008) 458–466.
[20] T. Thorsen, T. Holt, The potential for power production from salinity gradients by pressure retarded osmosis, J. Memb. Sci. 335(1–2) (2009) 103–110.
[21] R.L. McGinnis, J.R. McCutcheon, M. Elimelech, A novel ammonia-carbon dioxide osmotic heat engine for power generation, J. Membr. Sci. 305 (2007) 13–19.
[22] J.R. McCutcheon, M. Elimelech, Modeling water flux in forward osmosis: implications for improved membrane design, AIChE J. 53 (2007) 1736–1744.
[23] R.L. McGinnis, M. Elimelech, Energy requirements of ammonia-carbon dioxide forward osmosis desalination, Desalination 207 (2007) 370–382.
[24] J.R. McCutcheon, M. Elimelech, Influence of concentrative and dilutive internal concentration polarization on flux behavior in forward osmosis, J. Membr. Sci. 284 (2006) 237–247.
[25] G.T. Gray, J.R. McCutcheon, M. Elimelech, Internal concentration polarization in forward osmosis: role of membrane orientation, Desalination 197 (2006) 1–8.
[26] T.Y. Cath, A.E. Childress, M. Elimelech, Forward osmosis: principles, applications, and recent developments, J. Membr. Sci. 281 (2006) 70–87.
[27] E.R. Cornelissen, D. Harmsen, K.F. de Korte, C.J. Ruiken, J.J. Qin, H. Oo, L.P. Wessels, Membrane fouling and process performance of forward osmosis membranes on activated sludge, J. Membr. Sci. 319 (2008) 158–168
[28] J.N. Wienstein, F.B. Leitz, Electric power from differences in salinity: the dialytic battery, Science 191 (1976) 557–559.
[29] R.E. Lacey, Energy by reverse electrodialysis, Ocean Eng. 7 (1980) 1–47.
[30] R. Audinos, Electrodialyse inverse. Etude de l'énergie électrique obtenue à partir de deux solutions de salinités differentes, J. Power Sources 10 (1983) 203–217.
[31] J. Jagur-Grodzinski, R. Kramer, Novel process for direct conversion of free energy of mixing into electric power, Ind. Eng. Chem. Process Des. Dev. 25 (1986) 443–449.
[32] M. Turek, B. Bandura, Renewable energy by reverse electrodialysis, Desalination 205 (2007) 67–74.
[33] M. Turek, B. Bandura, P. Dydo, Power production from coal-mine brine utilizing reversed electrodialysis, Desalination 221 (2008) 462–466
[34] P. Długołęcki, K. Nijmeijer, S. Metz, M. Wessling, Current status of ion exchange membranes for power generation from salinity gradients, J. Membr. Sci. 319(1–2) (2007) 214–222.
[35] J.W. Post, H.V.M. Hamelers, C.J.N. Buisman, Energy recovery from controlled mixing salt and fresh water with a reverse electrodialysis system, Environ. Sci. Technol. 42(15) (2008) 5785–5790.
[36] J. Veerman, J.W. Post, M. Saakes, S.J. Metz, G.J. Harmsen, Reducing power losses caused by ionic shortcut currents in reverse electrodialysis stacks by a validated model, J. Membr. Sci. 310 (2008) 418–430.
[37] J. Veerman, M. Saakes, S.J. Metz, G.J. Harmsen, Reverse electrodialysis: performance of a stack with 50 cells on the mixing of sea and river water, J. Memb. Sci. 327(1–2) (2009) 136–144.

[38] F. Suda, T. Matsuo, D. Ushioda, Transient changes in the power output from the concentration difference cell (dialytic battery) between seawater and river water, Energy 32 (2007) 165–173.
[39] E. Brauns, Towards a world sustainable and simultaneous large-scale production of renewable energy and potable water through salinity gradient power by combining reversed electrodialysis and solar power? Desalination 219 (2008) 312–323.
[40] X. Tongwen, Y. Weihua, Fundamental studies of a new series of anion exchange membranes: membrane preparation and characterization, J. Membr. Sci. 190 (2001) 159–166.
[41] F.G. Wilhelm, I.G.M. Punt, N.F.A. van der Vegt, H. Strathmann, M. Wessling, Cation permeable membranes from blends of sulfonated poly (ether ether ketone) and poly (ether sulfone), J. Membr. Sci. 199 (2002) 167–176.
[42] P. Xing, G.P. Robertson, M.D. Guiver, S.D. Mikhailenko, K. Wang, S. Kaliaguine, Synthesis and characterization of sulfonated poly (ether ether ketone) for proton exchange membranes, J. Membr. Sci. 229 (2004) 95–106.
[43] R.K. Nagarale, G.S. Gohil, V.K. Shahi, R. Rangarajan, Preparation and electrochemical characterizations of cation-exchange membranes with different functional groups, Colloids Surf. A 251 (2004) 133–140.
[44] J. Schauer, L. Brozova, Z. Pientka, K. Bouzek, Heterogeneous ion-exchange polyethylene-based membranes with sulfonated poly(1,4-phenylene sulfide) particles, Desalination 200 (2006) 632–633.
[45] G.E. Molau, Heterogeneous ion-exchange membranes, J. Membr. Sci. 8 (1981) 309–330.
[46] F. Helfferich, Ion Exchange, McGraw Hill, New York, 1962.
[47] S. Koter, Influence of the layer fixed charge distribution on the performance of an ion-exchange membrane, J. Membr. Sci. 108 (1995) 177–183.
[48] Ralex, Ion exchange membranes, Product brochure Mega, a.s., Czech Republic.
[49] Selemion, Ion exchange membranes, Product brochure Asahi Glass Co., Ltd., Japan.
[50] Neosepta, Ion exchange membranes, Product brochure Tokuyama Co., Japan.
[51] Fumasep, Ion exchange membranes, Product brochure, FuMA-Tech GmbH, Germany.
[52] Y. Mizutani, Structure of ion exchange membranes, J. Membr. Sci. 49 (1990) 121–144.
[53] P.V. Vyas, B.G. Shah, G.S. Trivedi, P. Ray, S.K. Adhikary, R. Rangarajan, Characterization of heterogeneous anion-exchange membrane, J. Membr. Sci. 187 (2001) 39–46.
[54] K. Sutherland, Profile of the International Membrane Industry: Market Prospects to 2008, Elsevier, Amsterdam, 2003.
[55] M.H. Dirkse, W.P.K. van Loon, J.D. Stigter, J.W. Post, J. Veerman, G.P.A. Bot, Extending potential flow modelling of flat-sheet geometries as applied in membrane-based systems, J. Membr. Sci. 325 (2008) 537–545.
[56] J.P. Van der Hoek, D.O. Rijnbende, C.J.O. Lokin, P.A.C. Bonné, M.T. Loonen, M.H. Hofman, Electrodialysis as an alternative for reverse osmosis in an integrated membrane system, Desalination 117 (1998) 159–172.

CHAPTER 6

Ion-Exchange Membrane Processes in Water Treatment

H. Strathmann*

Institute of Chemical Engineering, University of Stuttgart, Stuttgart, Germany

Contents

1. INTRODUCTION

Electromembrane processes such as electrodialysis, continuous electrodeionization, and diffusion dialysis have experienced a steady growth since they were first applied on an industrial scale for the desalination of brackish water during the middle of the last century [1]. Desalination of brackish water is still the dominant application of electromembrane processes. But a

*Corresponding author.
E-mail address: Heiner.Strathmann@t-online.de

Sustainability Science and Engineering, Volume 2 ISSN 1871-2711, DOI 10.1016/S1871-2711(09)00206-2

number of new applications in the production of high-quality industrial process water and in the treatment of industrial effluents have been identified very recently [2,3]. The development of processes such as continuous electrodeionization, capacitive deionization, and electrodialysis with bipolar membranes has further extended the range of application of electromembrane processes beyond their traditional use in brackish water desalination. The term "electromembrane process" is used to describe an entire family of processes that can be quite different in their basic concept and their application. But they are all based on the coupling of a mass transport with an electrical current through an ion perm-selective membrane. Electromembrane processes are used to remove ionic components such as salts from electrolyte solutions or to produce certain compounds such as NaOH and Cl_2 from NaCl solutions, or to convert chemical into electrical energy as in the H_2/O_2 fuel cell. This chapter, however, is concentrated mainly on technically relevant processes related to water treatment while other processes used in electrochemical synthesis and energy conversion will only be briefly described in their fundamental principle.

1.1 Definition of terms and description of electromenbrane processes

Before discussing the various electromembrane processes and their applications in detail, the definition of some basic terms and the principle of these processes as well as the main components determining their performance shall be described briefly.

1.1.1 The ion-exchange membranes, and their structure and function

The key components in electromembrane processes are the ion-exchange membranes, which resemble ion-exchange resins in sheet form. There are two different types of ion-exchange membranes: (1) cation-exchange membranes, which contain negatively charged groups fixed to the polymer matrix, and (2) anion-exchange membranes, which contain positively charged groups fixed to the polymer matrix. In a cation-exchange membrane, the fixed negative charges are in electrical equilibrium with mobile cations in the interstices of the polymer as indicated in Fig. 1, which shows schematically the matrix of a cation-exchange membrane with fixed anions and mobile cations, the latter are referred to as counterions. The mobile anions are referred to as co-ions. They are more or less completely excluded from the polymer matrix because of their electrical charge, which is identical to that of the fixed ions. Due to the exclusion of the co-ions, a cation-exchange membrane permits predominantly the transfer of cations.

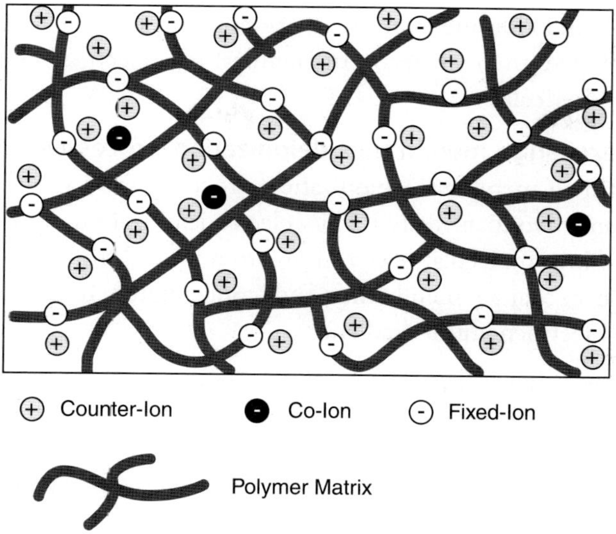

Figure 1 Schematic drawing illustrating the structure of a cation-exchange membrane.

Anion-exchange membranes carry positive fixed charges. Therefore, they are mainly permeable to anions. The properties of ion-exchange membranes are determined by different parameters such as the density of the polymer network, the hydrophobic or hydrophilic character of the matrix polymer, and the type and concentration of the fixed charges in the polymer. To which extent the co-ions are excluded from an ion-exchange membrane depends not only on membrane properties but also on the ion concentration of the feed solution.

The most desired properties of ion-exchange membranes are high permselectivity and low electrical resistance and good mechanical, chemical, and thermal stability. Many of today's available membranes meet most of these requirements. Especially, the cation-exchange membranes based on fluorocarbon polymers have quite satisfactory properties.

1.2 Principle of ion-exchange membrane processes

The processes that utilize ion-exchange membranes as a key component can conveniently be divided into three types: (1) electrodeionization processes, (2) electrosynthesis processes, and (3) electromembrane energy conversion processes. In the first type of processes, an electrical potential gradient is used to remove charged components such as dissociated salts from a solution. In the second type of processes, the transport of ions is combined with an electrochemical reaction producing certain chemicals such as bases and chlorine from the corresponding salts. The third type of processes involves the conversion of chemical into electrical energy as, for example, in fuel cells.

1.2.1 Ion-exchange membranes deionization processes

In ion-exchange membrane deionization processes such as electrodialysis, diffusion, and Donnan dialysis, or electrodeionization and capacitive deionization low-molecular-weight ions are removed from a feed solution through ion-exchange membranes and concentrated under the driving force of an electrochemical gradient.

Electrodialysis. The principle of electrodialysis is illustrated in Fig. 2, which shows a schematic diagram of an electrodialysis cell arrangement consisting of a series of anion- and cation-exchange membranes arranged in an alternating pattern between an anode and a cathode to form individual cells. If an ionic solution such as an aqueous salt solution is pumped through these cells and an electrical potential is established between the anode and cathode, the positively charged cations migrate toward the cathode and the negatively charged anions toward the anode. The cations pass through the negatively charged cation-exchange membrane but are retained by the positively charged anion-exchange membrane. Likewise, the negatively charged anions pass through the anion-exchange membrane, and are retained by the cation-exchange membrane. The overall result is an increase

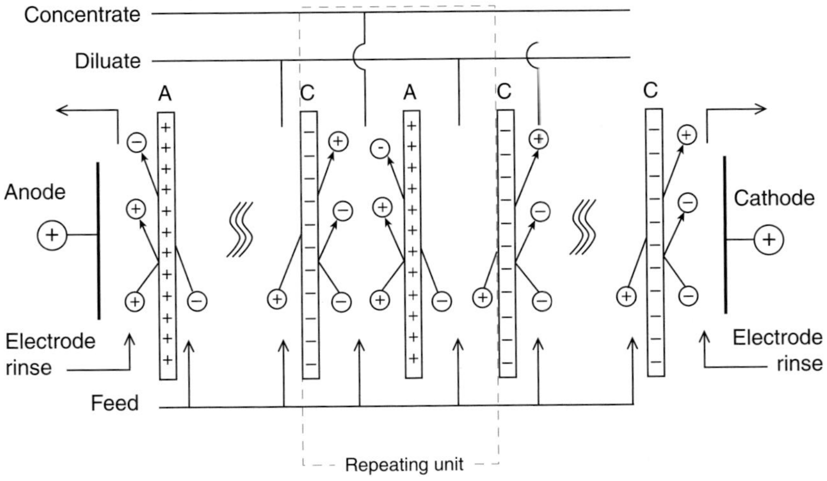

Figure 2 Schematic diagram illustrating the principle of desalination by electrodialysis in a stack with cation- and anion-exchange membranes in alternating series between two electrodes.

in the ion concentration in alternate compartments, while the other compartments simultaneously become depleted of ions. The depleted solution is referred to as the diluate and the concentrated solution as the brine. The space between two contiguous membranes occupied by the diluate and the brine and the two contiguous anion- and cation-exchange membranes make up a cell pair, which is a repeating unit in a so-called electrodialysis stack, which may have a few hundreds cell pairs between two electrodes [4].

Electrodialysis is used mainly today for desalination of brackish water and demineralization of solution in the food and drug industry as well as in the concentration of salts from seawater.

Electrodialysis with bipolar membranes. The conventional electrodialysis can be combined with bipolar membranes and utilized to produce acids and bases from the corresponding salts. In this process monopolar cation- and anion-exchange membranes are installed together with bipolar membranes in alternating series in an electrodialysis stack as illustrated in Fig. 3. A bipolar membrane consists of a laminate of a cation- and an anion-exchange layer. If an electrical potential difference is established across the membrane, charged species are removed from the interphase between the two ion-exchange layers. When this interphase contains only water, the transport of electrical charges is accomplished by protons and hydroxide ions, which are

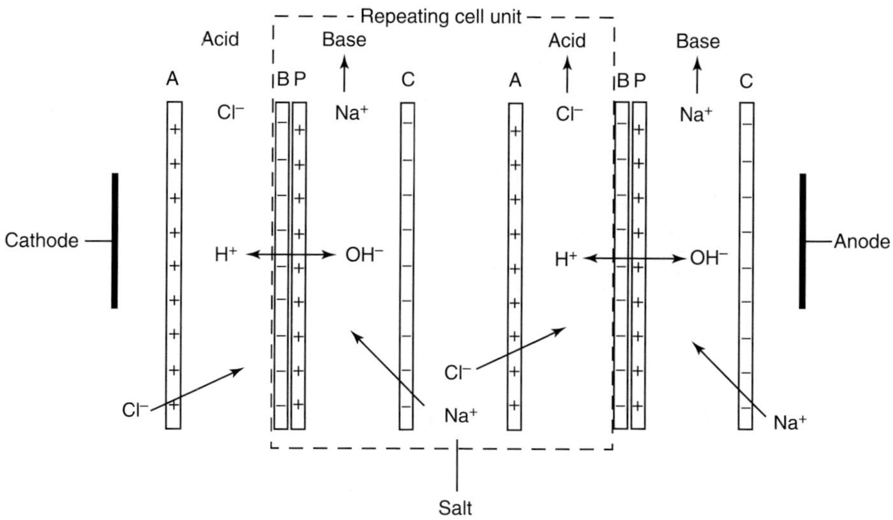

Figure 3 Schematic drawing illustrating the principle of electrodialytic production of acids and bases from the corresponding salts with bipolar membranes.

produced continuously in the bipolar membrane by water dissociation due to the driving force of an electrical potential gradient. The H^+ and OH^- ions, removed from the interphase of the bipolar membrane form with the salt ions of the feed solution an acid and a base in the two compartments between the two monopolar and the bipolar membranes. A cation-exchange, an anion-exchange, and a bipolar membrane form a repeating unit in the stack between two electrodes. Thus, a repeating unit is composed of three separate flow streams, that is, the salt containing feed solution flow stream and two product solution flow streams containing an acid and a base.

The utilization of electrodialysis with bipolar membranes is economically very attractive and has a multitude of interesting potential applications [5].

Continuous electrodeionization. Continuous electrodeionization is very similar to conventional electrodialysis. However, the cell of the diluate flow stream is filled with a mixed-bed ion-exchange resin. The principle of the process is illustrated in Fig. 4.

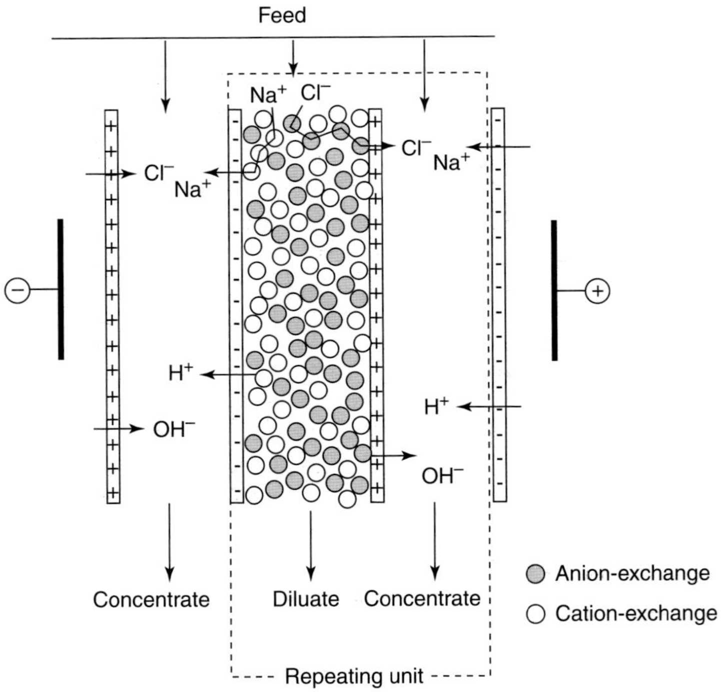

Figure 4 Schematic drawing illustrating the principle of the continuous electrodeionization process.

The mixed-bed ion-exchange resin in the diluate cell of an electrodialysis stack binds the ions of a feed solution. Due to an applied electrical field, the ions migrate through the ion-exchange bed toward the adjacent concentrate cells. The ion-exchange resin increases the conductivity in the diluate cell substantially, and at very low salt concentrations in the feed solution water is dissociated at the contact point of the cation- and anion-exchange resin beads generating protons and hydroxide ions, which further replace the salt ions in the resins. The result is completely deionized water as a product. Compared to the deionization by a conventional mixed-bed ion-exchange resin, continuous electrodeionization has several advantages since no chemicals are needed for the regeneration of the ion-exchange resins, which is time-consuming, labor-intensive, and generates a salt containing wastewater. Fig. 4 shows just one concept of an electrodeionization stack. In practicable applications, various stack concepts based on separate ion-exchange beds or bipolar membranes are used [6].

Diffusion dialysis. Diffusion dialysis is used mainly today to recover acids or bases from a mixture with salt ions. Its principle is illustrated in Fig. 5, which shows a schematic diagram of a typical diffusion dialysis cell arrangement consisting of a series of anion-exchange membranes arranged in parallel to form individual cells. If a feed solution containing a salt in a mixture with an acid is separated by an anion-exchange membrane from a

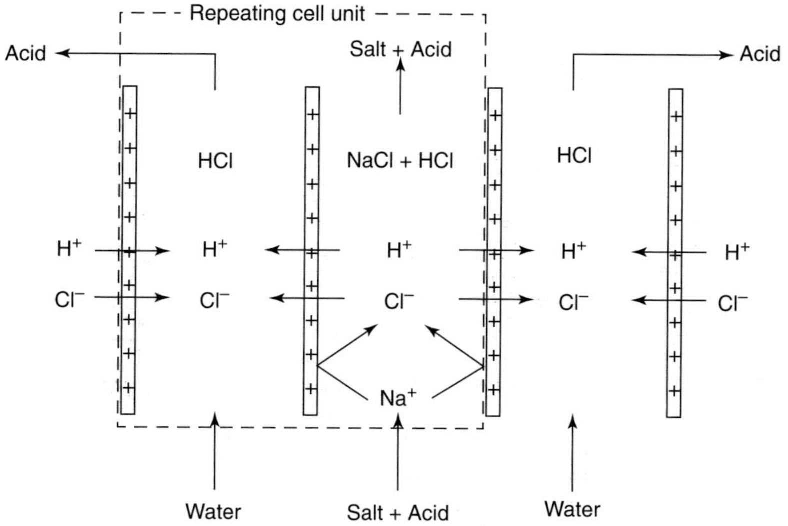

Figure 5 Schematic drawing illustrating the principle of diffusion dialysis used to recover an acid from a mixture with salt in a stack of anion-exchange membranes only.

compartment containing pure water as so-called stripping solution anions will diffuse from the feed solution through the ion-exchange membrane into the stripping solution due to a concentration difference, while the salt cations will be retained by the membrane. The protons, however, can pass the anion-exchange membrane in spite of their positive charge. Thus, the acid will be removed from the salt solution. Correspondingly, a base can be removed from mixtures with salts if cation-exchange membranes are used.

Diffusion dialysis is used to recover acids from pickling solutions in the metal surface treating industry [7]. However, its commercial relevance is still rather limited because of costs. Since the diffusion through the relatively thick ion-exchange membranes is a rather slow process, large membrane areas are required to remove a significant amount of ions from a feed solution, resulting in high investment costs for a given capacity plant.

Donnan dialysis. The principle of Donnan dialysis is shown in Fig. 6. Only cation- or anion-exchange membranes are installed in a stack. The driving force for the transport of ions is their concentration difference in the two phases separated by the membranes [8]. A typical application of Donnan dialysis is the removal of divalent ions such as Ca^{2+} from a feed stream by the exchange for monovalent ions such as Na^{+} in water softening

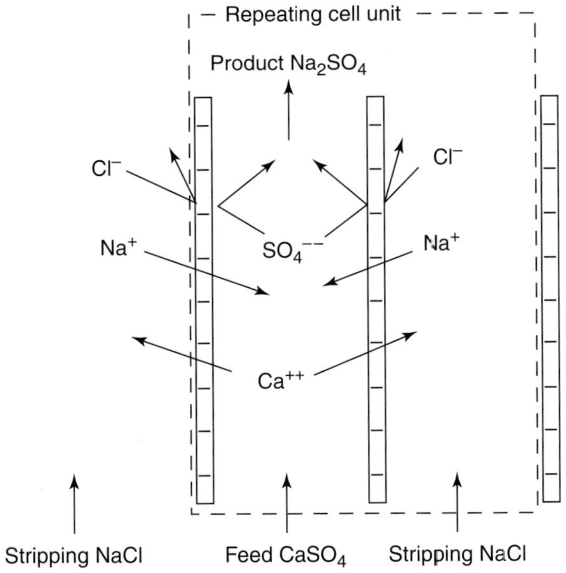

Figure 6 Schematic drawing illustrating the principle of the Donnan dialysis water softening process by the exchange of Na^{+} and Ca^{2+} ions in a stack with cation-exchange membranes only.

as illustrated in Fig. 6, which shows a feed solution containing $CaCl_2$ in relatively low concentration and a stripping solution containing NaCl in relatively high concentration flowing through alternating cells of a stack of cation-exchange membranes. Because of the concentration difference in the feed and the stripping solution Na^+ ions diffuse from the stripping solution through the cation-exchange membrane into the feed solution. Since the Cl^- ions cannot permeate the negatively charged cation-exchange membrane, an electrical potential is generated between the two solutions, which acts as driving force for the transport of Ca^{2+} ions from the feed to the stripping solution. Because of the required electroneutrality the identical charges are exchanged between the two solutions, that is, for two Na^+ ions diffusing from the stripping into the feed solution one Ca^{2+} ion is removed from the feed solution if the membrane is completely impermeable for Cl^- ions. The ion transport in Donnan dialysis is referred to as countercurrent transport.

In addition to water softening, there are several other interesting applications in wastewater treatment, but up to today, there is very little large-scale commercial use of Donnan dialysis.

Capacitive deionization. Capacitive deionization is an electrosorption process that can be used to remove ions from an aqueous solution by charge separation. The process is similar to conventional electrodialysis. But it also differs in a number of ways from electrodialysis as well. The main difference is that in capacitive deionization ions are removed from a solution without an oxidation/reduction reaction and the electrode compartments participate directly in the deionization and ion concentration process, that is, the anolyte and catholyte are contained within the porous electrodes and electrons are not transmuted by oxidation/reduction reactions but by electrostatic adsorption [9]. A cell of a capacitive deionization unit consists of two electrodes made out of activated carbon separated by a spacer that acts as a flow channel for an ion containing solution as illustrated in Fig. 7. The system resembles a "flow-through capacitor." If an electrical potential is applied between the electrodes, ions are removed from the solution and adsorbed at the surface of the charged electrodes. When the carbon electrodes are saturated with the charges, that is, the ions are released from the electrodes by reversing the potential, that is, the cathode becomes the anode and vice versa the anode becomes the cathode. Thus, capacitive deionization is a two-step process. In a first step as shown in Fig. 7a, ions are removed from a feed solution by electrosorption and migration in the feed solution under an electrical potential driving force, resulting in deionized

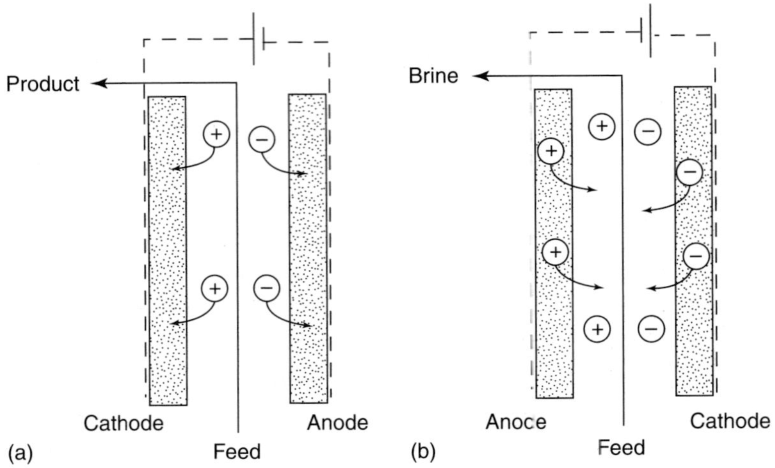

Figure 7 Schematic diagrams illustrating the capacitive deionization process: (a) sorption of ions from a feed solution at the porous carbon electrodes producing deionized product water and (b) desorption of ions from the porous carbon electrodes into the feed solution due to a change of polarity-producing concentrated brine.

product water. In a second step as shown in Fig. 7b, the adsorbed ions are released from the carbon electrodes and transported back into the feed solution by reversing the polarity-producing concentrated brine.

A key component in this process is the carbon electrode. Since the number of ions adsorbed at the electrodes is directly proportional to the available surface area the specific surface area, that is, the surface area per unit weight of the electrodes should be as high as possible. Activated carbon, carbon nanotubes, and especially carbon aerogels are the most promising materials. Their specific surface area is up to $1100\,m^2\,g^{-1}$. Another parameter that determines the energy consumption required to transport the ions from the feed solution to the electrodes is determined by the number of ions removed from the feed solution, that is, the concentration difference between the feed solution and the product and the applied voltage between the electrodes. Therefore, the resistance of the feed solution should be as low as possible. To avoid electrode reaction, which results in water dissociation and the production of hydrogen and oxygen or chlorine, the voltage drop at the electrodes should not exceed a certain value given by the water dissociation potential. Therefore, capacitive deionization cells are operated at a voltage drop of between 0.8 and 1.5 V. The deionization of a given feed solution and the regeneration of the capacitor is a function of time. The efficiency of the capacitive deionization is impaired by incomplete sorption and desorption of ions at the corresponding electrode especially at high ion concentrations due to the electrode pore solution concentration. Dissolved counterions in the pore solution are adsorbed on the electrode surface while the co-ions are expelled. Thus, counterions occupy capacitance within the electrode, which then is unavailable for the removal of ions from the feed solution. Co-ions expelled from the electrodes enter into the feedwater stream and increase the ion concentration in the purified product water in the deionization step. During the regeneration step, ions are desorbed from the electrodes and transported into the feed solution, increasing its concentration above its original value. However, when the voltage is reversed, ions are simultaneously adsorbed and repelled. This affects the upper limit of the concentration of the regeneration stream and reduces the ionic efficiency of capacitive deionization. The effect of the pore solution transport can be reduced significantly by placing a charged barrier between the feed solution and the electrodes as illustrated in Fig. 8, which shows schematically the ion transport in membrane-capacitive deionization.

During the deionization step, anions are prevented to diffuse into the product water by a cation-exchange membrane on the cathode- and an anion-exchange membrane on the anode as shown in Fig. 8a. In the

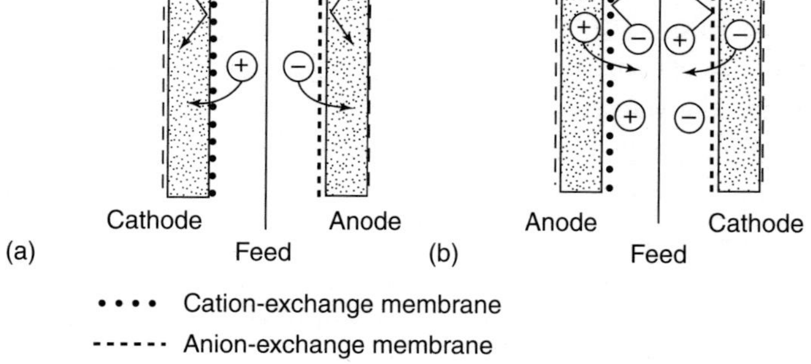

Figure 8 Schematic diagrams illustrating the capacitive deionization process with ion-exchange membranes between the feed solution and the porous carbon electrodes (a) shows adsorption, that is, the deionization step and (b) shows the desorption of ions due to a change of polarity-producing concentrated brine, that is, the regeneration step.

regeneration step under reverse polarity condition, the cation-exchange membrane prevents now the transport of anions toward the anode and the anion-exchange membrane the transport of cations toward the cathode as shown in Fig. 8b and thus avoids ion adsorption at the electrodes during the regeneration step. The consequence of introducing an ion-exchange membrane between the feed solution and the electrode is that more ions are adsorbed during the deionization step and more ions are desorbed and released during the regeneration step than in a capacitive deionization process without ion-exchange membranes between feed solution and electrodes. In an industrial-size capacitive deionization unit, a cathode, a feedflow channel, two ion-exchange membranes, and an anode are stacked as repeating units between two endplates.

1.2.2 Ion-exchange membranes in electrochemical synthesis and power generation

Ion-exchange membranes are also used in electrochemical synthesis of certain organic compounds and chemicals such as chlorine and caustic soda, or oxygen and hydrogen. They are also used in energy conversion systems, such as fuel cells and reverse electrodialysis systems. The applications are not subjects of this discussion, which is concentrated on water treatment only.

2. TRANSPORT OF IONS IN MEMBRANES AND SOLUTIONS

The transport rate of a component in a membrane and a solution is determined by its concentration, its mobility in a given environment, and the driving force or forces acting on the component. The concentration and mobility of a component are determined by its interaction with other components in its surrounding. The driving forces for the transport are gradients in the electrochemical potential. For applying an electrical potential in an electrolyte solution, two electron conductors must be in contact with an electrolyte. At the electrode/electrolyte interface, the electron conductance is converted to an ionic conductance by an electrochemical reaction. In electrolyte solutions, the electrostatic forces must always be balanced, that is, the number of positive and negative electrical charges must always be equal on a macroscopic electrolyte volume.

2.1 Electric current in electrolyte solutions and Ohm's law

When an electrical potential difference is established between two electrodes in contact with an electrolyte solution, the anions migrate toward the anode where they are oxidized by releasing electrons to the electrode in an electrochemical reaction. Likewise, cations migrate towards the cathode where they are reduced by receiving electrons from the electrode. Thus, the transport of ions in an electrolyte solution between electrodes results in a transport of electrical charges, that is, an electrical current, which can be described by the same mathematical relation as the transport of electrons in a metallic conductor, that is, by Ohm's law, which is given by

$$U = RI \tag{1}$$

where U is the electrical potential between two electron sources, for example, between two electrodes, I is the electrical current between the electron sources, and R the electrical resistance.

The resistance R is a function of the specific resistance of the material, the distance between the electron sources, and the cross-section area of the material through which the electric current passes. It is given by

$$R = \rho \frac{l}{q} \tag{2}$$

where R is the overall resistance, ρ the specific resistance, l the length, and q cross-section area of the conducting material.

The reversal of the resistance and of the specific resistance, respectively, is the conductivity and the specific conductivity, thus is

$$S = \frac{1}{R} \text{ and } \kappa = \frac{1}{\rho} \tag{3}$$

where S is the conductivity and κ the specific conductivity.

The conductivity of electrons in metal conductors, however, is generally 3–5 orders of magnitude higher than that of ions in electrolyte solutions. Furthermore, the conductivity of metals decreases with increasing temperature while the conductivity of electrolyte solutions increases with increasing temperature. The most important difference between electron and ion conductivity, however, is the fact that ion conductivity is always coupled with a transport of mass while, due to the very small mass of an electron, virtually no mass is transported in an electron conductor.

The conductivity of electrolyte solutions depends on the concentration and the charge number of the ions in the solution. It is expressed as the equivalent conductivity which is given by

$$\Lambda_{eq} = \frac{\kappa}{C(z_a v_a + z_c v_c)} \tag{4}$$

where Λ_{eq} is the equivalent conductivity, C is the molar concentration of the electrolyte in the solution, z_a and z_c are the charge numbers of the anion and cation, respectively, and v_a and v_c are the stoichiometric coefficients of the anion and cation, respectively.

The stoichiometric coefficient gives the number of anions and cations in a mole electrolyte and the charge number gives the number of charges related to an ion. For example, for NaCl v_c and v_a are identical and 1 and also z_a and z_c are 1. However, for $MgCl_2$ is $v_c = 1$, $v_a = 2$, and $z_c = 2$, $z_a = 1$.

The number of electrical charges carried by all the ions of an electrolyte under the driving force of an electrical potential gradient through a certain area is given by

$$J_e = \sum_i z_i u_i v_i C F \frac{\Delta\varphi}{l} = \sum_i z_i F J_i = \sum_i z_i v_i C \lambda_{eq} \frac{\Delta\varphi}{l} \tag{5}$$

where J_e is the flux of the electrical charges and J_i that of the individual ions; z, u, and v are the charge number, the ion mobility, and the stoichiometric coefficient, respectively; C is the concentration of the electrolyte; $\Delta\varphi$ and l are the potential difference and the distance between two points, respectively; F is the Faraday constant, which is $F = 96{,}485\ \mathrm{C\,eq^{-1}}$; and λ_{eq} is the equivalent conductivity.

Thus, the flux of electrical charges represents an electrical current, which is according to Ohm's law given by

$$I = \frac{U}{R} = \sum_i z_i F J_i A = \sum_i z_i v_i C \lambda_{eq} \frac{\Delta\varphi}{l} A = \kappa \frac{\Delta\varphi}{l} = iA \tag{6}$$

where I is the current, U the applied voltage, R the resistance, $\Delta\varphi$ the voltage difference between two points, l the distance between the two points, κ the conductivity, A the cross-section area of the conducting media, and i the current density.

2.2 Mass transport in ion-exchange membranes and electrolyte solutions

To describe the mass transport in an electrolyte solution or in an ion-exchange membrane, three independent fluxes must be considered, that is, the fluxes of the cations, the flux of anions, and the flux of the solvent [10]. The transport of ions is the result of an electrochemical potential gradient and the transport of the solvent through the membrane is a result of osmotic and electroosmotic effects.

2.2.1 Fluxes and driving force in ion transport

The driving force for the flux of a component in electromembrane processes is a gradient in its electrochemical potential, which is given at constant temperature by

$$d\tilde{\mu}_i = d\mu_i + d\varphi = \bar{V}_i dp + RT d \ln a_i + z_i F d\varphi \tag{7}$$

where $d\tilde{\mu}$, $d\mu$, $d \ln a$, dp, and $d\phi$ are the gradients of the electrochemical potential, the chemical potential, the activity, the hydrostatic pressure, and the electrical potential; F is the Faraday; R the gas constant; and T the temperature.

The mass transport in electromembrane processes at constant pressure and temperature can be described as a function of the driving force by a phenomenological equation, that is,

$$J_i = \sum_i L_{ik} \frac{d\eta_k}{dz} = \sum_i L_{ik} \left(RT \frac{d \ln a_i}{dz} + z_i F \frac{d\varphi}{dz} \right) \tag{8}$$

where L_{ik} is a phenomenological coefficient relating the driving forces to the corresponding fluxes, and the subscripts i and k refer to various components in the system [10].

Assuming an ideal solution in which the activity of a component is identical to its concentration and no kinetic coupling occurs between individual fluxes, then Eq. (8) becomes identical with the Nernst–Planck flux equation [11], which is given by

$$J_i = -D_i \left(\frac{dC_i}{dz} + \frac{z_i F C_i}{RT} \frac{d\varphi}{dz} \right) \tag{9}$$

where D_i is the diffusion coefficient of the component i, which is related to the phenomenological coefficient by

$$D_i = \frac{L_{ii} RT}{C_i} \tag{10}$$

The first term $D_i\,(dC_i/dz)$ in Eq. (9) represents the diffusion and the second term $D_i\,(z_i C_i F/RT)\,(d\varphi/dz)$ the migration of a component.

Thus, the Nernst–Planck equation is an approximation of the more general phenomenological equation.

2.2.2 Electrical current and ion fluxes

The electric current in an electrolyte solution is transported by ions only and described by

$$i = \frac{I}{A} = F \sum_{i} z_i J_i \tag{11}$$

where i is the current density, I the current, A the membrane surface, F the Faraday constant, J the flux, and z the charge number, the subscript i refers to cations and anions.

Introducing Eq. (9) into 11 and rearranging leads to

$$i = F \sum_{i} z_i J_i = F^2 \sum_{\mathrm{i}} z_{\mathrm{i}}^2 \frac{C_i D_i}{RT} \left(\frac{RT}{z_i C_i F} \frac{\mathrm{d}C_i}{\mathrm{d}z} + \frac{\mathrm{d}\varphi}{\mathrm{d}z} \right) \tag{12}$$

where i is the current density, C the concentration, F the Faraday constant, φ the electrical potential, z the charge number, D the ion diffusivity, R the gas constant, T the absolute temperature, and the subscript i refers to anions and cations.

The term $(RT/z_i C_i F)(\mathrm{d}C_i/\mathrm{d}z)$ has the dimension of an electrical potential gradient and represents the concentration potential, which is established between two electrolyte solutions of different concentrations.

2.2.3 The transport number and the membrane permselectivity

In an electrolyte solution, the current is carried by both ions. However, cations and anions usually carry different portions of the overall current. In ion-exchange membranes, the current is carried preferentially by the counterions.

The fraction of the current, which is carried by a certain ion, is expressed by the ion transport number, which is given by

$$T_i = \frac{z_i J_i}{\sum_{i} z_i J_i} \tag{13}$$

where T_i is the transport number of the component i, J_i is its flux, and z_i its charge number.

The transport number T_i indicates the fraction of the total current that is carried by the ion i, the sum of the transport number of all ions in a solution is 1.

The membrane permselectivity is an important parameter for determining the performance of a membrane in a certain ion-exchange membrane process. It describes the degree to which a membrane passes an ion of one charge and retains an ion of the opposite charge. The permselectivity of cation- and anion-exchange membranes can be defined by the following relations [8]:

$$\Psi^{\mathrm{cm}} = \frac{T_{\mathrm{c}}^{\mathrm{cm}} - T_{\mathrm{c}}}{T_{\mathrm{a}}} \quad \text{and} \quad \Psi^{\mathrm{am}} = \frac{T_{\mathrm{a}}^{\mathrm{am}} - T_{\mathrm{a}}}{T_{\mathrm{c}}} \tag{14}$$

where Ψ is the permselectivity of a membrane, T is the transport number, the superscripts cm and am refer to cation- and anion-exchange membranes, and the subscripts c and a refer to cation and anion, respectively.

An ideal perm-selective cation-exchange membrane would transmit positively charged ions only, that is, for a transport number of a counterion in a cation-exchange membrane is $T_{\mathrm{c}}^{\mathrm{cm}} = 1$ and the permselectivity of the membrane is $\Psi^{cm} = 1$. The permselectivity approaches zero when the transport number within the membrane is identical to that in the electrolyte solution, that is, for $T_{\mathrm{c}}^{\mathrm{cm}} = T_{\mathrm{c}}$ is $\Psi^{cm} = 0$. For the anion-exchange membrane, the corresponding relation holds.

The transport number of a certain ion in the membrane is proportional to its concentration in the membrane, which again is a function of its concentration in the solutions in equilibrium with the membrane phase due to the Donnan exclusion. Thus, the selectivity of ion-exchange membranes results from the exclusion of co-ions from the membrane phase.

The concentration of a co-ion in an ion-exchange membrane can be calculated from the Donnan equilibrium. For a monovalent salt and a diluate salt solution and assuming the activity coefficients of the salt in the membrane and the solution to be 1, the co-ion concentration in the membrane is given to a first approximation by [12]:

$$^{\mathrm{m}}C_{\mathrm{co}} = \frac{^{\mathrm{s}}C_{\mathrm{s}}^{2}}{C_{\mathrm{fix}}} \tag{15}$$

where C is the concentration, the subscripts co, s, and fix refer to co-ion, salt, and fixed ion of the membrane, the superscripts s and m refer to membrane and solution.

Eq. (15) indicates that the co-ion concentration in the membrane and with that the permselectivity of the membrane is decreasing with the salt concentration in the solution and will vanish when the salt concentration in the solution is identical to the fixed ion concentration of the membrane.

2.2.4 Membrane counterion permselectivity

The transport number of different counterions in an ion-exchange membrane can be quite different. The transport rates of ions through a membrane are determined by their concentration and mobility. The concentration of the counterions is always close to the concentration of the fixed charges of the membrane. The mobility of the ions depends mainly on the radius of the hydrated ions and the membrane structure. The mobility of different ions in an aqueous solution does not differ very much from each other. Exceptions are the H^+ and OH^- ions. Their mobility is about a factor of 5–8 higher than that of other ions. This exceptionally high mobility of protons and hydroxide ions is the result of molecular interaction of water dipoles with electrical charges. Protons form hydronium ions and are transported mostly via a so-called tunnel mechanism from one hydronium ion to the next water molecule. This explains not only the extraordinary high mobility of protons but it is also one of the reasons for the high permeability of anion-exchange membranes for protons while these membranes generally have a very low permeability for salt cations, which move with their hydrate shell through the solution by diffusion. The same mechanism holds also true for the transport of hydroxide ions and thus the permeability of hydroxide ions is much higher than that of other salt anions. Because protons and hydroxide ions are transported only to a small extent as individual ions surrounded by a hydration shell, they contribute very little to the electroosmotic transport of water, and their water transport number is always quite low.

The permselectivity of an ion-exchange membrane for different counterions is determined by the concentration and the mobility of the different ions in the membrane as indicated earlier. The concentration of the different counterions in the membrane is determined mainly by electrostatic effects referred to as "electroselectivity" [13]. The mobility depends on the size of the hydrated ion.

A typical counterion-exchange sequence of a cation-exchange membrane containing a SO_3 group as fixed charge is:

$$Ba^{2+} > Pb^{2+} > Sr^{2+} > Ca^{2+} > Mg^{2+} > Ag^{+} > K^{+} > NH_4^{+} > Na^{+} > Li^{+}$$

A similar counterion-exchange sequence is obtained for anions in an anion-exchange membrane containing quaternary ammonium groups as fixed charges:

$$I^- > Br^- > Cl^- > F^-$$

The permselectivity is the product of the ion-exchange selectivity and mobility selectivity. The mobility of different ions is determined mainly by steric effects, that is, the size of the ions and the cross-linking density of the membrane.

2.2.5 Water transport in electrodialysis

Water transport in electrodialysis from the diluate to the concentrate process stream can affect the process efficiency significantly. If a convective flux as a result of pressure differences between flow streams can be excluded, there are still two sources for the transport of water from the diluate to the concentrate solution. The first one is the result of osmotic pressure differences between the two solutions, and the second is due to electroosmosis, which results from the coupling of water to the ions being transported through the membrane due to the driving force of an electrical potential.

Each of the two fluxes may be dominant depending on the perm-selectivity of the ion-exchange membrane, the concentration gradient, and the current density. In a highly perm-selective membrane and with moderate differences in the salt concentration in the two solutions separated by the membrane the electroosmotic flux is dominating and generally much higher than the osmotic solvent flux. In electrodialysis, the water flux due to electroosmosis can be expressed by a solvent transport number which gives the number of water molecules transported by one ion.

The water transport number thus is

$$^{\mathrm{m}}T_{\mathrm{w}} = \frac{J_{\mathrm{w}}}{\sum_i J_i} \qquad (16)$$

where $^{\mathrm{m}}T_{\mathrm{w}}$ is the water transport number, J_{w} is the water flux, and J_i is the flux of ions through a given membrane.

The water transport number refers to the number of water molecules transferred by one ion through a given membrane. It depends on the membrane and on the electrolyte, that is, on the size of the ions, their valence, and their concentration in the solution. In aqueous salt solutions and commercial ion-exchange membranes, the water transport number is in the order of 4–8, that is, 1 mol of ions transports ca. 4–8 mol of water through a typical commercial ion-exchange membrane.

3. ION-EXCHANGE MEMBRANES, AND THEIR PROPERTIES AND PREPARATION

The key components in electrodialysis and related processes are the ion-exchange membranes. There are three different types of ion-exchange membranes: (1) cation-exchange membranes, which contain negatively charged groups fixed to the polymer matrix, (2) anion-exchange membranes, which contain positively charged groups fixed to the polymer matrix, and (3) bipolar membranes, which are composed of an anion- and a cation-exchange layer laminated together. The fixed charges of an ion-exchange membrane are in electrical equilibrium with mobile ions in the interstices of the polymer matrix as indicated in Fig. 1 which shows schematically the structure of a cation-exchange membrane with negative charges fixed to the polymer matrix. The most desired properties of ion-exchange membranes are high permselectivity, low electrical resistance, good mechanical and form stability, and high chemical and thermal stability. In addition to these properties, bipolar membranes should have high catalytic water dissociation rates.

The properties of ion-exchange membranes are determined by two parameters, that is, the basic material they are made from and the type and concentration of the fixed ionic moiety. The basic material determines to a large extent the mechanical, chemical, and thermal stability of the membrane. The type and the concentration of the fixed ionic charges determine the permselectivity and the electrical resistance of the membrane, but they also have a significant effect on the mechanical properties of the membrane and their swelling in solution.

The following moieties are used as fixed charges in cation-exchange membranes:

$$-SO_3^-,\ -COO^-,\ -PO_3^{2-},\ -PHO_2^-,\ -AsO_3^{2-},\ -SeO_3^-$$

In anion-exchange membranes, fixed charges may be the following moieties:

$$-\overset{+}{N}H_2R,\ -\overset{+}{N}HR_2,\ -\overset{+}{N}R_3,\ -\overset{+}{P}R_3,\ -\overset{+}{S}R_2$$

The sulfonic acid group is completely dissociated over nearly the entire pH range, while the carboxylic acid group is virtually undissociated in the pH range <3. The quaternary ammonium group again is completely dissociated over the entire pH range, while the secondary ammonium group is only weakly dissociated. Accordingly, ion-exchange membranes are referred to as being weakly or strongly acidic or basic in character.

3.1 Preparation of ion-exchange membranes

Ion-exchange membranes are ion-exchange resins in sheet form. There are, however, significant differences between ion-exchange resins and membranes as far as the mechanical properties and especially the swelling behavior are concerned. Ion-exchange resins are mechanically weak or tend to swell drastically in diluate electrolyte solutions [14]. The most common solution to this problem is the preparation of a membrane with a backing of a stable reinforcing material, which gives the necessary strength and dimensional stability. Two techniques are used today for the preparation of ion-exchange membranes. One leads to a more heterogeneous and the other to a more homogeneous structure. Both preparation procedures are described in great detail in the patent literature [15–17].

3.1.1 Preparation of heterogeneous ion-exchange membranes

Ion-exchange membranes with a heterogeneous structure consist of fine ion-exchange particles embedded in an inert binder polymer such as polyethylene, phenol resins, or polyvinylchloride. Heterogeneous ion-exchange membranes are characterized by the discontinuous phase of the ion-exchange material. The efficient transport of ions through a heterogeneous membrane requires either a contact between the ion-exchange particles or an ion-conducting solution between the particles. Heterogeneous ion-exchange membranes can easily be prepared by mixing an ion-exchange powder with a dry binder polymer and extrusion of sheets under the appropriate conditions of pressure and temperature or by dispersion of ion-exchange particles in a solution containing a dissolved film-forming binder polymer, casting the mixture into a film and then evaporating the solvent.

Heterogeneous ion-exchange membranes with useful low electrical resistances contain more than 65% by weight of the ion-exchange particles. Membranes that contain significantly less than 65 wt% ion-exchange particles have high electric resistance and membranes with significantly more resin particles have poor mechanical strength. Furthermore, heterogeneous membranes develop water filled interstices in the polymer matrix during the swelling process, which affects both the mechanical properties as well as the permselectivity.

The ion-exchange capacity of heterogeneous membranes is in the range of 1–2 eq kg^{-1} dry membrane and thus significantly lower than that of homogeneous membranes, which is between 2 and 3 eq kg^{-1} dry resin. In general, heterogeneous ion-exchange membranes have higher electrical resistances and lower permselectivity than homogeneous membranes.

3.1.2 Preparation of homogeneous ion-exchange membranes

Homogeneous ion-exchange membranes can be prepared by polymerization of monomers that contain a moiety that either is or can be made anionic or cationic, or by polymerization of a monomer that contains an anionic or a cationic moiety, or by introduction of anionic or cationic moieties into a polymer dissolved in a solvent by a chemical reaction, or grafting functional groups into a preformed polymer film.

A method of preparing both cation- and anion-exchange membranes, which is used for the preparation of commercial cation-exchange membranes, is the polymerization of styrene and divinylbenzene and its sulfonation according to the following reaction scheme [13]:

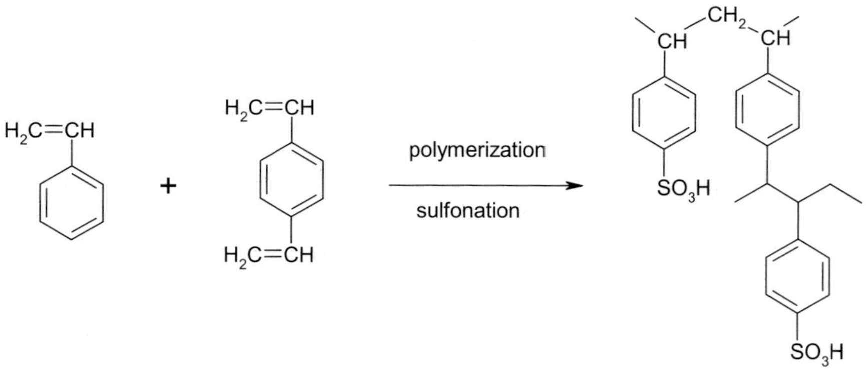

In a first step, styrene is partially polymerized and cross-linked with divinylbenzene and then in a second step sulfonated with concentrated sulfuric acid. The obtained membranes show high ion-exchange capacity and low electrical resistance. To increase the mechanical strength, the membrane is cast on a support screen.

A homogeneous anion-exchange membrane can be obtained by introducing a quaternary amine group into polystyrene by a chloromethylation procedure followed by an amination with a tertiary amine according to the following reaction scheme:

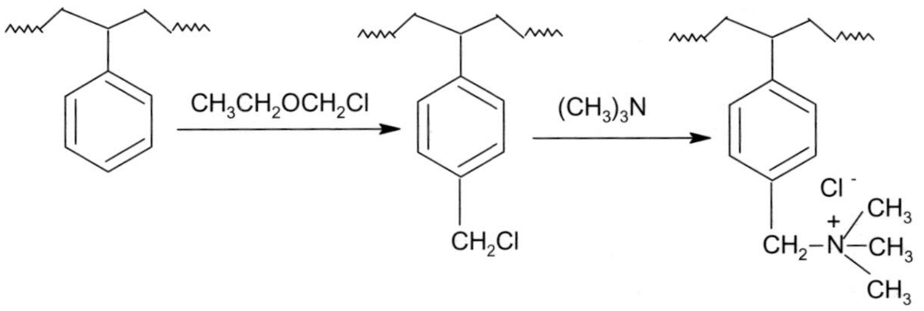

The membrane structures and their preparation described above are just two examples. There are many variations of the basic preparation procedure, resulting in slightly different products. Instead of styrene, often substituted styrene such as methylstyrene or phenylacetate is used and instead of divinylbenzene monomers such as divinylacetylene or butadiene are used [2].

More recently cation-exchange membranes with good mechanical and chemical stability and well controlled ion-exchange capacity are prepared by sulfonation of dissolved polysulfone [18]. The sulfonation is carried out with chlorosulfonic acid according to the following scheme:

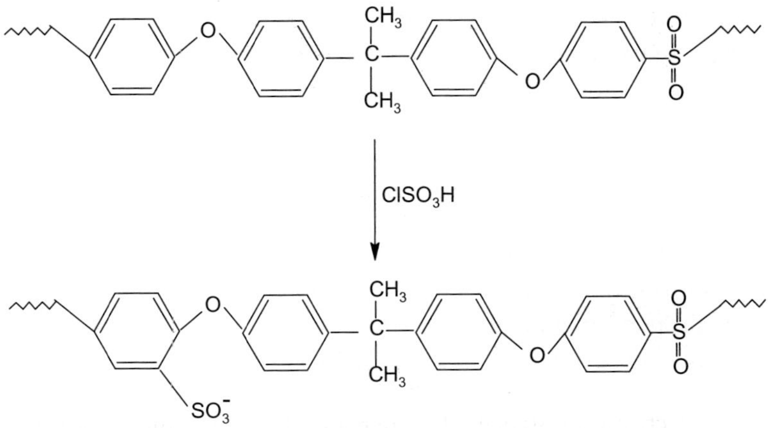

To obtain membranes with different ion-exchange capacity the sulfonated polysulfone can be mixed with unsulfonated polymer in a solvent such as *N*-methylpyrrolidone. By changing the ratio of the sulfonated to unsulfonated polymer, the fixed charge density can easily be adjusted to a desired value. The sulfonated polysulfone can be cast as a film on a screen. After the evaporation of the solvent, a reinforced membrane with excellent chemical and mechanical stabilities and good electrochemical properties is obtained.

Anion-exchange membranes based on polysulfone can be prepared by halomethylation of the backbone polymer and subsequent reaction with a tertiary amine.

For the preparation of cation-exchange membranes also, polyetheretherketone is used as the basic polymer. It can very easily be sulfonated with concentrated sulfuric acid according to the following scheme:

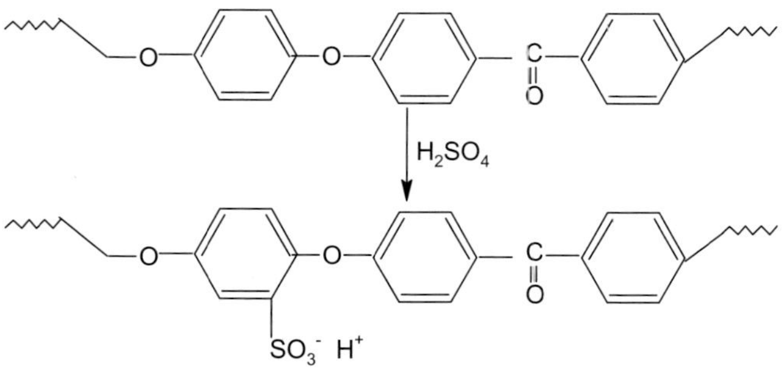

Sulfonation occurs on one polyetheretherketone block only and is thus very easy to control. To obtain membranes with different ion-exchange capacities, the sulfonated polyetheretherketone can be mixed with polyethersulfone in a solvent such as *N*-methylpyrrolidone. By changing the ratio of the sulfonated polyetheretherketone to polyethersulfone, the fixed charge density can easily be adjusted to a desired value.

3.1.3 Special property membranes

In addition to the monopolar membrane described above, a large number of special property membranes are used in various applications such as low-fouling anion-exchange membranes used in certain wastewater treatment applications or composite membranes with a thin layer of weakly dissociated carboxylic acid groups on the surface used in the chlorine/alkaline production, and bipolar membranes composed of laminate of an anion- and a cation-exchange layer used in the production of protons and hydroxide ions to convert a salt in the corresponding acids and bases. The preparation techniques are described in detail in the literature [2,19,20].

One of the technically and commercially most important cation-exchange membranes developed in recent years is based on perfluorocarbon polymers. Membranes of this type have extreme chemical and thermal stabilities and they are the key component in the chlorine/alkaline electrolysis as well as in most of today's fuel cells. They are prepared by copolymerization of tetrafluoroethylene with perfluorovinylether having a carboxylic or sulfonic acid group at the end of a side chain. There are several variations of a general basic structure commercially available today [21]. The various preparation techniques are described in detail in the patent literature.

Today's commercially available perfluorocarbon membranes have the following basic structure:

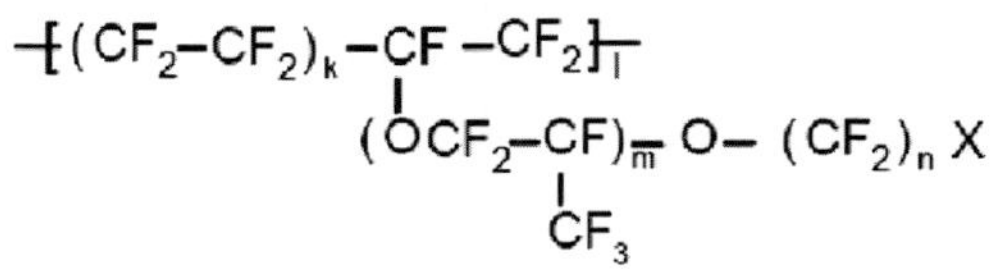

$k = 5\text{-}8,\ l = 600\text{-}1200,\ m = 1\text{-}2,\ n = 1\text{-}4,\ X = SO_3^-\ ,\ COO^-$

The synthesis of the perfluorocarbon membranes is rather complex and requires a multistep process. In addition to the various perfluorinated cation-exchange membranes also, perfluorinated anion-exchange membranes have been developed. The anion-exchange membranes have similar chemical and thermal properties as the cation-exchange membranes.

4. THE DESIGN OF ION-EXCHANGE MEMBRANE SEPARATION PROCESSES

The function of all ion-exchange membrane separation processes used in practical water treatment are based on the same general principle, that is, the selective transport of ions through ion-exchange membranes under the driving force of an electrochemical potential difference. The design of the various processes in their practical application can be very different. In all processes, the technical feasibility and economics in a given application is affected by the design of the process and certain hardware components.

4.1 Electrodialysis process and system design

The efficiency of electrodialysis is determined to a large extent by the properties of the membranes. But it is also affected by the process and system design, which determine the limiting current density, the current utilization, the concentration polarization, etc. Therefore, the process design has a significant effect on the overall efficiency and costs [22,23] in electrodialysis.

4.1.1 The electrodialysis stack and process parameters

A key element in electrodialysis is the so-called stack, which is a device to hold an array of membranes between the electrodes that the streams being processed are kept separated. A typical electrodialysis stack used in water desalination contains 100–300 cell pairs stacked between the electrodes. The electrode containing cells at both ends of a stack are often rinsed with a separate solution, which does not contain Cl^- ions to avoid chlorine formation.

The membranes in an electrodialysis cell are separated by spacer gaskets as indicated in Fig. 9, which shows schematically the design of a so-called sheet-flow electrodialysis stack. The spacer gasket consists of a screen, which supports the membranes and controls the flow distribution in the cell and a gasket that seals the cell to the outside and also contains the manifolds to distribute the process fluids in the different compartments. To minimize the resistance of the solution in the cell, the distance between two

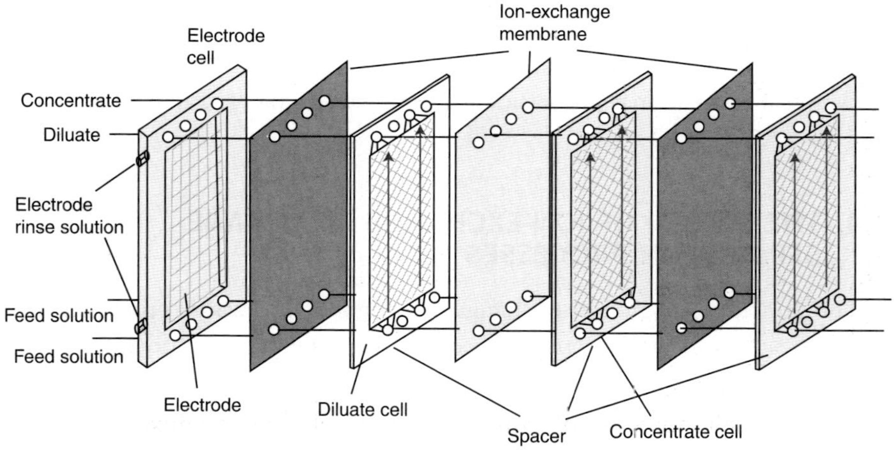

Figure 9 Exploded view of a sheet-flow type electrodialysis stack arrangement, indicating the individual cells and the spacer gaskets containing the manifold for the distribution of the different flow streams.

membranes is kept as small as possible and is in the range of 0.5–2 mm in industrial electrodialysis stacks. A proper electrodialysis stack design ensures uniform flow distribution and mixing of the solutions to minimize concentration polarization at the membrane surfaces at minimized pressure loss of the solution flow in the stack.

Concentration polarization and limiting current density. The limiting current density is the maximum current that may pass through a given cell pair area without detrimental effects. If the limiting current density is exceeded, the electric resistance in the diluate will increase and water dissociation may occur at the membrane surface, which affects current utilization and can lead to pH changes in the solutions.

The limiting current density is determined by concentration polarization at the membrane surface in the diluate containing compartment, which is determined by the diluate concentration, the compartment design, and the feedflow velocity. Concentration polarization in electrodialysis is also the result of differences in the transport number of ions in the solution and in the membrane. The transport number of a counterion in an ion-exchange membrane is generally close to 1 and that of the co-ion close to 0 while in the solution the transport numbers of anions and cations are not very different. At the surface of a cation-exchange membrane facing the diluate solution, the concentration of ions in the solution is reduced due to a transport number of the cations that is lower in the solution than in the membrane. Because of the electroneutrality requirements, the number of anions is reduced in the boundary layer by migration in the opposite direction. The net result is a reduction of the electrolyte concentration in the solution at the surface of the membrane and a concentration gradient is established in the solution between the membrane surface and the well-mixed bulk. This concentration gradient results in a diffusive electrolyte transport. A steady-state situation is obtained when the additional ions, which are needed to balance those removed from the interface due to the faster transport rate in the membrane, are supplied by the diffusive transport. The other side of the cation-exchange membrane faces the concentrate solution where the electrolyte concentration at the membrane surface is increased. The effect of concentration polarization is illustrated in Fig. 10, which shows the salt concentration profiles and the fluxes of cations and anions in the concentrate and diluate solution at the surface of a cation-exchange membrane.

The symbols J and C in Fig. 10 denote the fluxes and the concentration of ions, the superscripts mig and diff refer to migration and diffusion, the

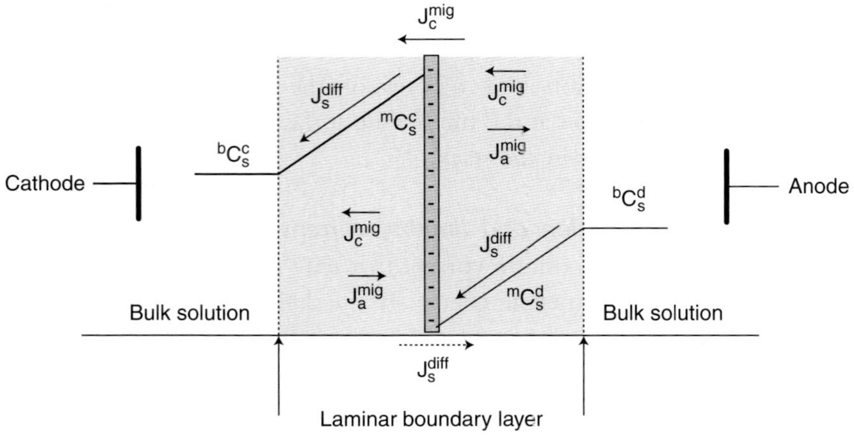

Figure 10 Schematic drawing illustrating the concentration profiles of a salt in the laminar boundary layer on both sides of a cation-exchange membrane and the flux of ions in the solutions and the membrane.

superscripts d and c refer to diluate and concentrate solution, and the superscripts b and m refer to bulk phase and membrane surface, respectively, the subscripts a, c, and s refer to anion, cation, and salt, respectively.

The concentration polarization occurring in electrodialysis can be calculated by a mass balance taking into account all fluxes in the boundary layer and the hydrodynamic conditions in the flow channel between the membranes. To a first approximation, the salt concentration at the membrane surface can be calculated by applying the so-called Nernst film model, which assumes that the bulk solution between the laminar boundary layers has a uniform entrance to the exit. In a practical electrodialysis stack, there will be entrance and exit effects and the idealized model hardly exists. Nevertheless, the Nernst model provides a very simple approach to the mathematical treatment of the concentration polarization, which results in an expression for the current density as a function of the bulk solution concentration, the transport number of the ions, the diffusion coefficient of the electrolyte and the thickness of the laminar boundary layer.

$$i = \frac{z_i F D_i}{T_i^{\mathrm{m}} - T_{i_i}^{\mathrm{s}}} \frac{\Delta C_i^{\mathrm{d}}}{Z_{\mathrm{b}}} \tag{17}$$

where i is the current density, T the transport number of the counterion, ΔC the concentration difference between the solution in the diluate at the membrane surface and in the bulk, D the diffusion coefficient, F the Faraday constant, z the charge number, Z_b the boundary layer thickness, the subscript i refers to cations or anions, and the superscripts d, m, and s refer to diluate, membrane, and solution, respectively.

When the flow conditions are kept constant, the boundary layer will be constant and the current density will reach a maximum value independent of the applied electrical potential gradient. This maximum current density, which is referred to as the limiting current density, is reached when the counterion concentration at the membrane surface becomes 0. Thus, $i = i_{\text{lim}}$ for ${}^{\text{m}}C_{\text{s}}^{\text{d}} \rightarrow 0$.

$$i_{\text{lim}} = \frac{z_i F D_{\text{s}}}{T_i^{\text{m}} - T_i^{\text{b}}} \frac{{}^{\text{b}}C_{\text{s}}^{\text{d}}}{Z_{\text{b}}} \tag{18}$$

where i_{lim} is the limiting current density and ${}^{\text{b}}C_{\text{s}}^{\text{d}}$ is the salt concentration of the diluate in the bulk solution. All other symbols have the same meaning as in Eq. (17).

The assumptions made in the Nernst film model are often not permissible in an electrodialysis stack. Therefore, the limiting current density in practical applications is generally experimentally determined and described as a function of the feedflow velocity in the electrodialysis stack by the following relation [23]:

$$i_{\text{lim}} = au^{\text{b}} F C_{\text{s}}^{\text{d}} \tag{19}$$

where C_{s}^{d} is the concentration of the solution in the diluate cell, u the linear flow velocity of the solution through the cells parallel to the membrane surface, F the Faraday constant, and a and b are characteristic constants for a given stack design and must be determined experimentally. This is done in practice by measuring the limiting current density in a given stack configuration at constant feed solution salt concentrations as a function of the feedflow velocity.

Current utilization. In practical applications, electrodialysis is affected by incomplete current utilization. The reasons for the incomplete current utilization are poor membrane permselectivity, parallel current through the stack manifold, and water transport by convection and due to osmosis and electroosmosis. In a well-designed stack with no pressure difference between diluate and the concentrate, the convective water transport is negligibly low and also the current through the manifold can be neglected.

Under these conditions, the overall current utilization is given by

$$\xi = n(\psi^{\mathrm{cm}} T_{\mathrm{a}}^{\mathrm{s}} + \psi^{\mathrm{am}} T_{\mathrm{c}}^{\mathrm{s}})[1 - (T_{\mathrm{w}}^{\mathrm{cm}} + T_{\mathrm{w}}^{\mathrm{am}})\bar{V}_{\mathrm{w}}(C_{\mathrm{s}}^{\mathrm{ind}} - C_{\mathrm{s}}^{\mathrm{outd}})] \quad (20)$$

where ξ is the current utilization; ψ is the membrane permselectivity; T is the transport number; n is the number of cell pairs in the stack; $\bar{V}_{\mathrm{w}}$ is the partial molar volume of water; and C is the concentration; a, c, s, and w refer to anion, cation, solution, and water, respectively; and the superscripts cm, am, ind, and outd refer to cation-exchange membrane, anion-exchange membrane, and diluate at the inlet and outlet of a stack, respectively.

Electrodialysis equipment and plant design. In most commercially used electrodialysis stacks, the membranes are arranged in parallel between two electrodes and are separated by spacers, which also hold the manifolds for the distribution of the individual flow channel as indicated in Fig. 11. There are two major concepts as far as the construction of the spacers is concerned. One is the so-called sheet-flow spacer concept, which is illustrated in Fig. 11a and the other is the so-called tortuous path concept, which is illustrated in Fig. 11b. The main difference between the sheet-flow and the tortuous path flow spacer is that in the sheet-flow spacer the compartments are vertically arranged and the process path is relatively short. The flow velocity of the solutions in the cells formed by two membranes and a spacer

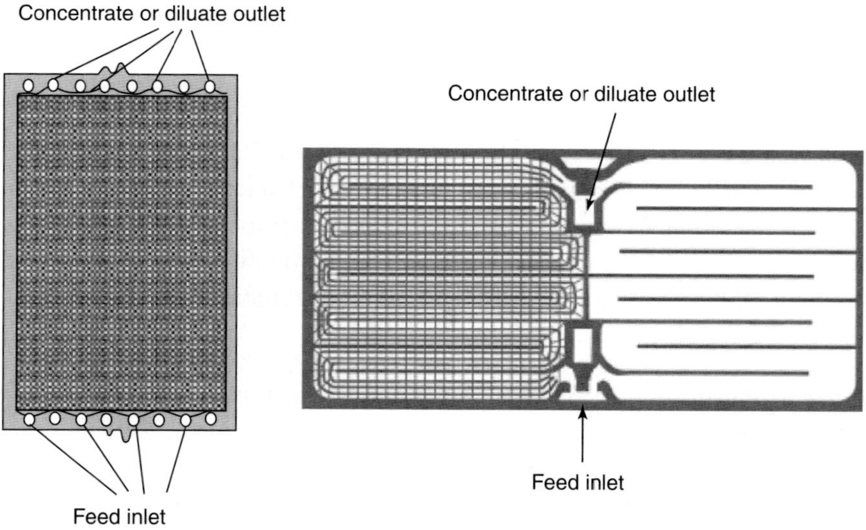

Figure 11 Schematic drawing illustrating the sheet-flow and a tortuous path spacer concept.

is between 2 and 4 cm s^{-1} and the pressure loss is 0.2–0.4 bar. In the tortuous path flow stack, the membrane spacers are horizontally arranged and have a long serpentine cut-out, which defines a long narrow channel for the fluid path. The feedflow velocity in the stack is relatively high, that is, between 6 and 12 cm s^{-1}, which provides a better control of concentration polarization and higher limiting current densities, but the pressure loss in the feedflow channels is quite high, that is, between 1 and 2 bar.

In the practical application of electrodialysis, there are two main process operation modes. The first one is referred to as the unidirectional electrodialysis and the second one electrodialysis reversal [24]. In a unidirectional-operated electrodialysis system, the electric field is permanently applied in one direction and the diluate and concentrate cells are also permanently fixed over the period of operation. Unidirectional-operated electrodialysis plants are rather sensitive to membrane fouling and scaling and require periodical rinsing of the stack with acid or detergent solutions. In desalination of brackish or surface waters, generally electrodialysis reversal is applied. In the electrodialysis reversal operating mode, the polarity of the electric field applied to the electrodialysis stack is reversed in certain time intervals. Simultaneously, the flow streams are reversed, that is, the diluate cell becomes the concentrate cell and vice versa with the result that matter being precipitated at the membrane surface will be redissolved and removed with the flow stream passing through the cell. The principle of the electrodialysis reversal is illustrated in Fig. 12.

Fig. 12 shows an electrodialysis cell unit formed by a cation- and anion-exchange membrane between two electrodes and a feed solution containing negatively charged large "fouling" components. If an electric field is applied, these components will migrate to the anion-exchange membrane and be deposited on its surface to form a so-called fouling layer, which affects the efficiency of the electrodialysis process. If the polarity is reversed, the negatively charged components will now migrate away from the anion-exchange membrane back into the feed stream and the membrane properties are restored. This procedure, which is referred to as "clean in place," is very effective not only for the removal of colloidal fouling materials but also for removing precipitated salts and is used today in almost all electrodialysis water desalination systems.

However, reversing the polarity of a stack has to be accompanied by a reversal of the flow streams. This requires a more sophisticated flow control. The flow scheme of an electrodialysis plant operated with reversed polarity is shown in Fig. 13. In the reverse polarity operating mode, the hydraulic

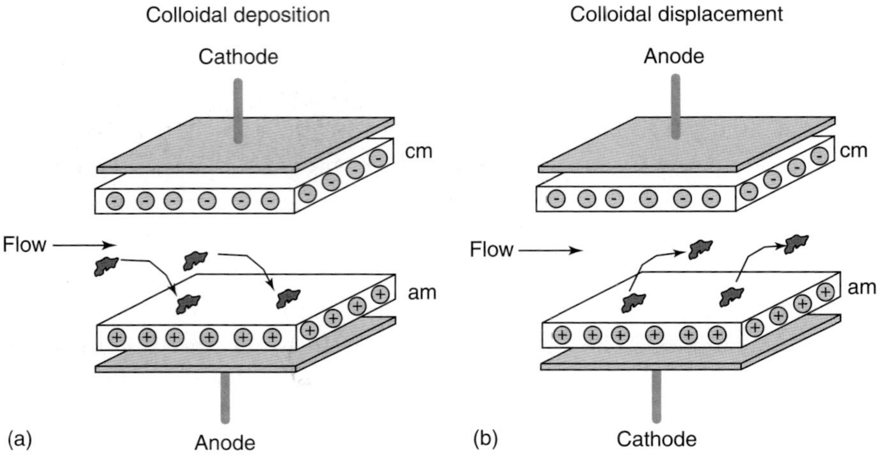

Figure 12 Schematic drawing illustrating the removal of deposited negatively charged colloidal components from the surface of an anion-exchange membrane by reversing the electric field in the electrodialysis reversal operating mode.

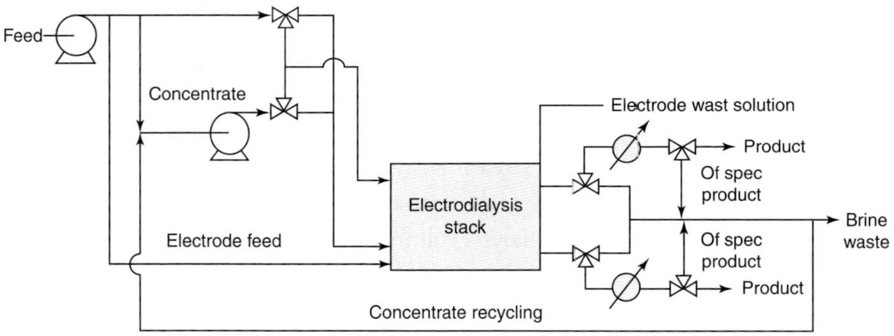

Figure 13 Flow scheme of electrodialysis reversal in a continuous operating mode with the feed solution also used as electrode rinse.

flow streams are reversed simultaneously, that is, the diluate cell will become the brine cell and vice versa. In this operating mode, the polarity of the current is changed at specific time intervals ranging from a few minutes to several hours.

While reversing the polarity and the flow streams, there is a brief period when the concentration of the desalted product exceeds the product quality specification. Therefore, the product water outlet has a concentration sensor, which controls an additional three-way valve. This valve diverts high concentrated product to waste and then, when the concentration

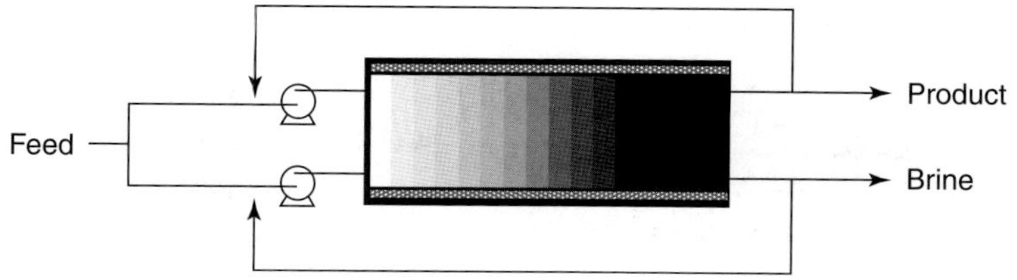

Figure 14 Flow scheme of an electrodialysis stack operated in a feed and bleed mode, that is, with partial recycling of the diluate and concentrate solutions.

returns to the specified quality, directs the flow to the product outlet. Thus, in electrodialysis reversal, there is always a certain amount of the product lost to the waste stream.

The degree of desalination that can be achieved in passing the feed solution through a stack is a function of the solution concentration, the applied current density, and the residence time of the solution in the stack. If the degree of desalination or concentration that can be achieved in a single path through the stack is insufficient, several stacks are operated in series or part of the diluate or concentrate is fed back to the feed solution as shown in Fig. 14.

In the so-called feed and bleed operating mode, both the brine and the product concentration can be determined independently and very high recovery rates can be obtained.

4.1.2 Electrodialysis process costs

The total costs in electrodialysis are the sum of fixed charges associated with the plant investment costs and the plant operating costs. Both the capital costs as well as the plant operating costs per unit product are a function of the feed solution and the required product and brine concentration. But they are also strongly affected by the plant capacity and the overall process design [23].

Investment-related costs. The investment costs are determined mainly by the required membrane area for a certain plant capacity. Other items such as pumps and process control equipment are considered as a fraction of the required membrane area. The required membrane area for a given capacity plant can be calculated from the current required to remove a certain number of ions from the feed solution. Thus, the total current required for the desalination process is proportional to the concentration difference between the feed and diluate solution, the total volume flow of the diluate through the stack, and the Faraday constant. It is inversely proportional to the number of cell pairs in the stack and the current utilization. The total current passing through the stack is given by

$$I = \frac{Q_{st}^{d} F(C^{ind} - C^{outd})}{N_{cell}\xi} = \frac{A_{st} i}{N_{cell}} = \frac{Q_{cell}^{d} F(C^{ind} - C^{outd})}{\xi} = A_{cell} i \quad (21)$$

Thus

$$N_{cell} Q_{cell} = Q_{st} \text{ and } N_{cell} A_{cell} = A_{st} \text{ and } A_{st} = \frac{Q_{st}^{d} F(C^{ind} - C^{outd})}{i\xi} \quad (22)$$

where I and i represent the electric current and the current densities, A is the cell pair area, N_{cell} refers to the number of cell pairs in the stack, Q is the volume flow, C is the concentration expressed in equivalent per volume, F is the Faraday constant, and ξ the current utilization. The subscripts st and cell refer to the stack and cell pair, and the superscripts outd and ind refer to diluate at stack outlet and inlet.

The voltage drop across the stack is constant over the entire cell length of a stack while the resistance changes from the feed inlet to the product outlet due to an increase of the resistance of the diluate as a result of the concentration change. Therefore, the current density also decreases from the feed entrance to the diluate exit. The current density is related to the resistance and the voltage by

$$U_{st} = N_{cell} i \bar{R} A_{cell} \quad (23)$$

where is U_{st} the voltage drop across the stack, A_{cell} is the cell pair area, and $\bar{R}$ is the average resistance of a cell pair.

The average resistance $\bar{R}$ can be calculated from the average concentrations in the diluate and concentrate cells $\bar{C}^{d}$ and $\bar{C}^{c}$ [23] and is given by

$$\bar{R} = N_{cell} \frac{1}{A_{cell}} \left[\frac{\Delta \ln(C^{ind}/C^{inc})(C^{outc}/C^{outd})}{\Lambda_{eq}(C^{ind} - C^{outd})} + r^{am} + r^{cm} \right] \quad (24)$$

where $\bar{R}$ is the average resistance, N_{cell} and A_{cell} are the number of cell pairs in a stack and the C^{ind} and C^{outd} are the salt concentrations of the diluate at the inlet and outlet of the cells, C^{inc} and C^{outc} are the salt concentrations of the concentrate at the inlet and outlet, Λ_{eq} is equivalent conductivity of the solutions in the cells, Δ is the cell thickness, and r^{am} and r^{cm} are the membrane area resistances.

The voltage drop in an electrodialysis stack is the result of the resistances of the membranes and the solutions, the concentration potential difference between the concentrate and diluate streams, and the voltage drop related to the electrode reaction. Most electrodialysis stacks used in practical applications consist of several hundred geometrically identical cells, which are operated in co-current flow, and it can be assumed that the concentration potentials as well as the electrode reactions can be neglected and that in the concentration range of interest, the equivalent conductivity is independent of the concentration [23]. Since the voltage drop is proportional to the current density, which should not exceed the limiting current density, there is a maximum voltage drop that may be applied. The relation between maximum voltage drop and the limiting current density is given by

$$U_{\max} = N_{\text{cell}} i_{\lim} \bar{R} A_{\text{cell}} = N_{\text{cell}} i_{\lim} \times \left[\frac{\Delta \ln(C^{\text{ind}}/C^{\text{inc}})(C^{\text{outc}}/C^{\text{outd}})}{\Lambda_{\text{eq}}(C^{\text{ind}} - C^{\text{outd}})} + r^{\text{am}} + r^{\text{cm}} \right] \tag{25}$$

where $U_{\max}$ is the maximum applied voltage across the stack and $i_{\lim}$ is the limiting current density. All other symbols are identical to that of Eq. (24).

The membrane area required for a certain plant capacity as function of the feed and product concentrations is obtained by combination, and rearranging of Eqs. (21)–(25) gives

$$A_{\text{st}} = \frac{\ln(C^{\text{ind}}/C^{\text{inc}})(C^{\text{outc}}/C^{\text{outd}}) + [\Lambda_{\text{eq}}(r^{\text{am}} + r^{\text{cm}})(C^{\text{ind}} - C^{\text{outd}})/\Delta]}{(C^{\text{outd}}/C^{\text{outc}}) + 1 + (\Lambda_{\text{eq}} C^{\text{outd}}/\Delta)(r^{\text{am}} + r^{\text{cm}})} \times \frac{Q^{\text{d}}_{\text{st}} F C^{\text{outd}}}{i_{\lim} \xi} \tag{26}$$

where A_{st} is the total membrane area in a stack and N_{cell} is the number of cell pairs in a stack. All other symbols are the same as the ones in the Eqs. (24) and (25).

The total investment-related costs depend on the price of the membranes and their useful life under operating conditions, which is in practical application 5–8 years, and on the price of the additional plant components and their life.

Operating costs. The operating costs are composed of labor cost, the cost of maintenance of the plant, and the energy cost. The labor and maintenance costs are proportional to the size of the plant and calculated as a percentage of the investment-related costs. The energy required in an electrodialysis process is an additive of two terms: (1) the electrical energy to transfer the ionic components from one solution through membranes into another solution and (2) the energy required to pump the solutions through the electrodialysis unit. The energy consumption due to electrode reactions can generally be neglected since in a modern electrodialysis stack more than 200 cell pairs are placed between the two electrodes. The energy required for operating the process control devices can generally also be neglected.

The total energy required in electrodialysis for the actual desalination process is given by the current passing through the electrodialysis stack multiplied with the total voltage drop encountered between the electrodes:

$$E_{\text{des}} = I_{\text{st}} U_{\text{st}} t \tag{27}$$

where E_{des} is the energy consumed in a stack for the transfer of ions from a feed to a concentrate solution, I_{st} the current passing through the stack, U_{st} the voltage applied across the stack, that is, between the electrodes, and t the time of operation.

The total current through the stack is given by Eq. (21) and the voltage across the stack is given by Eq. (24). Introducing the two equations into Eq. (26) and dividing by the produced diluate gives the desalination energy per volume product:

$$\begin{aligned} E_{\text{de,spc}} = {} & \frac{N_{\text{cell}} t}{A_{\text{cell}} V_{\text{pro}}} \left[\frac{\Delta \ln(C^{\text{ind}}/C^{\text{inc}})(C^{\text{outc}}/C^{\text{outd}})}{\Lambda(C^{\text{ind}} - C^{\text{outd}})} + r^{\text{am}} + r^{\text{cm}} \right] \\ & \times \left[\frac{Q^{\text{d}}_{\text{cell}} F(C^{\text{ind}}_{\text{s}} - C^{\text{outd}}_{\text{s}})}{\xi} \right]^2 \end{aligned} \tag{28}$$

where $E_{\text{de,spc}}$ is the specific desalination energy, t is the time of operation, C^{ind} and C^{inc} are the equivalent concentrations of the diluate and the concentrate at the cell inlet, C^{outd} and C^{outc} are the concentrations of the diluate and the concentrate at the cell outlet, Λ is the equivalent conductivity of the salt solution, r^{am} and r^{cm} are the area resistances of the anion- and cation-exchange membrane, Δ is the cell thickness, ξ is the current utilization, $Q^{\text{d}}_{\text{cell}}$ is the diluate flow rate in a cell, A_{cell} is the cell pair area, N_{cell} is the number of cell pairs in a stack, and V_{pro} is a volume product water.

Eq. (28) shows that the energy dissipation due to the resistance of the solutions and membranes increases with the current density. The electrical energy for a given resistance is proportional to the square of the current, whereas the salt transfer is directly proportional to the current. Hence, the power necessary for the production of a given amount of product increases with the current density. However, the required membrane area for a given capacity is decreasing with the current density as illustrated in Fig. 15, which shows the total costs of desalination, the membrane costs, and current density-related costs as a function of the current density. The figure shows that at a certain current density, the total desalination cost reaches a minimum. However, in a practical application, the operating current density must be lower than the limiting current density.

The operation of an electrodialysis unit requires one or more pumps to circulate the diluate, the concentrate, and the electrode rinse solution through the stack. The energy required for pumping these solutions is determined by the volumes of the solutions to be pumped and the pressure drop. It can be expressed by

$$E_{\text{p,spec}} = \frac{E_{\text{p}}}{Q^{\text{d}}t} = k_{\text{eff}} \frac{(Q^{\text{d}}\Delta p^{\text{d}} + Q^{\text{c}}\Delta p^{\text{c}} + Q^{\text{e}}\Delta p^{\text{e}})}{Q^{\text{d}}} \tag{29}$$

where $E_{\text{p,spec}}$ is the total energy for pumping the diluate, the concentrate, and the electrode rinse solution through the stack per unit diluate water; k_{eff} is an efficiency term for the pumps; and Q^{d}, Q^{c}, and Q^{e} are the volume

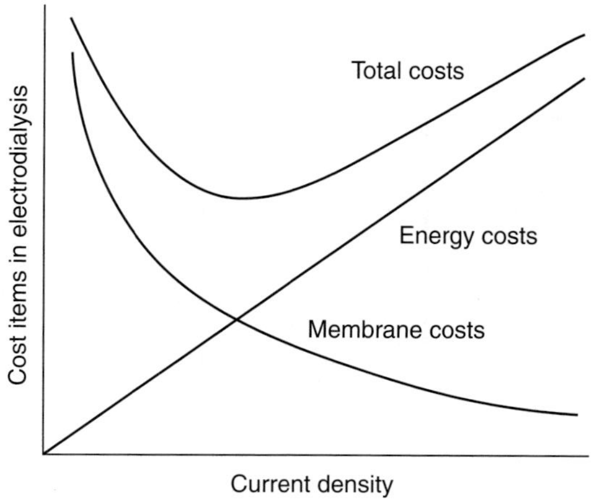

Figure 15 Schematic diagram illustrating the various cost items in electrodialysis as function of the applied current density.

flow rates of the diluate, the concentrate, and the electrode rinse solution through the stack.

The energy consumption due to the pressure loss in the electrode rinse solution can be neglected in most practical applications because the volume of the electrode rinse solution is very small compared to the volumes of the diluate and concentrate.

The pressure losses in the various cells are determined by the solution flow velocities and the cell design. The energy requirements for circulating the solution through the system may become a significant or even dominant portion of the total energy consumption for solutions with rather low salt concentration.

4.2 Electrodialysis with bipolar membranes

Electrodialysis with bipolar membranes utilizes water dissociation of bipolar membranes in an electrodialysis stack for the production of acids and bases from the corresponding salts. The process and the stack design are in many aspects similar to conventional electrodialysis. The key element in electrodialysis with bipolar membranes is the bipolar membrane. Its performance determines to a very large extent the technical feasibility and economics of the process [25].

4.2.1 The bipolar membrane, and its structure and function

The function of the bipolar membrane is illustrated in Fig. 16a, which shows a bipolar membrane consisting of an anion- and a cation-exchange

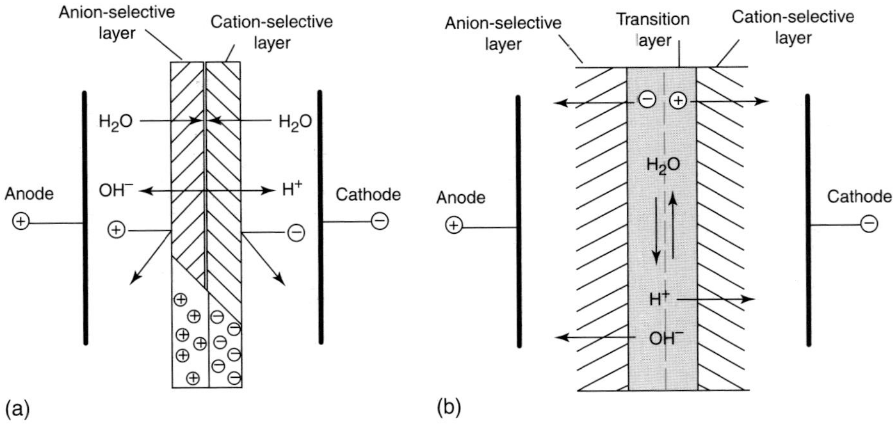

Figure 16 Schematic diagram illustrating the function of a bipolar membrane showing (a) a bipolar membrane and (b) the 4–5 nm thick transition region at the interphase of the two cation- and anion-exchange layers.

layer arranged in parallel between two electrodes. If a potential difference is established between the electrodes, all charged components will be removed from the interphase between the two ion-exchange layers. If only water is left in the solution between the membranes, further transport of electrical charges can be accomplished only by protons and hydroxyl ions, which are regenerated due to the water dissociation in a very thin, that is, 4–5 nm thick, transition region between the cation- and anion-exchange layers of the bipolar membrane as shown in Fig. 16b. The water dissociation equilibrium is given by

$$2H_2O \Leftrightarrow H_3O^+ + OH^-$$

The energy required for water dissociation can be calculated from the Nernst equation for a concentration chain between solutions of different pH values [28]. It is given by

$$\Delta G = F\Delta\varphi = 2.3RT\Delta \mathrm{pH} \tag{30}$$

where ΔG is the Gibbs free energy, F is the Faraday constant, R is the gas constant, T is the absolute temperature, and ΔpH and $\Delta\varphi$ are the pH and the voltage differences between the two solutions separated by the bipolar membrane. For 1 mol L^{-1} acid and base solutions in the two phases separated by the bipolar membrane, ΔG is 0.022 kWh mol^{-1} and $\Delta\varphi$ is 0.828 V at 25 °C.

The transport rate of H^+ and OH^- ions from the transition region into the outer phases cannot exceed the rate of their generation. However, the generation rate of H^+ and OH^- ions in a bipolar membrane is drastically increased compared to the rate obtained in water due to a catalytic reaction [26–28]. Therefore, very high production rates of acids and bases can be achieved in bipolar membranes.

4.2.2 System and process design of electrodialysis with bipolar membranes

The design of an electrodialysis process with bipolar membranes is closely related to that of a conventional electrodialysis desalination process. The main difference is in the stack construction and additional energy requirements for water dissociation. Furthermore, the mono- and bipolar membranes as well as other hardware components must have excellent chemical stability in strong acids and bases.

Stack design in bipolar membrane electrodialysis. A typical arrangement of an electrodialysis stack with bipolar membranes is illustrated in Fig. 3, which shows the production of an acid and a base from the corresponding salt in a repeating cell unit, which consists of three individual cells containing the salt solution, the acid and the base, and three membranes, that is, a cation-exchange, an anion-exchange, and a bipolar membrane. In industrial-size stacks, 50–100 repeating cell units may be placed between two electrodes.

The main difference between an electrodialysis desalination stack and a stack with bipolar membranes used for the production of acids and bases is the manifold for the distribution of the different flow streams. Since in most practical applications, high acid and base concentrations are required, the stack is usually operated in a feed and bleed concept as shown Fig. 17.

4.2.3 Electrodialysis with bipolar membrane process costs

The determination of the costs for the production of acids and bases from the corresponding salts follows the same general procedure as applied for the costs analysis in electrodialysis desalination. The contributions to the overall costs are the sum of the investment-related costs and the operating costs.

Investment costs in electrodialysis with bipolar membranes. The investment costs are directly related to the required membrane area for a certain plant capacity and can be expressed as a certain percentage of the total required membrane area for a given capacity plant, which can be calculated from the current density by

$$A_{\text{unit}} = \frac{Q_{\text{pro}} F C_{\text{pro}}}{i\xi} \quad (31)$$

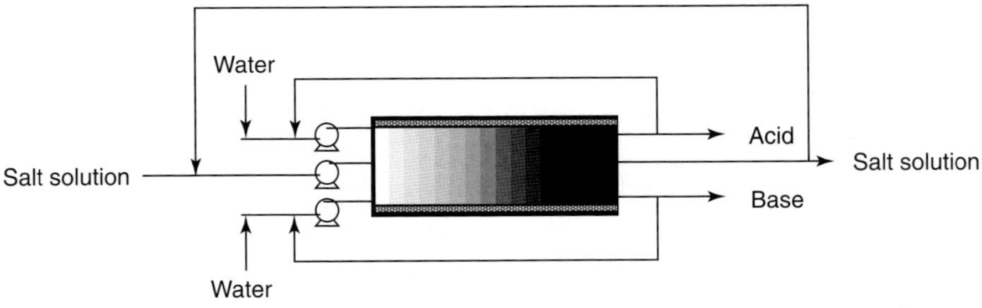

Figure 17 Schematic diagram indicating the production of acids and bases from the corresponding salt in a stack with feed and bleed operation.

where A_{unit} is the required cell unit area containing a bipolar membrane, and a cation- and an anion-exchange membrane; i the current density; Q_{pro} the product volume flow; F the Faraday constant; ξ the current utilization; and C_{pro} the concentration of the product.

Operating costs in electrodialysis with bipolar membranes. The operating costs in electrodialysis with bipolar membranes are strongly determined by the energy requirements, which are composed of the energy required for the water dissociation in the bipolar membrane and the energy necessary to transfer the salt ions from the feed solution, and protons and hydroxide ions from the transition region of the bipolar membrane into the acid and base solutions. The total energy for the production of an acid and a base from the corresponding salt is as in electrodialysis desalination given by the total current passing through the stack and the voltage drop across the stack. The voltage drop across the stack is the result of the electrical resistance of the membranes, that is, that of the cation- and anion-exchange membranes and the bipolar membranes and the resistances of the acid, the base and the salt containing flow streams in the stack. In addition to the voltage drop required to overcome the various electrical resistances of the stack, additional voltage drop is required to provide the energy for the water dissociation which is given by Eq. (30). Assuming that the three cells of a cell unit in the stack have the same geometry and flow conditions, the total energy consumption in an electrodialysis stack is given analog to the energy in conventional electrodialysis expressed by [29]

$$E_{spe,pro} = \frac{N_{unit}A_{unit}}{V_{pro}}\left(\sum_i \frac{\Delta}{\Lambda_i \bar{C}_i} + r^{am} + r^{cm} + r^{bm} + \frac{2.3RT\Delta pH}{Fi}\right) \times \left(\frac{Q_{pro}F(C_{pro}^{out} - C_{pro}^{in})}{N_{unit}A_{unit}\xi}\right)^2 t \qquad (32)$$

where $E_{spc,pro}$ is the energy needed for the production of a certain amount of acid and base; i is the current density passing through the stack; N_{unit} is the number of cell units in a stack; A_{unit} is the cell unit area; C and $\bar{C}$ are the concentration and the average concentration in a cell; Δ is the thickness of the individual cells; Λ is the equivalent conductivity; r is the area resistance; ξ is the current utilization; R is the gas constant; T is the absolute temperature; F is the Faraday constant; ΔpH is the difference in the pH value between the acid and base; the subscript pro refers to product and the subscript i refers to salt, acid, and base; the superscripts am, cm, and bm refer to the cation-exchange, the anion-exchange, and the bipolar membrane; the superscripts out and in refer to cell outlet and inlet; Q is the flow of the acid or base through the stack; and t is the time.

The average concentrations of the acid, the base, and the salt in the bulk solutions are the integral average of the solutions given by

$$\bar{C}_i = \frac{\ln(C_i^{\text{out}}/C_i^{\text{in}})}{C_i^{\text{out}} - C_i^{\text{in}}} \tag{33}$$

Operation of an electrodialysis unit with bipolar membranes requires three pumps to circulate the salt solution, the acid, and the base through the stack.

The energy required for pumping flow streams through the stack is given by

$$E_{\text{p,spec}} = \frac{E_{\text{p}}}{Q_{\text{pro}}t} = k_{\text{eff}} \frac{\sum_{\text{fs}} Q^{\text{fs}} \Delta p^{\text{fs}}}{Q_{\text{pro}}} \tag{34}$$

where $E_{\text{p,,spec}}$ is the total energy for pumping the solutions through the stack per unit product, k_{eff} is an efficiency term for the pumps, Q is the volume flow rate, the superscript fs refers to the different flow streams, and the subscripts p and pro refer to pumping and product, respectively.

The total costs of the electrodialytic water dissociation with bipolar membranes are the sum of fixed charges associated with the amortization of the plant investment costs, the energy costs, and of the operating costs, which include maintenance costs and all pre- and posttreatment procedures.

Problems in practical application of bipolar membrane electrodialysis. In addition to the precipitation of multivalent ions in the base containing flow stream and the stability of the ion-exchange membranes in strong acids and bases, a serious problem is the contamination of the products by salt ions, which permeate the bipolar membrane, which in general is not completely perm-selective. Furthermore, the current utilization is affected by the leakage of H^+ and OH^- ions through the monopolar anion- and cation-exchange membranes. Especially, when high concentrations of acids and bases are required, the salt contamination is generally quite high and the current utilization low. This is a major limitation for the practical application of bipolar membranes for the effective production of acids and bases from the corresponding salt solutions.

4.3 Continuous electrodeionization

Continuous electrodeionization is widely used at present for the preparation of high-quality deionized water for the electronic industry or in analytical laboratories [30]. Compared to the deionization by conventional ion-exchange techniques, the continuous electrodeionization has the advantage that no chemicals are needed for the regeneration of the ion-exchange resins, which is time-consuming, labor-intensive, and generates a salt containing wastewater. There are some variations of the basic design as far as the distribution of the ion-exchange resin is concerned, and recently bipolar membranes are also used in the process [6].

4.3.1 System components and process design aspects

The process design in electrodeionization is very similar to that of conventional electrodialysis. The main difference is that in a continuous electrodeionization stack, the diluate cells and sometimes also the concentrate cells are filled with ion-exchange resins. There are different concepts used for the distribution of the cation- and anion-exchange resins in the cell. Two more frequently applied stack designs are illustrated in Fig. 18a and b.

Fig. 18a shows a conventional electrodialysis stack in which the diluate cell is filled with a mixed-bed ion-exchange resin. Both cations and anions are adsorbed by the ion-exchange resins and then transported by an electrical potential gradient through the corresponding ion-exchange resin toward the adjacent concentrate cells facing the cathode and the anode, respectively. Since the ion conductivity in the ion-exchange resin is several orders of magnitude higher than in the deionized water, the stack can be operated economically at relatively high current density compared to conventional electrodialysis.

However, the use of a mixed-bed ion-exchange resin in continuous electrodeionization results in a rather poor removal of weak acids and bases such as boric or silicic acid. Much better removal of weakly dissociated electrolytes can be obtained in a system in which the cation- and anion-exchange resins are placed in separate beds with a bipolar membrane placed in between as illustrated in Fig. 18b, which shows a diluate cell filled with a cation-exchange resin facing toward the cathode separated by a bipolar membrane from a diluate cell facing the anode. A cation-exchange membrane, a cation-exchange resin, a bipolar membrane, an anion-exchange resin, an anion-exchange resin, and a concentrate cell form a repeating unit between two electrodes.

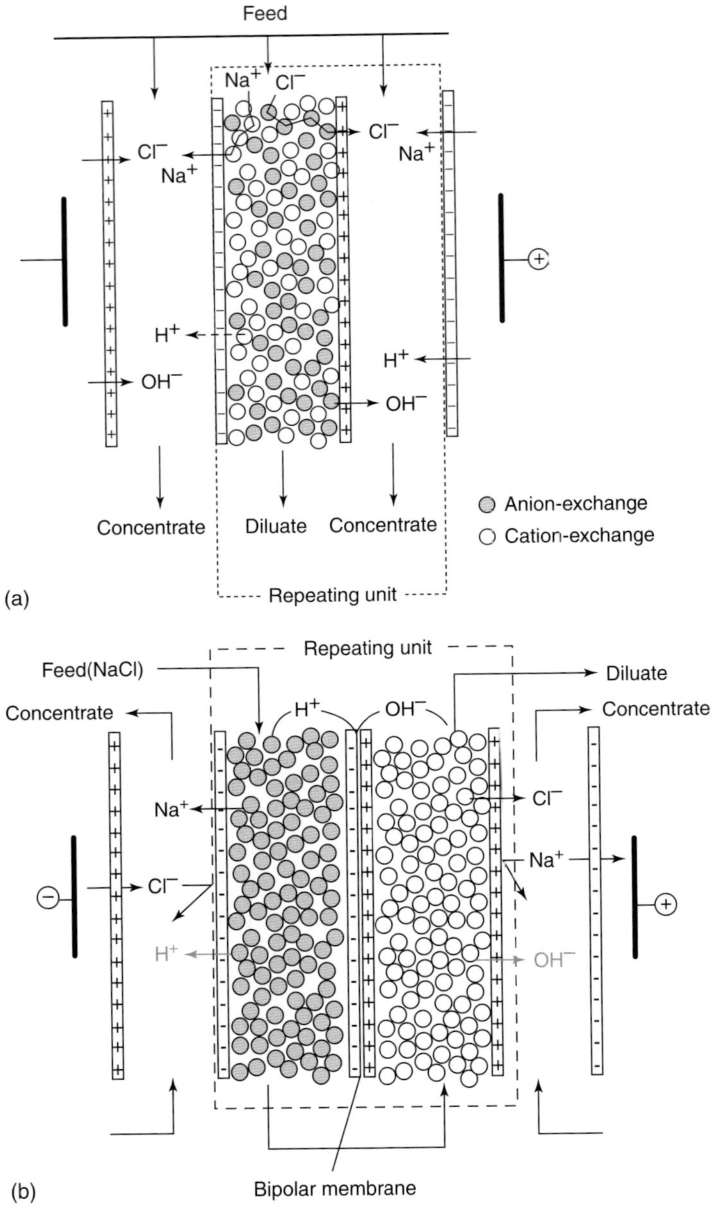

Figure 18 Schematic drawing illustrating different stack concepts used in continuous electrodeionization: (a) a conventional stack with diluate cells filled with a mixed-bed ion-exchange resin and (b) a stack with cation-exchange and anion-exchange resins in different diluate cells and regeneration of the ion-exchange resins by H^+ and OH^- ions generated in a bipolar membrane.

The main difference between the electrodeionization system with the mixed-bed ion-exchange resins and the system with separate beds is that in mixed-bed electrodeionization systems, anions and cations are simultaneously removed from the feed while the solution leaving the diluate cell is neutral. In the electrodeionization system with separate ion-exchange beds and bipolar membranes, the cations will first be exchanged by the protons generated in the bipolar membrane with the result that the solution leaving the cation-exchange bed is acidic. This solution is then passed through the cell with the anion-exchange resin where the anions are exchanged by the OH^- ions generated in the bipolar membrane and the solution is neutralized, and at the exit of the anion-exchange-filled cell, the solution is also neutral.

4.3.2 Operational problems in practical application of electrodeionization

The main problem of an electrodeionization system with a mixed-bed ion-exchange resin is the incomplete removal of weak acids or bases. But also electrodeionization systems with separate resin beds are affected by uneven flow distribution in the resin bed, which leads to poor utilization of the ion-exchange resins. The fouling of the ion-exchange resins by organic components such as humic acids and bacterial growth on the surface of the resin is another problem that requires a very thorough pretreatment of the feed solution to guarantee a long-term stability of the system.

4.4 Other electromembrane separation processes

In addition to the processes discussed so far, there are two more electromembrane separation processes in which the driving force is not an externally applied electrical potential but a concentration gradient. The processes are referred to as diffusion dialysis and Donnan dialysis. Diffusion dialysis utilizes anion- or cation-exchange membranes only to separate acids and bases from mixtures with salts. Donnan dialysis can be used to exchange ions between two solutions separated by an ion-exchange membrane. Both processes have so far gained only limited practical relevance [3].

5. APPLICATIONS OF ION-EXCHANGE MEMBRANE SEPARATION PROCESSES

Ion-exchange membrane separation processes are used today in a large variety of applications from water desalination by conventional electrodialysis to the production of acids and bases by electrodialysis with bipolar membranes, or the elimination of toxic components from industrial effluents. The technical and commercial relevance of the various processes is very different. Some processes such as electrodialysis are used on a large industrial scale for the desalination of water or the removal of certain electrolytes from food and chemical products. Other processes such as bipolar membrane water dissociation or the continuous electrodeionization are used on a much smaller scale for the production of ultrapure water or for the purification of certain food and pharmaceutical components.

5.1 Applications of electrodialysis

Conventional electrodialysis is today commercially by far the most relevant ion-exchange membrane separation process. Electrodialysis was first developed for the desalination of brackish water to produce potable water. In this application, electrodialysis is replaced today to some extent by other membrane processes such as reverse osmosis and nanofiltration. Nevertheless, water desalination is still the most important large-scale application of electrodialysis. But other applications of electrodialysis in the food, the pharmaceutical and chemical industry as well as in wastewater treatment are gaining increasing importance. Another large-scale application of electrodialysis is the preconcentration of seawater for the production of table salt. Some of the more important large-scale industrial applications of conventional electrodialysis and the stack and process design used in this application as well as the major limitations are listed in Table 1.

5.1.1 Brackish water desalination by electrodialysis

The production of potable water from brackish water is presently the largest single application of electrodialysis. In brackish water desalination, electrodialysis competes directly with reverse osmosis. However, in a certain range of feedwater salt composition, electrodialysis has a clear economic advantage over other desalination processes. Electrodialysis is mainly used in small- to medium-sized plants with capacities of less than a few 100 m^3 per day to more than 20,000 m^3 per day with a brackish water salinity of 1000–5000 $mg\,L^{-1}$ total dissolved solids. Since in electrodialysis both the energy consumption and the required membrane area increase with increasing feedwater concentration, reverse osmosis is considered to have an economic advantage for the desalination of water with total dissolved salts in excess of 10,000 $mg\,L^{-1}$. The advantages of electrodialysis compared to reverse osmosis are the following: high water recovery rates, long useful life of membranes, operation at elevated temperatures up to 50 °C, and less membrane fouling or scaling due to process reversal. The disadvantage of electrodialysis compared to reverse osmosis is that neutral toxic components such as viruses or bacteria are not removed from a feed stream. Thus, the

Table 1 Industrial applications of conventional electrodialysis

Industrial applications	Stack and process design	Status of application	Limitations	Key problems
Brackish water desalination	Sheet flow, tortuous path stack, reverse polarity	Commercial	Concentration of feed and costs	Scaling, costs
Boiler feedwater production	Sheet flow, tortuous path stack, reverse polarity	Commercial	Product water quality and costs	Costs
Waste and process water treatment	Sheet flow stack, unidirectional	Commercial	Membrane properties	Membrane fouling
Demineralization of food products	Sheet flow or tortuous path stack, unidirectional	Commercial pilot phase	Membrane selectivity and costs	Membrane fouling, product loss
Table salt production	Sheet flow stack, unidirectional	Commercial	Costs	Membrane fouling
Concentration of reverse osmosis brine	Sheet flow stack, unidirectional	Pilot phase	Costs	Waste disposal

Figure 19 Industrial-size electrodialysis reversal desalination plant. Photograph, courtesy of Ionics Incorporated.

product water may require a posttreatment procedure when used as potable water. In spite of the fact that electrodialysis reversal is significantly less sensitive to membrane fouling than reverse osmosis, some pretreatment of the feedwater is required. Especially, the iron and manganese ion concentrations must be kept below 0.3 and $0.05\,\mathrm{mg\,L^{-1}}$. A typical large-scale electrodialysis water desalination plant is shown in the photograph of Fig. 19. The plant is built by Ionics Inc., Watertown, MA, USA, using the Aquamite EDR system.

5.1.2 Production of industrial water by electrodialysis

Depending on its application, industrial process water must meet certain quality standards in terms of total dissolved solids and colloidal material. Traditionally, precipitation, filtration, and ion-exchange are used in the production of industrial water. Today, these processes are replaced or increasingly used in combination with microfiltration, reverse osmosis, and electrodialysis. Major applications of electrodialysis in industrial water processing include re-demineralization of boiler feed and desalination of contaminated industrial water for reuse.

5.1.3 Electrodialysis wastewater treatment

During recent years, there has been a substantial increase in water costs. Also, the discharge of certain industrial wastewaters is often problematic and costly because of high salt concentrations or toxic water constituents. This has resulted in a trend to reuse industrial wastewater. A typical application of industrial water reuse is the recycling of cooling tower blow-down water. Electrodialysis is particularly suited for this purpose since high recovery rates up to 95% and high brine concentrations up to 100,000 mg L^{-1} can be achieved, which results not only in lower feedwater costs but also in a reduction in wastewater discharge. Furthermore, ion-exchange membranes can be operated at temperatures up to 50 °C, which is in the range of most cooling systems.

The recovery of water is not always the main objective in the treatment of industrial effluents. Very often toxic or valuable components such as heavy metal ions are removed to avoid pollution of the environment and save disposal chemical costs by recycling valuable materials. A large number of interesting applications for electrodialysis are in the galvanic industry and in metal surface treatment processes. A typical example is the recovery and concentration of nickel from the still rinse water and the recovery of acids such as HCl, HNO_3, HF, etc. from spent pickling solutions. There are many more interesting applications of electrodialysis in wastewater treatment. Some of these applications require only relatively small electrodialysis units, as, for example, the treatment of small amounts of effluents from chemical processes containing highly toxic compounds. In other applications, rather large quantities of water must be treated. This is the case in the paper and pulp industry. Here electrodialysis has been successfully used to remove NaCl selectively from the chemical recovery cycle of Kraft Pulp production.

5.1.4 Food processing by electrodialysis

In the food and beverage industry but also in biotechnology and the pharmaceutical industry, electrodialysis has found a number of applications. Some of these applications can be considered as state-of-the-art processes such as the deionization of whey. Other applications such as the removal of salts from protein and sugar solutions or salts and organic acids such as lactic acid and certain amino acids from fermentation have been studied on a pilot plant scale. However, only a few applications have been commercialized so far. Many applications of conventional ion-exchange resins in the food industry can be replaced by electrodialysis providing better economics, simpler operation, and less waste material, which must be deposited.

5.1.5 Electrodialysis in hybrid processes and the preconcentration of salts

The use of electrodialysis in desalination is most economic in a certain range of concentration as far as the feed solution, the diluate, and the concentrate is concerned. To utilize the most effective range of operation of electrodialysis in desalination, a combination with other separation processes has proved to be very effective. Of very special interest, however, is the concentration of NaCl from seawater prior to evaporation for the production of table salt in Japan, which has no native salt deposit. Thus, for the production of table salt and as raw material for the chlorine alkaline production, the use of electrodialysis as a preconcentration step prior to evaporation of seawater results in substantial energy savings. Large plants with a capacity of 20,000 to more than 200,000 tons of salt per year are now in operation in Japan. Preconcentration of diluate solutions by electrodialysis is also of interest for the production of certain salts such as KBr or KI from raw water sources and for the treatment of certain industrial effluents to recover valuable or toxic wastewater constituents. In the production of ultrapure water, the combination of electrodialysis with ion-exchange is very effective and has finally resulted in the development of the continuous electrodeionization process and to the continuous regeneration of ion-exchange resins. Another application, which is presently studied on a laboratory scale, is the integration of ion-exchange membranes in the so-called membrane reactors. In many chemical and biochemical reactions, the reaction products or the reaction by-products inhibit the reaction when a certain concentration is exceeded. This often limits the achievable product concentration and requires additional separation and concentration steps. A continuous removal of the reaction inhibiting components often makes a continuous more economic production possible.

A substantial effort has also been concentrated on reversing electrodialysis to recover energy from mixing a concentrated salt solution such as seawater with river or surface water in an electrodialysis stack. Although the process is technically feasible, it is presently not economic.

5.2 Applications of bipolar membrane electrodialysis

Since bipolar membranes became available as commercial products in 1977 [5], a very large number of potential applications has been identified and has been studied extensively on a laboratory or pilot plant scale [31]. However, in spite of the obvious technical and economical advantages of the technology, large-scale industrial plants are still quite rare. The main reasons for the reluctant use of bipolar membrane electrodialysis are shortcomings of the available bipolar and monopolar membranes, which result in a short useful membrane life, poor current utilization, and high product contamination. Nevertheless, there are a number of smaller scale applications in the chemical process industry, biotechnology, food processing, and wastewater treatment. Some of the potential applications of electrodialysis are summarized in Table 2.

Table 2 Potential applications of electrodialysis with bipolar membranes, and their state of development and possible advantages and experienced problems

Application	Development state of process	Potential advantages	Problems related to application
Production of mineral acids and bases from salts	Pilot plant operation	Lower energy consumption	Contamination of products, poor membrane stability
Recovering of organic acids from fermentation processes	Commercial plants	Integrated process, lower costs	Unsatisfactory membrane stability and fouling
pH-control in chemical processes	Laboratory tests	Less chemicals and less salt production	Application experience, process costs
Removal of SO_2 from flue gas	Extensive pilot plant tests	Decreased salt production	Long-term membrane stability
Recycling of HF and HNO_3 from steel pickling solutions	Commercial plants	Recovered acids and decreased salt disposal	Relatively complex process, high investment costs
Ion-exchange resin regeneration	Pilot plant tests	Decreased salt disposal	High investment costs
High-purity water production	Laboratory tests	Better removal of weak acids and bases	No long-term experience

5.2.1 Production of acids and bases by bipolar membrane electrodialysis

The largest potential application of bipolar membrane electrodialysis is the production of acids and bases from the corresponding salts. Presently, caustic soda is produced as a coproduct with chlorine by electrolysis. Utilizing bipolar membrane electrodialysis and producing caustic soda and an acid instead of chlorine from the corresponding salts is an interesting alternative to the conventional chlorine/alkaline electrolysis because of lower energy consumption. However, the process is impaired by poor membrane stability and insufficient permselectivity at high ion concentrations, resulting in substantial product salt contamination, low current utilization, and short membrane life under operating conditions. Problem-free operation of the bipolar membrane electrodialysis also requires a substantial pretreatment of the salt solution. The overall result of extensive laboratory tests is that presently the production of mineral acids and bases by bipolar membrane electrodialysis does not meet the product quality requirements under economic conditions.

However, the situation is quite different when acids or bases must be recovered from salts obtained in chemical reactions or neutralization processes. In these cases, the requirements for the concentration and the purity of the recovered acids or bases are not as stringent as in the production of high-quality commercial products and electrodialysis with bipolar membrane can be applied economically.

5.2.2 Applications of bipolar membranes in wastewater treatment

Recovering acids and bases from their salts generated in neutralization reactions to minimize waste disposal is one of the most promising applications of bipolar membrane electrodialysis. One of the more promising applications is the recovery of acids such as hydrofluoric and nitric acids from an effluent stream containing potassium fluoride and nitrite generated by neutralization of a steel pickling bath. The process is illustrated in the simplified flow diagram of Fig. 20.

The spent pickling acid is neutralized with potassium hydroxide. The solution is then filtered and the precipitated heavy metal hydroxides are removed. The neutral potassium fluoride and nitrite containing solution is fed to the bipolar membrane electrodialysis unit in which the salts are converted to the corresponding acids and potassium hydroxide. Potassium hydroxide is recycled to the neutralization tank and the acids to the pickling bath. The depleted salt solution from the bipolar membrane electrodialysis unit is concentrated in a conventional electrodialysis system and recycled

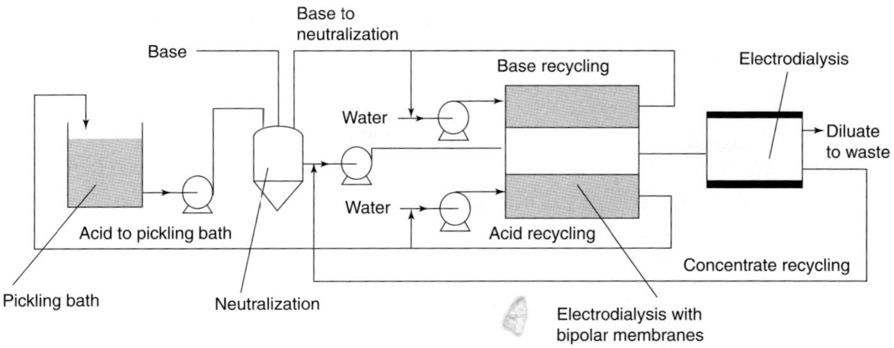

Figure 20 Simplified flow diagram of acid recovery and recycling from steel pickling neutralization bath.

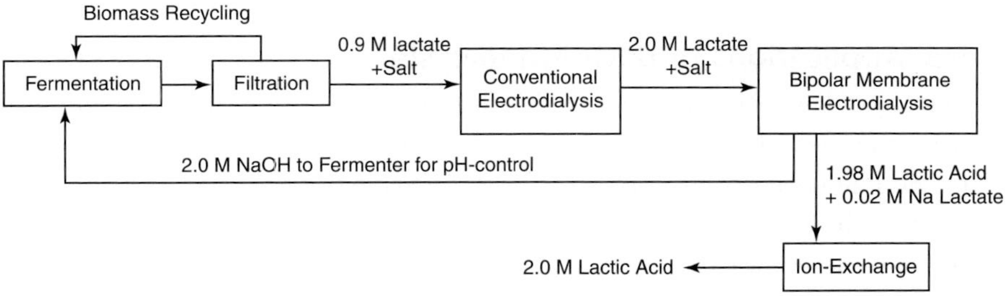

Figure 21 Simplified flow diagram illustrating the production process with integrated electrodialysis.

directly to the bipolar membrane unit. The diluate is used for rinsing and cleaning of the filter.

The treatment of alkaline or acid scrubbers that are used to remove components, which are harmful to the environment such as NO_x, SO_2, or NH_3 from waste air streams, is another interesting application for bipolar membrane electrodialysis. In alkaline and acid scrubbers, large amounts of acids or bases are consumed and salts are produced, which are often contaminated with toxic materials such as heavy metal ions or organic compounds.

5.2.3 Applications of bipolar membrane electrodialysis in biotechnology

A very promising application of electrodialysis with bipolar membranes is the recovery of organic acids from fermentation processes. The process is illustrated in Fig. 21, which shows the production of lactic acid by continuous fermentation with an integrated product recovery process based on bipolar membrane electrodialysis.

Conventionally, lactic acid is produced in a batch process and the separation and purification of the lactic acid is achieved mainly by ion-exchange, resulting in a large volume of wastewater from regeneration salts. In the production process with integrated electrodialysis, which is shown in Fig. 21, a minimum of ion-exchange resin is needed in a final purification step. The concentration of the lactate salt is achieved by conventional electrodialysis and the conversion of the lactate into lactic acid by bipolar membrane electrodialysis. The simultaneously produced base is recycled to the fermenter to control the pH value.

Other typical applications of bipolar membrane electrodialysis in biotechnology are the recovery of gluconic acid from sodium gluconate and the production of ascorbic acid from sodium ascorbate.

5.3 Applications of diffusion dialysis

Diffusion dialysis is used today mainly in the treatment of industrial effluents from the metal processing industry. There are several potential applications in the chemical process industry for the purification of acids or bases or in the food industry for deacidification of fruit juice. But these applications are presently of minor commercial relevance.

A typical application of diffusion dialysis is illustrated in Fig. 22, which shows a simplified flow scheme of the recovery of acids such as HF, HNO_3, and H_2SO_4 from spent pickling baths, which is a major commercial application. A number of diffusion dialysis plants are in operation for several years in Japan, the USA, and Europe. The recovery of HF and HNO_3 from spent pickling solutions seems to be particularly attractive because these acids are relatively expensive and cause severe pollution problems. But the recovery of HCl from aluminum etching rinse water or H_2SO_4 from steel pickling processes is also technically and economically feasible.

5.4 Applications of continuous electrodeionization

The demand by industry for high-purity water in the electronic industry but also in analytical laboratories and power stations is rapidly increasing. Generally, well or surface water is purified in a series of processes that include water softening, microfiltration, reverse osmosis, ultrafiltration, UV sterilization, and mixed-bed ion-exchange, which is used as a final polishing process, producing water with a conductivity of less than $0.06\,\mu S\,cm^{-1}$. While processes such as reverse osmosis, micro- and ultrafiltration, or UV sterilization can be operated in a continuous mode, the mixed-bed

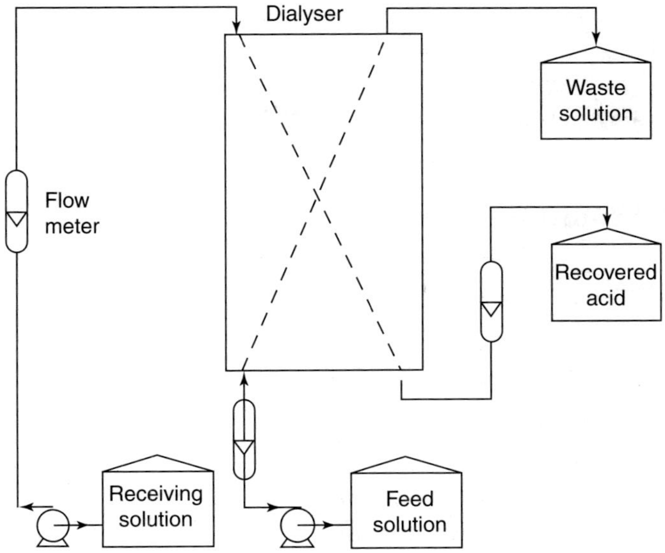

Figure 22 Schematic diagram illustrating the recovery of acids from a pickling solution containing an acids/salt mixture.

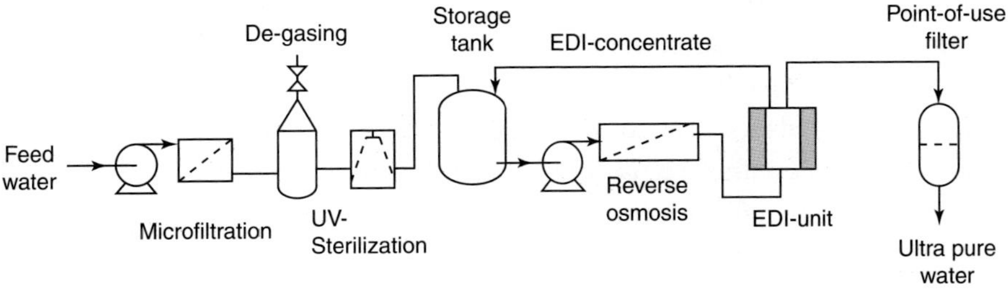

Figure 23 Ultrapure water production line used for the boiler feedwater preparation.

ion-exchanger, which is necessary to reach the required low conductivity, must be regenerated at certain time intervals. This regeneration is not only labor-intensive and costly but it requires extremely long rinse down times to remove traces of regeneration chemicals. By replacing the mixed-bed ion-exchanger by a continuous deionization using separate ion-exchange beds and bipolar membranes, the ultrapure water production can be substantially simplified, yielding consistently high-quality water in a completely continuous process as illustrated in Fig. 23.

The advantages of the ultrapure water production with an integrated electrodeionization unit compared to the use of a mixed-bed ion-exchanger are a simpler process, no necessity of regeneration chemicals, less raw water consumption, and a substantial reduction in costs.

LIST OF SYMBOLS

Roman letters

a	constant [–]
a	activity [mol m^{-3}]
A	area [m^2]
C	concentration [mol m^{-3}]
$\bar{C}$	average concentration [mol m^{-3}]
D	diffusion coefficient [$m^2\,s^{-1}$]
E	energy [A V s]
F	Faraday constant [A s eq^{-1}]
G	Gibbs free energy [J]
i	current density [A m^{-2}]
I	current [A]
i_{lim}	limiting current density [A m^{-2}]
J	flux [mol $m^{-2}\,s^{-1}$]
J_e	flux of electrical charges [A m^{-2}]
k	coefficient [various]
L	coefficient [$mol^2\,N^{-1}\,m^{-2}\,s^{-1}$]
l	length [m]
n	number [–]
N	number [–]
p	pressure [Pa]
q	area [m^2]
Q	volume flow rate [$m^3\,s^{-1}$]
r	area resistance [$\Omega\,m^2$]
R	electrical resistance [Ω]
$\bar{R}$	average electrical resistance [Ω]
R	gas constant [J $mol^{-1}\,K^{-1}$]
S	conductivity [Ω^{-1}]
T	temperature [K]
t	time [s]

T	transport number [−]
U	electrical potential [V]
u	ion mobility [$m^2 s^{-1} V^{-1}$]
$\bar{V}$	partial molar volume [$m^3 mol^{-1}$]
V	volume [m^3]
z	directional coordinate [m]
z	charge number [$eq\,mol^{-1}$]
Z_b	boundary layer thickness [m]

Greek letters

Ψ	membrane permselectivity [−]
Δ	cell thickness [m]
Δ	difference [−]
η	electrochemical potential [$A\,V\,s\,mol^{-1}$]
φ	electrical potential [V]
κ	specific conductivity [$\Omega^{-1} m^{-1}$]
Λ_{eq}	equivalent conductivity [$m^2 \Omega^{-1} eq^{-1}$]
μ	chemical potential [$J\,mol^{-1}$]
ν	stoichiometric coefficient [−]
ρ	specific resistance [Ω m]
ξ	current utilization [−]

Subscripts

a	anion
c	cation
cell	cell or cell pair
co	co-ion
eff	efficiency
des	desalination
e	electric charge
fix	fixed ion
i	component
k	component
lim	limited value
max	maximum value
min	minimum value
p	pump

pro	product
s	salt
spec	specific
st	stack
tot	total
unit	cell element
w	water

Superscripts

am	anion-exchange membrane
b	bulk solution
b	constant
bm	bipolar membrane
c	concentrate
cm	cation-exchange membrane
e	electrode rinse
d	diluate
diff	diffusion
fs	flow stream
in	inlet
m	membrane
mig	migration
out	outlet
pro	product
s	solution

REFERENCES

[1] J.R. Wilson (Ed.), Demineralization by Electrodialysis, Butterworth Scientific Publications, London, 1960.
[2] T. Sata, Recent trends in ion-exchange research, Pure Appl. Chem. 58 (1986) 1613.
[3] H. Strathmann, Ion-exchange membrane separation processes, Elsevier, Amsterdam, 2004.
[4] L.H. Schaffer, M.S. Mintz, Electrodialysis, in: K.S. Spiegler (Ed.), Principles of Desalination, Academic Press, New York, 1966, pp. 3–20.
[5] K.J. Liu, F.P. Chlanda, K.J. Nagasubramanian, Use of bipolar membranes for generation of acid and base: an engineering and economic analysis, J. Membr. Sci. 2 (1977) 109.
[6] A. Grabowskij, Z. Guiquing, H. Strathmann, G. Eigenberger, The production of high purity water by continuous electrodeionization with bipolar membranes, J. Membr. Sci. 281 (2006) 297.

[7] Y. Kobuchi, H. Mottomura, Y. Noma, F. Hanada, Application of ion-exchange membranes to the recovery of acids by diffusion dialysis, J. Membr. Sci. 27 (1986) 173.
[8] K.S. Spiegler, Electrochemical operations, in: F.C. Nachod, J. Schubert (Eds.), Ion-Exchange Technology, Academic Press, New York, 1956.
[9] J.C. Farmer, D.V. Fix, G.V. Mack, R.W. Pekola, J.F. Poco, Capacitive deionization of NaCl and $NaNO_3$ solutions with carbon aerogel electrodes, J. Electrochem. Soc. 143 (1996) 159.
[10] K.S. Spiegler, Transport processes in ionic membranes, Trans. Faraday Soc. 54 (1958) 1408.
[11] O. Kedem, A. Katchalsky, A physical interpretation of the phenomenological coefficients of membrane permeability, J. Gen. Physiol. 45 (1961) 143.
[12] H. Strathmann, Electrodialysis and related processes, in: R.D. Nobel, S.A. Stern (Eds.), Membrane Separation Technology, Elsevier, Amsterdam, 1995, pp. 213–281.
[13] F. Helfferich, Ion-Exchange, McGraw-Hill, London, 1962.
[14] G. Pourcelly, Conductivity and selectivity of ion exchange membranes: structure-correlations, Desalination 147 (2002) 359.
[15] D.S. Flett, Ion-Exchange Membranes, E. Horwood Ltd, Chichester, UK, 1983.
[16] F. Bergsma, Ch.A. Kruissink, Ion-exchange membranes, Fortschr. Hochpolym.-Forsch. 21 (1961) 307.
[17] A. Eisenberg, H.L. Yeager, Perfluorinated Ionomer Membranes, , ACS Symposium Series 180American Chemical Society, Washington, DC, 1982
[18] P. Zschocke, D. D. Quellmalz, Novel ion exchange membranes based on an aromatic polyethersulfone, J. Membr. Sci. 22 (1985) 325.
[19] R. Simons, Preparation of a high performance bipolar membrane, J. Membr. Sci. 78 (1993) 13.
[20] F.W. Wilhelm, Bipolar membrane preparation, in: A.J.B. Kemperman (Ed.), Bipolar Membrane Technology, Twente University Press, Enschede, The Netherlands, 2000.
[21] W. Grot, Perfluorinated cation exchange polymers, Chem. Ing. Tech. 47 (1975) 617.
[22] E.L. Huffmann, R.E. Lacey, Engineering and economic considerations in electro-membrane processing, in: R.E. Lacey, S. Loeb (Eds.), Industrial Processing with Membranes, John Wiley & Sons, New York, 1972, pp. 39–55.
[23] H.J. Lee, F. Safert, H. Strathmann, S.H. Moon, Designing of an electrodialysis desalination plant, Desalination 142 (2002) 267.
[24] W.E. Katz, The electrodialysis reversal (EDR) process, Desalination 28 (1979) 31.
[25] K.N. Mani, Electrodialysis water splitting technology, J. Membr. Sci. 58 (1991) 117.
[26] H. Strathmann, J.J. Krol, H.J. Rapp, G. Eigenberger, Limiting current density and water dissociation in bipolar membranes, J. Membr. Sci. 125 (1997) 123.
[27] S. Mafé, P. Ramirez, A. Alcaraz, V. Aguilella, Ion transport and water splitting in bipolar membranes: theoretical background, in: A.J.B. Kemperman (Ed.), Handbook of Bipolar Membrane Technology, Twente University Press, Enschede, The Netherlands, 2000.
[28] H.J. Rapp, Die Elektrodialyse mit bipolaren Membranen – Theorie und Anwendung, PhD Thesis, University of Stuttgart, Germany, 1965.
[29] H. Strathmann, G.H. Koops, Process economics of electrodialytic water dissociation for the production of acids and bases, in: A.J.B. Kemperman (Ed.), Handbook Bipolar Membrane Technology, Twente University Press, Enschede, The Netherlands, 2000.
[30] Ganzi, G.C. (1988). Electrodeionization for high purity water production, in: K.K. Sirkar, D.R. Lloyd (Eds.), New Membrane Materials and Processes for Separation, AIChE Symposium Series, American Chemical Society, Vol. 84, Washington, DC, p. 73.
[31] G.P. Pourcelly, C. Gavach, Electrodialysis water splitting – applications of electrodialysis with bipolar membranes, in: A.J.B. Kemperman (Ed.), Handbook Bipolar Membrane Technology, Twente University Press, Enschede, The Netherlands, 2000.

PART Three

Water Recycling Technologies

CHAPTER 7

Micropollutants in Water Recycling: A Case Study of N-Nitrosodimethylamine (NDMA) Exposure from Water versus Food

Andrea I. Schäfer[1,*], William Mitch[2], Sophie Walewijk[3], Albert Munoz[4], Emma Teuten[1] *and* Martin Reinhard[3]
[1]School of Engineering, University of Edinburgh, Edinburgh EH9 3JL, UK
[2]Department of Chemical Engineering, Yale University, New Haven, CT 06520, USA
[3]Department of Civil and Environmental Engineering, Stanford University, Stanford, CA 94305-4020, USA
[4]Environmental Engineering, University of Wollongong, NSW 2125, Australia

Contents

*Corresponding author.
E-mail address: andrea.schaefer@ed.ac.uk

Sustainability Science and Engineering, Volume 2 ISSN 1871-2711, DOI 10.1016/S1871-2711(09)00207-4

1. INTRODUCTION

N-Nitrosodimethylamine (NDMA) is an environmental contaminant that has been shown to initiate cancer though methylation of DNA in animal studies. While it has not been definitively identified as a human carcinogen, incidents of certain types of cancer were, as an example, significantly higher in rubber workers with high exposure to NDMA than in workers with lower exposure. NDMA is classified as a "probable human carcinogen" by the International Agency for Research on Cancer [1].

Human exposure to NDMA is typically assumed to occur through exposure to food, cosmetics, cigarette smoke, and industrial exposure. Little attention was given to the presence of NDMA in drinking water until the early 1990s, following its detection in the water supply in Ohsweken, Ontario, Canada [2]. Subsequent routine monitoring indicated a persistent, fluctuating presence. Following the detection of NDMA in drinking water wells near rocket engine testing facilities in California, a statewide survey of drinking water facilities indicated that NDMA can form as a disinfection by-product [3]. More widespread analyses at drinking water facilities, predominantly in North America, have revealed that NDMA is often present above the detection limit at facilities practicing chloramination, albeit with mean concentrations typically below 10 ng/L [4]. Surveys across the world are currently in progress and NDMA is found in many locations.

Despite a recent analysis by Fristachi and Rice [5] suggesting that $<1\%$ of the dietary contribution of NDMA originates from water, there has been a surge of interest in NDMA in drinking water. Meta-analyses have indicated that long-term consumption of chlorinated drinking water is associated with an increased risk of cancers of the bladder, colon, rectum, esophagus, and breast [6–8]. The exact agent responsible for this increased cancer risk is uncertain, although carcinogenic nitrosamines cannot be ruled out and the application of the precautionary principle is advisable.

While generally viewed as a contaminant of concern, concentrations of NDMA in drinking water are regulated in few countries. Only the governments of California, USA, and Ontario, Canada have set tolerable limits of 10 and 9 ng/L, respectively. The World Health Organization (WHO) is currently considering a guideline value of 100 ng/L, which is the calculated health-based value (HBV) associated with an upper-bound excess lifetime cancer risk of 10^{-5} [9]. A consequence of regulation is the requirement to remove NDMA from drinking water supplies. Current discussions accept that 10 ng/L represents a 10^{-5} lifetime cancer risk.

This chapter summarizes the human sources of NDMA, putting recent exposure into a historical context. Sources and concentrations of NDMA in drinking water are described and the potential impact of NDMA in drinking water on human health is discussed. The discussions are intended to give a thorough case study of the issues surrounding xenobiotics as well as an assessment of the relative risk of such compounds through water intake.

2. CHARACTERISTICS, FORMATION, AND SOURCES OF NDMA

2.1 Chemical and physical properties of NDMA

NDMA belongs to the family of compounds known as nitrosamines and is the simplest dialkyl nitrosamine. Table 1 presents the physicochemical properties of NDMA.

2.2 Formation pathways of NDMA

There are three formation pathways of greatest relevance to food and drinking water:

- Acid-nitrite or NO_x nitrosation in food products
- Dichloramine–oxygen pathway during drinking water chloramination
- Free chlorine–nitrite pathway during chlorination in of nitrite-rich waters

Other pathways have been identified, including the ozonation of a specific fungicide degradation product [16] and formation associated with

Table 1 Physicochemical properties of NDMA

Physical/chemical property	Unit	Value	Reference
Formula		$C_2H_6N_2O$	[10]
Chemical structure		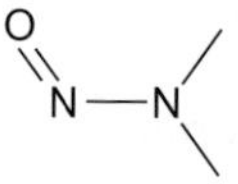	[11]
Molecular weight	g/mol	74.08	[12]
Vapour pressure (at 25 °C)	Pa	1080	[12]
Henry's law constant (at 25 °C)	Pa m^3/mol	3.34	[13]
Octanol water partition coefficient	log K_{ow}	−0.57	[14]
Organic carbon water partition coefficient	log K_{oc}	1.07	[15]

radicals generated during breakpoint chlorination [17]. However, these are likely of limited importance.

2.2.1 Acid–nitrite or NO_x nitrosation in food products

The nitrosation mechanism has been used to explain endogenous nitrosamine formation during ingestion of meats preserved with nitrite [18]. The nitrosyl cation ($^+N{=}O$) forms from acidification of nitrite in the stomach:

$$2H^+ + NO_2^- \rightarrow H_2NO_2^+ \rightarrow H_2O + {}^+N = O$$

The nitrosyl cation is partially stabilized by interaction with anions, including nitrite and chloride:

$${}^+N = O + NO_2^- \rightarrow N_2O_3 \text{ or } {}^+N = O + Cl^- \rightarrow ClNO$$

These partially stabilized forms of nitrosyl cations are highly potent nitrosating agents, which transfer nitrosyl cations to the lone pairs of amines in the rate-limiting step for nitrosamine formation [19]:

$$N_2O_3 + (CH_3)_2NH \rightarrow (CH_3)_2N - N = O + H^+ + NO_2^-$$

Formation rates were maximized near pH 3.4 [18]. Elevated anion concentrations, including the saliva constituent thiocyanate, were found to enhance formation rates [20]. Due to the pH dependency of the reaction, this formation mechanism would not be important for drinking water disinfection, even in the presence of elevated anion concentrations.

Nitric oxide (NO) and nitrogen dioxide (NO_2) in air pollution can form N_2O_3. Dissolution of N_2O_3 into amine-containing solutions can rapidly form nitrosamines.

2.2.2 Dichloramine–oxygen pathway during chloramination

Early research recognized the association between chloramination (used in treatment as well as distribution systems in various countries) and nitrosamine formation. However, attempts were made to link this formation with monochloramine, the predominant chloramine species under typical chloramination conditions [21,22]. Briefly, this research suggested that monochloramine reacted with dimethylamine to form NDMA via an unsymmetrical dimethylhydrazine (UDMH) intermediate over the course of days at overall yields of $<3\%$.

However, noting that chloramination of dimethylamine formed much more NDMA than did chloramination of UDMH, more recent research has indicated that nearly all nitrosamine formation could be explained by a reaction of dichloramine with organic amine precursors [23]. This pathway posits a reaction between dichloramine and unchlorinated dimethylamine to form a chlorinated UDMH intermediate. Oxidation of the intermediate by chloramines to form uncharacterized products competes with oxidation with dissolved oxygen to form NDMA. Under typical chloramination conditions, monochloramine is the predominant species, yet dichloramine is always present according to the equilibrium:

$$2NH_2Cl + H^+ \leftrightarrow NHCl_2 + NH_4^+$$

The dichloramine model was robust over a variety of conditions and could even account for NDMA formation when preformed monochloramine was applied to dimethylamine solely from the traces of dichloramine that formed from disproportionation of monochloramine.

2.2.3 Free chlorine–nitrite pathway

Previous research suggested that chlorination in the presence of nitrite would form a $ClNO_2$ intermediate [24]. Reaction of $ClNO_2$ with nitrite would lead to the formation of the nitrating and nitrosating agent, dinitrogen tetraoxide (N_2O_4). NDMA and dimethylnitramine formed simultaneously and rapidly during chlorination in the presence of nitrite and dimethylamine [17].

However, this pathway is less relevant to drinking water systems as free chlorine and nitrite rarely coexist in these systems. In nitrifying sections of distribution systems, the practice of boosting the free chlorine dose may promote nitrosamine formation via this pathway. On the other hand, both dimethylnitramine and NDMA were observed in chlorinated pools, with dimethylnitramine concentrations correlating with nitrite concentrations [25].

2.3 Precursors for and occurrence and fate of NDMA in drinking water and wastewater facilities

The specific identity of NDMA precursors has not been identified. However, while organic nitrogen is elevated in water impacted by wastewater and algal blooms [26], NDMA precursors are particularly associated with municipal wastewater-impacted water supplies [27,28]. Dimethylamine was reduced to <2 nM concentrations via biological wastewater treatment, such that it was not an important precursor [28]. Tertiary amines would be degraded to secondary amines via all three of the most important NDMA formation pathways [29]. However, although trimethylamine, like dimethylamine, is a urine constituent, trimethylamine was also removed readily by biological wastewater treatment.

As a class, tertiary amines with dimethylamine functional groups were determined to be potent NDMA precursors [28]. However, specific tertiary amines with dimethylamine functional groups that may be responsible for NDMA formation within wastewater-impacted drinking water supplies have not been identified.

More recently, ozonation has been associated with significant levels of NDMA formation [30]. Reported values before and after ozonation were 16–290 and 14–280 ng/L, respectively. Some sources were attributed to industrial effluents, which resulted in an extremely high concentration of 10,000 ng/L in one case.

Although their permanent positive charges renders them unlikely to react with oxidants, quaternary amine polymers used as coagulants [28,31] and for anion exchange resins [32] during water treatment have been noted as NDMA precursors. Whether the NDMA formation results from reactions with quaternary amine functional groups or tertiary or secondary amine impurities in the treatment polymers is unclear.

In contrast to the long history of NDMA measurements in food, concentrations of NDMA in drinking waters have only been measured since the 1990s. Some reported values are shown in Table 2. While the mean concentrations are low, concentrations have exceeded 100 ng/L following disinfection. However, in most situations, utilities cognizant of

Table 2 Concentrations of NDMA reported in drinking water

Location	Date	[NDMA] (ng/L)		Treatment method	Reference
		Range	Mean		
Water Factory 21, California, USA	2001	nd to 200	7±16	UV	[33]
Ohsweken, Ontario, Canada	1994	nd to 52	3±8	UV	[34]
Alberta, Canada	2004	nd to 12±4		Chloramination	[4]
Alberta, Canada	2004	nd to 100±9		Chlorination	[4]

elevated NDMA concentrations have employed remedial techniques to reduce NDMA concentrations supplied to the consumer. Dillon et al. [35] have surveyed a number of UK drinking waters and detected concentrations of up to 5.8 ng/L. Those numbers are consistent with river water concentrations detected in Japan [36].

In a recent survey, untreated wastewater was shown to contain a median NDMA concentration of 80 ng/L [37]. NDMA occurrence in wastewater is particularly significant in cases where water of wastewater origin percolates into aquifers used as drinking water supply or where wastewater is treated such that it can be directly injected into potable aquifers [3,38–40]. For example, 50% of the water produced by Orange County Sanitation District Plant 1 augments potable water supplies [40]. Other cases are found in Virginia and Texas, and in Namibia and South Africa [38].

2.4 Origin and occurrence of NDMA in food, drinks, and cigarette smoke

High concentrations of nitrosamines were widely reported in foods in the 1970s, particularly in cured meats and fish, cheese, beer, and dried milk [41]. Since NDMA is a potent liver carcinogen, this was good reason for concern. Subsequently, the mechanisms for NDMA formation in these foods were determined. Those are nitrosation of naturally occurring amine by sodium nitrite (a preservative added to fish, meat, and cheese), or by gaseous nitrogen oxides (formed during flue drying of milk and malt used in brewing). Once the routes to NDMA formation in food were understood, steps were taken to reduce its presence. Consequently, the NDMA levels reported in these foodstuffs has been significantly reduced. As an example, Table 3 compares early reports of NDMA concentrations in food with those measured more recently. Reports from the 1990s to 2000s show more than an order of magnitude less NDMA than studies from the 1970s to 1980s. Preparation methods are important, for example, cooking meat at high temperatures (frying or baking) gives a significant increase in NDMA [43].

Synergistic effects of dietary components may increase the likelihood of tumors arising from NDMA. Alcohol and diets rich in fats have been shown to increase the occurrence of liver tumors in laboratory animals fed NDMA [53,54]. Furthermore, cadmium in the diet has been shown to promote NDMA carcinogenesis [55].

Many studies indicate that endogenous NDMA formation from amines and nitrites/nitrates in the stomach is important, and the contribution of endogenous NDMA formation was estimated to be significant by Fristachi and Rice [5]. This occurs because nitrates are reduced to nitrites in the oral

Table 3 Concentrations of NDMA reported in foods

Product	Unit	Year	NDMA concentration		Reference
			Range	Mean	
Cured fish	ng/g	1971	nd to 26		[42]
		2001–2005		0.54–1.99	[43]
Cured meat	ng/g	1975	nd to 35		[44]
		2004		7.3 ± 0.93	[45]
Bacon	ng/g	1973	nd to 30		[46]
		1993–1994	nd to 3		[47]
Cheese	ng/g	1978	nd to 68		[48]
		1995	nd to 0.84	0.28	[47, 49]
Dried milk	ng/g	1981	0.45 to 4.2	1.69 ± 0.17	[50]
		1995	nd to 0.18		[51]
Beer	ng/L	1978–1979	nd to 78,000		[52]
		2000–2006	nd to 660		[52]

cavity, and many amines in foods are rapidly nitrosated in the presence of nitrite under the acidic conditions pertaining to the stomach. Nitrosamines can also form through bacterial nitrosation of amines. However, it is very difficult to accurately estimate endogenous NDMA formation. This has consequences on the intake of nitrites and nitrates in drinking water also, for which health effects are currently unknown.

In vitro experiments frequently use excessively high nitrite concentrations, making extrapolation to realistic physiological concentrations complicated. Investigation of the endogenous NDMA formation *in vivo* is particularly difficult due to its rapid metabolism. Furthermore, coingestion of foods rich in some antioxidants, such as strawberries, garlic, and green tea, significantly inhibit nitrosation under gastric conditions [56,57]. For example, human excretion of NDMA was 26 times lower following green tea ingestion. These factors make predicting endogenous formation of NDMA in humans difficult, and we will not address it further herein.

3. HUMAN EXPOSURE TO NDMA

The daily tolerable limit for NDMA intake has been identified to be 4.0–9.3 ng/(kg day), using a rat liver tumor study and a range of modifying factors. This translates to 280–650 ng/day for a 70 kg person [58]. For estimations below, a person of 70 kg has been assumed.

3.1 Exposure to NDMA through water

The main source of NDMA in drinking water is from its unwanted formation of a disinfection by-product. Many disinfection methods, including chlorination, chloramination, and ozonolysis, can result in the formation of NDMA. However, the highest concentrations of NDMA are generated during chloramination (up to 100 ng/L), in which the oxidant chloramine nitrosates numerous amines present in water supplies. Recent evidence on ozonation also indicates high levels of NMDA formation.

Exposure of humans to NDMA from drinking water varies considerably depending on the concentrations in the water supply. While it was recently suggested that on average $<1\%$ of the dietary contribution of NDMA originates from its presence in water, some communities are likely exposed to much higher amounts due to high NDMA concentrations in some drinking water supplies, particularly those that treat the water with chloramines.

Considering the recommended consumption of 2 L of water per day, the maximum concentrations shown in Table 2 (200 ng/L, which is twice the proposed WHO guideline value) indicate that exposure to NDMA from drinking water would be up to 5.7 ng/(kg day) in situations where these concentrations have not been mitigated by further treatment processes. If the WHO guidelines are used for estimations then a maximum of 2.9 ng/(kg day) can be expected from water.

In addition to exposure to NDMA in water through ingestion, dermal sorption from water is another possible uptake mechanism. The rate of penetration of NDMA through the skin has been estimated as 10^{-4} cm/h. Assuming a skin surface area of 1.8 m^2, a 10-min shower in water containing 200 ng/L NDMA would expose a person to just 0.06 ng of NDMA. High concentrations of NDMA have also been measured in swimming pools and hot tubs [25]. Even when considering the maximum NDMA concentrations reported by Walse and Mitch [25], a 20-min exposure is predicted to result in exposure to 0.03 and 0.3 ng NDMA uptake through the skin for swimming and bathing in hot tubs, respectively.

Hence the dermal uptake of NDMA from water is likely negligible compared to other sources, such as food (or drinking water). Uptake of

NDMA through open wounds, or inhalation through volatilization, may increase the exposure dose, but this is unlikely to make it significant compared to other exposure routes.

3.2 Exposure to NDMA through food, drinks, and smoking

The estimated daily exposure to NDMA from selected food and drinks compared with water and cigarettes at the example of Australian data is outlined in Table 4. The total daily intake varies from 5.7–44.4 ng/(kg day) which means for a 70 kg person an intake of 400–3100 ng per day. This is higher than the daily tolerable limit specified above. Fristachi and Rice [5] estimated a lower mean daily intake from food of 110 ng per person [1.6 ng/(kg day)]. Uncertainties remain and improvement in production processes is likely to reduce those figures further. The main contributors are food products, namely bacon/ham, cheese, and poultry. Average concentrations of NDMA are in the order of 1 μg/kg for meat products [60], which can be used to estimate exposure based on consumption. Munoz has summarized reported values from various sources including detailed assessment of food types extensively [61]. Figures vary widely depending on nationality (see Table 5) given different dietary habits. The relative toxicity of NDMA consumed in food compared to that from water is uncertain, although an article published in 1956 reported that NDMA administered in the basal diet was more potent than that in drinking water.

Alcoholic drinks can contribute significantly to NDMA intake. For example, Glória et al. [71] found NDMA in beer at concentrations of 0.05–0.55 μg/kg, originating from the fermentation process. Assuming a daily beer consumption of about 0.3 L/day, the range of beer-related NDMA intake is 0.3–3.3 ng/(kg day), which has decreased significantly in recent years due to improvements in the brewing process. In fact, Tricker et al. [59] indicated that 31% of the intake of men is from beer, while this figure used to be 65%. In whiskey, NDMA originates from the kilning of malt and concentrations reported on occasion exceeded the "action level" of 5 μg/kg [72] with a median of 0.3 μg/kg, values declining in new (rather than extensively aged) products due to improved production processes.

Smoking is a further source of NDMA, among other carcinogens. An intake range of 5–76 ng per cigarette has been reported [59]. For an average 20 cigarettes per day smoker, daily NDMA intake from smoking would be 1.4–21.7 ng/(kg day). Hence, cigarettes (and some drinks) are potentially a significantly greater source of NDMA than other sources for some individuals.

Table 4 Estimation of mean and maximum daily intake of NDMA by Australians based on consumer statistics in 2000 and NDMA food concentrations from literature [59]

Type of product	Amount per capita per day consumed	Units	NDMA load [ng/(kg day)]		Mean contribution (%)	
			Mean	Maximum	Mean	Maximum
Food						
Cheese	29.6	g/day	1.4	28.7	24.0	60.4
Sausage products	26.0	g/day	0.2	3.5	3.4	7.4
Bacon, ham	26.0	g/day	2.5	6.4	42.9	13.5
Fish	22.2	g/day	0.3	2.5	5.1	5.3
Poultry	84.4	g/day	1.1	3.0	18.9	6.3
Powdered milk	27.4	g/day	0.2	0.3	3.4	0.6
Food subtotal			5.7	44.4		
Drinks & cigarettes						
Water	2.0	L/day	0.02	2.80	0.3	5.9
Beer	0.3	L/day	0.03	0.18	0.5	0.4
Cigarettes	7.4	Cigarettes per day	0.08	0.15	1.4	0.3
Drinks & cigarettes subtotal			0.13	3.13		
		Total	5.8	47.1		

Table 5 Estimation of daily intake of NDMA in various countries

Country	NDMA[a] [ng/(kg day)]	Major NDMA source (% contribution)	Reference
United Kingdom	7.6	Cured meats (81%)[b]	[48]
	8.6	Beer, cured meats	[62]
Netherlands	5.4	Beer (71%)	[63]
	1.4	Not evaluated[c]	[64]
Germany	15.7 (8.1)	Beer (65%), cured meats (10%)	[65]
	7.6 (5)	Beer (40%), cured meats (18%)	[66]
	4 (2.4)	Beer (31%), cured meats (36%)	[65]
Japan	25.7	Dried fish (91%)	[67]
	7.1	Beer (30%), fish products (68%)	[68]
Sweden	1.7	Beer (32%), cured meats (61%)	[69]
Finland[d]	1.4	Beer (31%), smoked fish (25%)	[70]
France	2.7	Alcohol (33.5%)	[60]

Source: Adapted from Biaudet et al. [60] and Tricker et al. [59].
[a] Data for women in parenthesis.
[b] Beer was not included in this study.
[c] Determined by a 24-h duplicate diet analysis.
[d] Based on limited data.

3.3 Exposure to NDMA through other sources

NDMA has been reported in a range of cosmetics and personal care products with concentrations of up to 24 ng/g [73]. However, only 1–4% of the NDMA in cosmetic preparations was estimated to penetrate through the skin [74]. At the maximum concentration reported by Spiegelhalder and Preussmann, 10 g of product would be expected to result in 2–10 ng of uptake of NDMA through the skin. Based on this limited amount of data cosmetics and toiletries are probably only a negligible source of NDMA for average consumers, although monitoring would be a wise precaution. No data is available on the formation of harmful substances on the skin following application and exposure to sunlight.

A significant contributor to NDMA-related health problems is workplace exposure. Industrial exposure affects workers, particularly in the rubber industry, where vulcanization and salt bath processes are applied. Concentrations reach up to 100 μg/m^3 in the air. This compares to concentrations of up to 0.07 μg/m^3 measured in a smoky bar [75], indicating that occupational sources can be a significant source of NDMA exposure, albeit limited to a high occupational risk group. A correlation between risk of occupational NDMA exposure and certain types of cancer has been identified.

3.4 Comparison of NDMA intake from various sources

A comparison of NDMA intakes from various sources outlines a number of key points. First, intake from food is substantial and in mean values constitutes 98% of intake (see Fig. 1). Second, this intake varies significantly with location, occupation, lifestyle, and dietary habits. Hence, a vegan with no consumption of cheese or meat products who does not drink alcohol or smoke will have the main NDMA intake originate from water, while for a person consuming high cheese and meat diet and a moderate-to-high alcohol, the contribution of NDMA intake from water would be negligible. Bearing in mind that the contribution of water is 5.9% (see Table 4) when considering maximum NDMA load in food and water at the WHO guideline value, it becomes clear that water can indeed make a significant contribution.

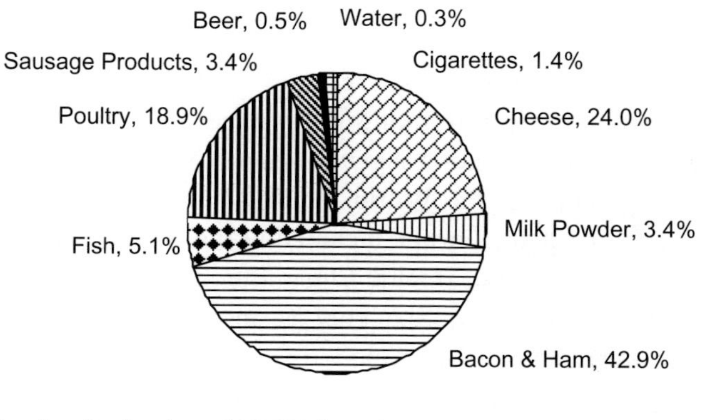

Figure 1 Intake distribution of NDMA based on mean intake distribution.

Fig. 1 summarizes the intake of NDMA through water, cigarettes, and a variety of foods as detailed in Table 4. Exposure routes are based on mean values and the average Australian diet. Clearly in this case water is not a significant contributor.

Given that the awareness of NDMA has made significant contributions to reduction of NDMA by improved manufacturing processes, it is expected that awareness and regulation of NDMA in drinking water will have a measurable health impact.

4. REGULATION OF NDMA IN DRINKING WATER

There are a number of regulatory and policy approaches regarding the issue of pollutants, which occur in wide ranges of contaminant concentrations, and ingestion habits. Mean values for NDMA used to calculate mean daily intakes based on average consumption values are somewhat meaningless when applied to individuals within a population. This is particularly true for NDMA, for which ingestion of the main dietary sources (cured meats and beer) varies from nil to multiples of the average amounts, depending on personal habits. Furthermore, looking at the relative contributions of NDMA from different sources is only marginally meaningful. It is the absolute amounts that are consumed that need to be addressed, and looking at ways to ensure that these are minimized and water plays an increasing role in those.

Given that NDMA can form during water and wastewater treatment, and that in some severe cases drinking water may provide as much or more dietary NDMA as food, some degree of formal monitoring of NDMA in drinking water must be required, particularly when disinfection techniques promoting formation are practiced.

Currently, NDMA in drinking water is regulated in very few countries worldwide, although it is now limited to 9 and 10 ng/L by the state and provincial governments in California and Ontario, respectively. These values exceed the 0.7 ng/L drinking water concentration suggested by the U.S. Environmental Protection Agency (EPA) to result in a 10^{-6} lifetime cancer risk [76]. Further, the World Health Organization is considering a guideline value of 100 ng/L to be included in the second addendum to the 3rd edition of the World Health Organization (WHO) Guidelines for Drinking Water Quality [35]. This is in agreement with German regulations for carcinogenic substances that have some level of uncertainty associated with their evaluation, which in the case of NDMA resulted at 100 ng/L [77]. There is presently no standard in the United Kingdom or the European Union. Since NDMA is likely a human carcinogen, its presence as a contaminant in drinking water is of concern, and consideration should be given to its wider regulation as well as the feasibility of monitoring and treatment options to prevent the occurrence of NDMA in water supplies.

5. CONTROL OF NDMA IN WATER

Two fundamentally different approaches to NDMA control have been pursued for drinking water treatment. One approach seeks to minimize NDMA formation while the other approach attempts to remove NDMA after its formation, but upstream of the drinking water distributions system.

5.1 Minimization of NDMA formation

As the predominant mechanism responsible for NDMA formation during disinfection involves a reaction between dichloramine and unchlorinated organic nitrogen precursors, removal or deactivation of organic nitrogen precursors, or minimization of dichloramine, would reduce NDMA formation. Chloraminating utilities employing cationic coagulation polymers, such as polyDADMAC, are attempting to reduce NDMA formation by reducing their polymer dosage [78]; however, these attempts must be balanced by the need to ensure proper coagulation.

In the case of wastewater-impacted source waters, the organic nitrogen precursors may be deactivated by a period of free chlorination [79] or ozonation [80]. As the dichloramine-associated formation mechanism requires an initial nucleophilic attack by the lone electron pair of organic amines on dichloramine, pretreatment with a strong oxidant, such as free chlorine or ozone, would oxidize the lone pair, preventing the formation. Interestingly, ozone was not effective at destroying NDMA itself.

Lastly, previous research indicated that dichloramine formation during chloramination, and the associated NDMA formation, could be reduced by altering the method by which the chlorine and ammonia reagents are added [79]. Chloramine formation reactions are fast compared to the timescale of reagent mixing. When chlorine is added downstream of ammonia, the chlorine to ammonia molar ratio can exceed 1 at the point of chlorine addition prior to complete mixing of chlorine into the flow stream. These conditions promote dichloramine formation at the point of chlorine addition. When chlorine is added upstream of ammonia, the opposite conditions prevail, and monochloramine formation is favored.

Moreover, as noted above, the short contact time with free chlorine before ammonia addition would aid in deactivating organic nitrogen precursors.

In certain waters, such as nonnitrified secondary municipal wastewaters in some wastewater recycling operations, ammonia is present in the influent water. In these situations, chloramines could be preformed under conditions promoting monochloramine formation (i.e., high pH with chlorine added prior to ammonia), and then applied to the process stream. This process was successfully pilot-tested at a wastewater recycling facility [81].

5.2 Membrane filtration

Historically, reverse osmosis (RO) membranes were designed to remove salt from seawater [82]. Over the next decades, a range of specialized membranes have become available for a variety of purposes, including nanofiltration (NF) membranes, which were designed for use as water softeners [83]. Today, RO and NF membranes are widely used in water treatment because they are able to remove various contaminants other than salts, including harmful trace organics, viruses, and dissolved organic matter. Sedlak found that NDMA precursors were 98% removed by RO [40,84]. However, RO and NF membranes are not always effective at removing small, neutral, and hydrophilic compounds such as NDMA. RO and NF only partially remove NDMA [3,39,40,84–87].

The more porous microfiltration (MF) and ultrafiltration (UF) membranes are unable to remove NDMA; however, they can reject its precursors with moderate efficiency (50%) [40].

To predict NDMA retention by RO and NF understanding the retention mechanisms is required. Although the actual rejection mechanisms for RO and NF are complex and not yet fully understood, key factors have been identified. These factors can be categorized as size exclusion, charge exclusion, and solute–membrane affinity [88], and they are related to the properties of the following – solute, membrane, solution (liquid matrix), and operating conditions [89,90]:

- Solute properties: molecular mass (MW) or molecular diameter, acid dissociation constant (pK_a), polarity, and hydrophobicity ($\log K_{ow}$).
- Membrane properties: molecular weight cutoff (MWCO), surface charge (zeta potential), and roughness.
- Water chemistry: pH, ionic strength, solute–solute interactions with other substances, in particular organic matter and colloidal matter.
- Operating conditions: pressure, flow rate, and recovery.
- Fouling status of the membrane.

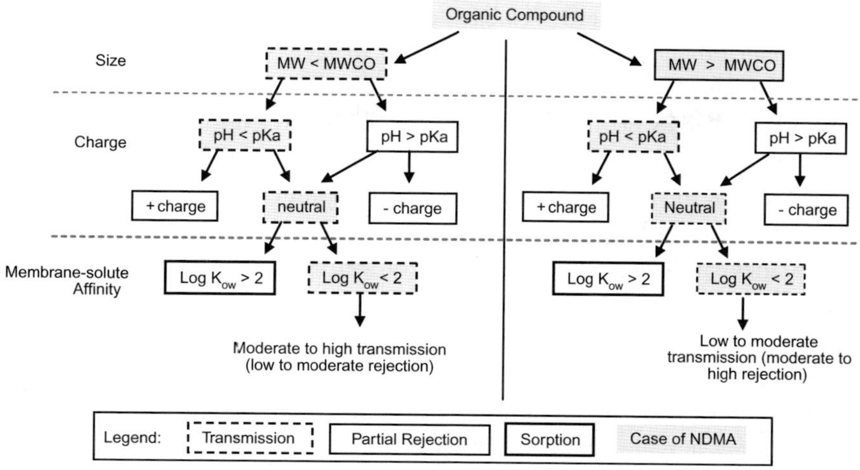

Figure 2 Predicting NDMA removal using a solute–matrix–membrane interaction scheme. Adapted from Ref. [89]. It is assumed that the membrane is negatively charged. The gray boxes highlight the case of NDMA. Left-hand side represents NF and "loose" RO; right-hand side represents "tight" RO.

Fig. 2 shows a schematic indicating solute–matrix–membrane interactions and their impact on organics rejection by membranes. The gray path follows the case for a negatively charged membrane and NDMA at pH values encountered during wastewater treatment and in environmental waters. Depending on the membrane, NDMA is smaller or larger than the MWCO.

Size exclusion depends on the MW and chemical structure of the solute as it relates to the membrane MWCO and/or pore size. The MWCO is defined as the molar mass above which more than 90% of a given compound is rejected. Compounds used for MWCO determinations are typically aqueous sugar or polyethylene glycol solutions. However, today there is no generally accepted industry standard [91], and MWCO ratings are not always comparable. As a first approximation, any solute that is larger than the MWCO will be efficiently rejected. NDMA has a MW of 73 Da, which is smaller than the MWCO of NF membranes (typically 200–500 Da). Although RO membranes are usually considered nonporous (i.e., no MWCO), some manufacturers report MWCO of approximately 100 Da for their RO membranes (Koch Membranes 2008). Because the MWCO of RO membranes (nonexistent to 100 Da) is closer to NDMA's MW (73 Da) than that of NF membranes (200–500 Da), tight RO membranes would be expected to better remove NDMA than NF membranes.

Holding other properties constant, a compound that is more charged is better rejected by NF and RO. The pK_a of NDMA is less than 1 [92], which renders this amine compound uncharged at ambient pH. Therefore, membrane charges do not contribute to NDMA rejection.

Finally, NDMA is expected to sorb poorly on the membrane and instead stay in the water phase due to its hydrophilicity (log $K_{ow} = -0.57$). For these reasons, NDMA is predicted to be poorly rejected by NF membranes and poorly to moderately rejected by RO membranes, depending on their "looseness," that is, MWCO (Fig. 2).

The available data show that RO and NF membranes do not achieve complete NDMA removal. Table 6 summarizes the available literature on NDMA removal by membranes in field and laboratory experiments. The data indicate a wide range of NDMA rejection values (10–70%), likely caused by different membrane–solute–matrix interactions. For example, in her laboratory studies, Steinle-Darling showed a decrease in NDMA rejection due to membrane fouling and water chemistry [87]. This is in agreement with other studies showing that fouling decreases rejection of small, uncharged contaminants [96,97].

To date, there is very little published data on NDMA rejection by NF membranes. An example is a study by Bellona et al. who determined the rejection of Filmtec NF90 [39,89] to be 42–47% NDMA. NF90 appears to be a borderline case because, based on the membrane's salt rejection and surface chemistry [98], NF90 could be classified as an RO membrane.

Advances in material science, such as membrane coatings, show promises in improving NDMA rejection by membranes. For example, a polyether polyamide block copolymer (PEBAX) coating increased NDMA rejection by LFC3 and BW-30 RO membranes by 6% and 15%, respectively, to 76% in both cases [87]. Further optimization of membranes may lead to greater removal efficiencies of small pollutants and greater reliance on membrane technology for organics removal.

Currently regulatory limits may be achieved by treating water with advanced oxidation (UV) or by blending it with water containing lower levels of NDMA.

5.3 UV treatment

UV treatment is the most common NDMA removal technique. However, treatment can employ UV fluences near 1000 mJ/cm^2, nearly an order of magnitude above those employed for disinfection [3]. NDMA exhibits a strong absorption band at 227 nm ($\varepsilon = 7000\ M^{-1}\ cm^{-1}$) associated with a $\pi \rightarrow \pi^\star$ transition. Laboratory studies indicate that low- and medium-pressure

Table 6 Literature studies on NDMA rejection by RO and tight NF membranes

Rejection of NDMA (%)	Membrane	MWCO[a]	Inlet NDMA concentration (ng/L)	Outlet NDMA concentration (ng/L)	Source
Laboratory study – distilled water matrix					
54	RO (ESPA3)	200[1]	200	92	[87]
70	RO (LFC3)	N/A	200	60	[87]
61	RO (BW-30)	N/A	200	78	[87]
Field studies at water reuse facilities (pilot and full scales) – pre-treated secondary wastewater matrix					
24–56	RO (ESPA2)	250–500[2]	18–57	14–41	[86]
32	RO (ESPA2)	250–500[2]	30	20	[39, 85]
40–65[b]	RO (ESPA2)[3]	250–500[2]	25–100	12–50	[40]
42–50[b]	RO	N/A	50–70	30–45	[40]
10–70[b]	RO	N/A	200–330	100–170	[40]
50	RO	N/A	N/A	N/A	[3]
~50	RO (TFC)	N/A	20–170 (wastewater concentration)		[84]
30	ULPRO (TFC-HR)	N/A	60	40	[39, 85]
50–55	ULPRO (TMG-10)	100[4]	20–40	10–20	[39, 85]
42–47	NF (NF90)	300[5]; 200[6]	10–20	4–12	[39, 85]

N/A means data not available.
[a] MWCO sources are (1) Fonseca et al. [93], the authors classified this membrane as NF. (2) Estimation in Kim et al. [94]. (3) Plumlee et al. [86]. (4) Drewes et al. [95]. (5) Kim et al. [94]. (6) DOW-Filmtec.
[b] Numbers were approximated from graphs.

lamps exhibit comparable photonic efficiencies for NDMA degradation [99]. Addition of 100 mg/L hydrogen peroxide (H_2O_2) to promote hydroxyl radical (OH*) formation results in no significant benefit regarding NDMA degradation. However, some wastewater recycling facilities employ a UV/ H_2O_2 advanced oxidation process (AOP) downstream of RO membranes, because of the combination of UV for NDMA removal and OH* for destruction of other contaminants that may pass through the membranes.

6. CONCLUSIONS

NDMA occurs in many foods, cigarettes, drinks, certain industrial processes, and some water supplies. While the regulation of NDMA in water supply can be described as an emerging issue with significant difficulties with regard to the implementation of monitoring procedures, the main aim of this chapter has been to present the available knowledge on NDMA in water regarding sources, occurrence, fate, effects, risk assessment, and risk management opportunities.

An increased awareness about the presence of traditionally high NDMA concentrations in some foods, and the understanding of how it was formed, has effectively led to large reductions in its occurrence. With the growing concern about NDMA in drinking water, in particular with increasing awareness and exploitation of water recycling opportunities, hopefully the next decade will see an increase in monitoring, elevated removal efficiency by treatment processes, and a decrease in its presence and consequently risk of exposure from water supplies. The avoidance of NDMA formation during treatment will play an important role in this process. Given the difficulty in removing NDMA in water treatment, this compound is an excellent model compound to deal with the implications of xenobiotics occurrence in water supplies.

REFERENCES

[1] World Health Organization International Agency for Research on Cancer, IARC Monographs on the Evaluation of Carcinogenic Risks to Humans Overall Evaluations of Carcinogenicity: An Updating of IARC Monographs, Vols. 1–42, Lyon, 1987.

[2] D.B. Jobb, R.B. Hunsinger, O. Meresz, V.Y. Taguchi, A Study of the Occurrence and Inhibition of Formation of *N*-Nitrosodimethylamine (NDMA) in the Ohsweken Water Supply, in: Proceedings of 5th National Conference on Drinking Water on Disinfection Dilemma – Microbiological Control versus By-products, September 13–15, 1992, Winnipeg, Canada, 1993, pp. 241–252.

[3] W.A. Mitch, J.O. Sharp, R.R. Trussell, R.L. Valentine, L. Alvarez-Cohen, D.L. Sedlak, *N*-Nitrosodimethylamine (NDMA) as a drinking water contaminant: a review, Environ. Eng. Sci. 20 (2003) 389–404.
[4] J.W.A. Charrois, J.M. Boyd, K.L. Froese, S.E. Hrudey, Occurrence of *N*-nitrosamines in Alberta public drinking-water distribution systems, J. Environ. Sci. Eng. 6 (2007) 103–114.
[5] A. Fristachi, G. Rice, Estimation of the total daily oral intake of NDMA attributable to drinking water, J. Water Health 5 (2007) 341–355.
[6] T.J. Doyle, W. Zheng, J.R. Cerhan, C.P. Hong, T.A. Sellers, L.H. Kushi, A.R. Folsom, The association of drinking water source and chlorination by-products with cancer incidence among postmenopausal women in Iowa: a prospective cohort study, Am. J. Public Health 87 (1997) 1168–1176.
[7] M. Koivusalo, E. Pukkala, T. Vartiainen, J.J. Jaakkola, T. Hakulinen, Drinking water chlorination and cancer a historical cohort study in Finland, Cancer Causes Control 8 (1997) 192–200.
[8] C.M. Villanueva, F. Fernández, N. Malats, J.O. Grimalt, M. Kogevinas, Meta-analysis of studies on individual consumption of chlorinated drinking water and bladder cancer, J. Epidemiol. Community Health 57 (2005) 166–173.
[9] World Health Organization, *N*-Nitrosodimethylamine in Drinking –Water, Geneva, 2006.
[10] United States Environmental Protection Agency, Ambient Quality Criteria for Nitrosamines, Office of Water Regulations: Standards Criteria and Standards Division, Washington, DC, 1980.
[11] Ontario Ministry of the Environment., Scientific Criteria Document for Multimedia Standard Development No. 01-90; *N*-Nitrosodimethylamine, Hazardous Contaminants Coordination Branch, Toronto, Ont., 1991.
[12] R.C. Weast, CRC Handbook of Chemistry and Physics, CRC Press Inc., Boca Raton, FL, 1983.
[13] S.S. Mirvish, I. Issenberg, H.C. Sornson, Air-water and ether-water distribution of *N*-nitroso compound: implications for laboratory safety, analytic methodology, and carcinogenicity for the rat esophagus, nose and liver, J. Natl. Cancer Inst. 56 (1976) 1125–1129.
[14] C. Hansch, A.J. Leo, Medchem Project Issue No. 26, Pomona College, Claremont, CA, 1985.
[15] R.G. Thomas, Volatilization from water, in: W.J. Lyman, W.F. Reehl, D.H. Rosenblatt (Eds.), Handbook of Chemical Property Estimation Methods, McGraw-Hill, New York, 1982, pp. 15–27.
[16] C.K. Schmidt, H.J. Brauch, *N,N*-Dimethylsulfamide as precursor for *N*-nitrosodimethylamine (NDMA) formation upon ozonation and its fate during water treatment, Environ. Sci. Technol. 42 (2008) 6340–6346.
[17] I.M. Schreiber, W. Mitch, Enhanced nitrogenous disinfection byproduct formation near the breakpoint: implications for nitrification control, Environ. Sci. Technol. 41 (2007) 7039–7046.
[18] S.S. Mirvish, Kinetics of nitrosamide formation from alkylureas, *N*-alkylurethans, and alkylguanidines: possible implications for the etiology of known gastric cancer, J. Natl. Cancer Inst. 46 (1971) 1183–1193.
[19] J. Casado, G. González-AlaTorre, C. Izquierdo, C. Brunner, The nitrosation of *N*-alkylureas: evidence for a proton-transfer mechanism, Int. J. Chem. Kinet. 28 (1996) 307.
[20] T.Y. Fan, S.R. Tannenbaum, Factors influencing the rate of formation of nitrosomorpholine from morpholine and nitrite: acceleration by thiocyanate and other anions, J. Agric. Food Chem. 21 (1973) 237–240.

[21] J. Choi, R.L. Valentine, Formation of *N*-nitrosodimethylamine (NDMA) from reaction of monochloroamine: a new disinfection by-product, Water Res. 36 (2002) 817–824.

[22] W.A. Mitch, D.L. Sedlak, Formation of *N*-nitrosodimethylamine (NDMA) from dimethylamine during chlorination, Environ. Sci. Technol. 36 (2002) 588–595.

[23] I.M. Schreiber, W.A. Mitch, Nitrosamine formation pathway revisited: the importance of dichloramine and dissolved oxygen, Environ. Sci. Technol. 40 (2006) 6007–6014.

[24] J. Choi, R.L. Valentine, *N*-Nitrosodimethylamine formation by free-chlorine-enhanced nitrosation of dimethylamine, Environ. Sci. Technol. 37 (2003) 4871–4876.

[25] S.S. Walse, W.A. Mitch, Nitrosamine carcinogens also swim in chlorinated pools, Environ. Sci. Technol. 42 (2008) 1032–1037.

[26] P. Westerhoff, H. Mash, Dissolved organic nitrogen in drinking water supplies: a review, J. Water SRT – Aqua 51 (2002) 415–448.

[27] A.C. Gerecke, D.L. Sedlak, Precursors of *N*-nitrosodimethylamine in natural waters, Environ. Sci. Technol. 37 (2003) 1331–1336.

[28] W.A. Mitch, D.L. Sedlak, Characterization and fate of *N*-nitrosodimethylamine precursors in municipal wastewater treatment plants, Environ. Sci. Technol. 38 (2004) 1445–1454.

[29] W.A. Mitch, I.M. Schreiber, Degradation of tertiary alkylamines during chlorination/chloramination: implications for formation of aldehydes, nitriles, halonitroalkanes and nitrosamines, Environ. Sci. Technol. 42 (2008) 4811–4817.

[30] K. Kosaka, M. Asami, Y. Konno, M. Oya, S. Kunikane, Identification of antiyellowing agents as precursors of *N*-nitrosodimethylamine production on ozonation from sewage treatment plant influent, Environ. Sci. Technol. 407 (11) (2009) 3540–3545.

[31] S.-H. Park, S. Wei, B. Mizaikoff, A.E. Taylor, C. Favero, C.-H. Huang, Degradation of amine-based water treatment polymers during chloramination as *N*-nitrosodimethylamine (NDMA) precursors, Environ. Sci. Technol. 43 (2009) 1360–1366.

[32] J.M. Kemper, P. Westerhoff, A. Dotson, W.A. Mitch, Nitrosamine, dimethylnitramine, and chloropicrin formation during strong-base anion-exchange treatment, Environ. Sci. Technol. 43 (2009) 466–472.

[33] Orange County Water District, Water Factory 21, Orange Country, CA, Orange County Water District, 2001.

[34] Ontario Ministry of the Environment and Energy, Removal of *N*-Nitrosodimethylamine from the Ohsweken (Six Nations) Water Supply, Toronto, Ont., 1994.

[35] G. Dillon, S. Blake, P. Rumsby, L. Rockett, T. Hall, P. Jackson, A. Rawlinson, NDMA – Concentrations in drinking water and factors affecting its formation (CSA7240/WT02049/DWI 70/2/210), Defra 7348 Final Report, March 2008.

[36] M. Asami, M. Oya, K. Kosaka, A nationwide survey of NDMA in raw and drinking water in Japan, Sci. Total Environ. 407 (2009) 3540–3545.

[37] D.L. Sedlak, R.A. Deeb, E.L. Hawley, W.A. Mitch, T.D. Turbin, S. Mowbray, S. Carr, Sources and fate of nitrosodimethylamine and its precursors in municipal wastewater treatment plants, Water Environ. Res. 77 (2005) 32–39.

[38] J. Anderson, The environmental benefits of water recycling and reuse, Water Sci. Technol.: Water Supply 3 (2003) 1.

[39] J. Drewes, C. Bellona, P. Xu, G. Amy, G. Filteau, G. Oelker, Comparing nanofiltration and reverse osmosis for treating recycled water, AWWARF Report 91212, 2008.

[40] D.L. Sedlak, M. Kavanaugh, Removal and Destruction of NDMA and NDMA Precursors during Wastewater Treatment, Final Report to the Water Reuse Foundation, Project 01-002, 2005.

[41] W. Lijinsky, *N*-Nitroso compounds in the diet, Mutat. Res. 443 (1999) 129–138.

[42] T. Fazio, J.N. Damico, J.W. Howard, R.H. White, J.O. Watts, Gas chromatographic determination and mass spectrometric confirmation of *N*-nitrosodimethylamine in smoke-processed marine fish, J. Agric. Food Chem. 19 (1971) 250–253.
[43] S. Yurchenko, U. Mölder, Volatile *N*-nitrosamines in various fish products, Food Chem. 100 (2007) 1713–1721.
[44] W.P. Sen, J.R. Iyengar, B.D. Donaldson, T. Panalaks, W.F. Miles, The analysis and occurrence of volatile nitrosamines in cured meat products, International Agency for Research on Cancer (IARC) Sci. Publ. 9 (1975) 49–52.
[45] M.W. Byun, H.J. Ahn, J.H. Kim, J.W. Lee, H.S. Yook, S.B. Han, Determination of volatile *N*-nitrosamines in irradiated fermented sausage by gas chromatography coupled to a thermal energy analyzer, J. Chromatogr. A 1054 (2004) 403–407.
[46] N.P. Sen, B. Donaldson, J.R. Iyengar, T. Panalaks, Nitrosopyrrolidine and dimethylnitrosamine in bacon, Nature 241 (1973) 473–474.
[47] M.B.A. Glória, J.F. Barbour, R.A. Scanlan, Volatile nitrosamines in fried bacon, J. Agric. Food Chem. 45 (1997) 1816–1818.
[48] T.A. Gough, K.S. Webb, R.F. Coleman, Estimate of the volatile nitrosamine content in U.K. food, Nature 272 (1978) 161.
[49] A. Dellisanti, G. Cerutti, L. Airoldi, Volatile *N*-nitrosamines in selected Italian cheeses, Bull. Environ. Contam. Toxicol. 57 (1996) 16–21.
[50] L. Lakritz, J.W. Pensabene, Survey of fluid and nonfat dry milks for *N*-nitrosamines, J. Dairy Sci. 64 (1981) 371–374.
[51] C.P. Oliveira, M. Beatriz, A. Gloria, J.F. Barbour, R.A. Scanlan, Nitrate, nitrite, and volatile nitrosamines in whey-containing food products, J. Agric. Food Chem. 43 (1995) 967–969.
[52] D.W. Lachenmeier, D. Fügel, Reduction of nitrosamines in beer: review of a success story, Brewing Sci. 297 (May/June) (2007) 84–89.
[53] L.M. Anderson, J.P. Carter, D.L. Logsdon, C.L. Driver, R.M. Kovatch, Characterization of ethanol's enhancement of tumorigenesis by *N*-nitrosodimethylamine in mice, Carcinogenesis 13 (1992) 2107–2111.
[54] E. Hietanen, H. Bartsch, J.C. Bereziat, M. Ahotupa, A.M. Camus, J.R. Cabral, M. Laitinen, Quantity and saturation degree of dietary fats as modulators of oxidative stress and chemically-induced liver tumors in rats, Int. J. Cancer 46 (1990) 640–647.
[55] G.G. Wade, R. Mandel, H.J.-P. Ryser, Marked synergism of dimethylnitrosamine carcinogenesis in rats exposed to cadmium, Cancer Res. 47 (1987) 6606–6613.
[56] S.Y. Choi, M.J. Chung, S.-J. Lee, J.H. Shin, N.K. Sung, *N*-Nitrosamine inhibition by strawberry, garlic, kale, and the effects of nitrite-scavenging and *N*-nitrosamine formation by functional compounds in strawberry and garlic, Food Control 18 (2007) 485–491.
[57] S.Y. Choi, M.J. Chung, N.J. Sung, Volatile *N*-nitrosamine inhibition after intake Korean green tea and Maesil (Prunus mume SIEB. et ZACC.) extracts with an amine-rich diet in subjects ingesting nitrate, Food Chem. Toxicol. 43 (2002) 949–957.
[58] D.J. Fitzgerald, N.I. Robinson, Development of a tolerable daily intake for *N*-nitrosodimethylamine using a modified benchmark dose methodology, J. Toxicol. Environ. Health Part A 70 (2007) 1670–1678.
[59] A.R. Tricker, D. Ditrich, R. Preussmann, *N*-Nitroso compounds in cigarette tobacco and their occurrence in mainstream tobacco smoke, Carcinogenesis 12 (1991) 257–261.
[60] H. Biaudet, T. Mavelle, G. Debry, Mean daily intake of *N*-nitrosodimethylamine from foods and beverages in France in 1987–1992, Food Chem. Toxicol. 32 (1994) 417–421.
[61] A. Munoz, A Comparison of *N*-Nitrosodimethylamine Human Health Risks from Water and Other Sources, MEng Thesis, University of Wollongong, Australia, 2005.

[62] Ministry of Agriculture Fisheries and Food (MAFF), Nitrate, nitrite and *N*-nitroso compounds in foods, Food Surveillance Paper, 20, 1987.
[63] R.W. Stephany, P.L. Schuller, Daily dietary intakes of nitrate, nitrite and volatile *N*-nitrosamines in the Netherlands using the duplicate portion sampling technique, Oncology 37 (1980) 203–210.
[64] G. Ellen, E. Egmond, J.W.V. Loon, E.T. Sahertian, K. Tolsma, Dietary intakes of some essential metals and non-essential trace elements, nitrate, nitrite and *N*-nitrosamines, by Dutch adults: estimated via 24-hour duplicate portion study, Food Addit. Contam. 7 (1990) 207–221.
[65] B. Spiegelhalder, G. Eisenbrand, R. Preussmann, Occurrence of volatile nitrosamines in food: a survey of the West German market, IARC Sci. Publ. 31 (1980) 467–479.
[66] B. Spiegelhalder, R.C. Preussmann, Occupational nitrosamine exposure. 1. Rubber and tyre industry, Carcinogenesis 4 (1983) 1147.
[67] T. Maki, Y. Tamura, Y. Shimamura, Y. Naoi, Estimate of the volatile nitrosamine content of Japanese food, Bull. Environ. Contam. Toxicol. 25 (1980) 257–261.
[68] M. Yamamoto, R. Iwata, H. Ishiwata, T. Yamada, A. Tanimura, Determination of volatile nitrosamine levels in foods and estimation of their daily intake in Japan, Food Chem. Toxicol. 22 (1984) 61–64.
[69] B.G. Österdahl, Volatile nitrosamines in foods on the Swedish market and estimation of their daily intake, Food Addit. Contam. 5 (1988) 587–595.
[70] P.L. Pentilla, L. Räsänen, S. Kimppa, Nitrate, nitrite and *N*-nitroso compounds in Finnish foods and estimation of the dietary intakes, Food Sci. Technol. 190 (1990) 336–340.
[71] M.B.A. Glória, J.F. Barbour, R.A. Scanlan, *N*-Nitrosodimethylamine in Brazilian, U.S. domestic, and U.S. imported beers, J. Agric. Food Chem. 1997 (1997) 814–816.
[72] Ministry of Agriculture Fisheries and Food (MAFF), UK, NDMA and ATNC in retail malt whisky, Food Surveillance Information Sheet, Number 57, February 1995, http://archive.food.gov.uk/maff/archive/food/infsheet/1995/no57/57whisky.htm, retrieved 06.17.2009.
[73] B. Spiegelhalder, R. Preussmann, Contamination of toiletries and cosmetic products with volatile and nonvolatile *N*-nitroso carcinogens, J. Cancer Res. Clin. Oncol. 108 (1984) 160–163.
[74] K.R. Brain, K.A. Walters, V.J. James, W.E. Dressler, D. Howes, C.K. Kelling, S.J. Moloney, S.D. Gettings, Percutaneous penetration of dimethylnitrosamine through human skin in vitro: application from cosmetic vehicles, Food Chem. Toxicol. 33 (1995) 315–322.
[75] G. Stehlik, O. Richter, H. Altmann, Concentration of dimethylnitrosamine in the air of smoke-filled rooms, Ecotoxicol. Environ. Saf. 6 (1982) 495–500.
[76] United States Environmental Protection Agency, Fact Sheet: Emerging Contaminant – *N*-Nitroso-dimethylamine (NDMA), 2008.
[77] Umweltbundesamt., Bewertung der Anwesenheit teil- oder nicht bewertbarer Stoffe im Trinkwasser aus gesundheitlicher Sicht, Bundesgesundheitsbl-Gesundheitsforsch-Gesundheitsschutz 46 (2003) S249–S251.
[78] A. Wilczak, A. Assadi-Rad, H.H. Lai, L.L. Hoover, J.F. Smith, R. Berger, F. Rodigari, J.W. Beland, L.J. Lazzelle, E.G. Kinicannon, H. Baker, C.T. Heaney, Formation of NDMA in chloraminated water coagulated with DADMAC cationic polymer, J. Am. Water Works Assoc. 95 (2003) 94–106.
[79] I.M. Schreiber, W.A. Mitch, Influence of the order of reagent addition on NDMA formation during chloramination, Environ. Sci. Technol. 38 (2005) 1445–1454.

[80] C. Lee, C. Schmidt, J. Yoon, U. VonGunten, Oxidation of *N*-nitrosodimethylamine (NDMA) precursors with ozone and chlorine dioxide: kinetics and effect on NDMA formation potential, Environ. Sci. Technol. 41 (2007) 2056–2063.
[81] W.A. Mitch, G.L. Oelker, E.L. Hawley, R.A. Deeb, D.L. Sedlak, Minimization of NDMA formation during chlorine disinfection of municipal wastewater by application of pre-formed chloramines, Environ. Eng. Sci. 22 (2005) 882–890.
[82] J. Glater, The early history of reverse osmosis membrane development, Desalination 117 (1998) 297–309.
[83] Verliefde., Priority organic micropollutants in water sources in Flanders and the Netherlands and assessment of removal possibilities with nanofiltration, Environ. Pollut. 146 (2007) 281–289.
[84] R. Deeb, E. Hawley, D. Sedlak, J. Loveland, M. Kavanaugh, Removal and destruction of NDMA and precursors during wastewater treatment, Water Environmental Federation's Annual Technical Exhibition and Conference, 2006.
[85] C. Bellona, J.E. Drewes, G. Oelker, J. Luna, G. Filteau, G. Amy, Comparing nanofiltration and reverse osmosis for drinking water augmentation, J. Am. Water Works Assoc. 100 (2008) 102.
[86] M.H. Plumlee, M. López-Mesas, A. Heidlberger, K.P. Ishida, M. Reinhard, *N*-Nitrosodimethylamine (NDMA) removal by reverse osmosis and UV treatment and analysis via LC-MS/MS, Water Res. 42 (2008) 347–355.
[87] E. Steinle-Darling, M. Zedda, M.H. Plumlee, H.F. Ridgway, M. Reinhard, Evaluating the impacts of membrane type, coating, fouling, chemical properties and water chemistry on reverse osmosis rejection of seven nitrosoalklyamines, including NDMA, Water Res. 41 (2007) 3959–3967.
[88] J.G. Wijmans, R.W. Baker, The solution–diffusion model: a review, J. Membr. Sci. 107 (1995) 1–21.
[89] C. Bellona, J.E. Drewes, P. Xu, G. Amy, Factors affecting the rejection of organic solutes during NF/RO treatment: a literature review, Water Res. 38 (2004) 2795.
[90] L.D. Nghiem, A.I. Schäfer, M. Elimelech, Removal of natural hormones by nanofiltration membranes: measurement, modeling, and mechanisms, Environ. Sci. Technol. 38 (2004) 1888–1896.
[91] Novasep, http://www.novasep.com/misc/glossary.asp?defId = 139&lookfor = & search = M, retrieved 04.03.2009.
[92] C. Lee, Y. Lee, C. Schmidt, J. Yoon, U. VonGunten, Oxidation of suspected *N*-nitrosodimethylamine (NDMA) precursors by ferrate (VI): kinetics and effect on the NDMA formation potential of natural waters, Water Res. 42 (2008) 433–441.
[93] A.C. Fonseca, R.S. Summers, A.R. Greenberg, M.T. Hernandez, Extracellular polysaccharides, soluble microbial products, and natural organic matter impact on nanofiltration membranes flux decline, Environ. Sci. Technol. 41 (2007) 2491–2497.
[94] J. Kim, F.A. DiGiano, R.D. Reardon, Autopsy of high-pressure membranes to compare effectiveness of MF and UF pretreatment in water reclamation, Water Res. 42 (2008) 697–707.
[95] J.E. Drewes, C. Bellona, M. Oedekoven, P. Xu, T.-U. K, G. Amy, Rejection of wastewater-derived micropollutants in high-pressure membrane applications leading to indirect potable reuse, Environ. Prog. 24 (2005) 400–409.
[96] H.Y. Ng, M. Elimelech, Influence of colloidal fouling on rejection of trace organic contaminants by reverse osmosis, J. Membr. Sci. 244 (2004) 215–226.
[97] X. Zhu, M. Elimelech, Colloidal fouling of reverse osmosis membranes: measurements and fouling mechanisms, Environ. Sci. Technol. 31 (1997) 3654–3662.

[98] C.Y. Tang, Y.-N. Kwon, J.O. Leckie, Fouling of reverse osmosis and nanofiltration membranes by humic acid: effects of solution composition and hydrodynamic conditions, J. Membr. Sci. 290 (2007) 86–94.
[99] C.M. Sharpless, K.L. Linden, Experimental and model comparisons of low- and medium-pressure Hg lamps for the direct and H_2O_2 assisted UV photodegradation of *N*-nitrosodimethylamine in simulated drinking water, Environ. Sci. Technol. 37 (2003) 1933–1940.

CHAPTER 8

TiO_2-Based Advanced Oxidation Nanotechnologies for Water Purification and Reuse

Hyeok Choi[1], Souhail R. Al-Abed[2], Dionysios D. Dionysiou[3,*], Elias Stathatos[4] *and* **Panagiotis Lianos[5]**

[1]Department of Civil Engineering, University of Texas at Arlington, 416 Yates Drive, Arlington, TX 76019-0308, USA
[2]National Risk Management Research Laboratory, US Environmental Protection Agency, 26W. Martin Luther King Dr., Cincinnati, OH 45268, USA
[3]Department of Civil and Environmental Engineering, University of Cincinnati, Cincinnati, OH 45221-0071, USA
[4]Department of Electrical Engineering, Technological-Educational Institute of Patras, 26334 Patras, Greece
[5]Department of Engineering Science, University of Patras, 26500 Patras, Greece

Contents

*Corresponding author.
E-mail address: dionysios.d.dionysiou@uc.edu

Sustainability Science and Engineering, Volume 2 ISSN 1871-2711, DOI 10.1016/S1871-2711(09)00208-6

1. ROLE OF ADVANCED OXIDATION TECHNOLOGIES FOR INNOVATIVE WATER TREATMENT

The so-called advanced oxidation technologies (AOTs) are among the most emerging chemical oxidation processes and are anticipated to play a crucial role in water treatment as stand-alone processes or posttreatment options in combination with conventional technologies in the near future [1]. AOTs refer to a set of chemical treatment processes designed to decompose organic and inorganic materials in water by oxidation route. The technologies are particularly useful to destroy biologically toxic and non-degradable chemicals such as aromatics, pesticides, and volatile compounds.

As summarized in Table 1, AOTs, as a powerful tool for water and wastewater treatment, include chemical oxidation, Fenton and photo-Fenton processes, ultraviolet (UV)-based processes, photocatalytic redox processes, supercritical water oxidation, sonolysis, and electron beams and

Table 1 Advanced oxidation technologies

Process	Chemicals or equipment used
Chemical oxidation	$O_3+H_2O_2$
Fenton and photo-Fenton processes	$Fe^{2+}+H_2O_2$, $Fe^{2+}+H_2O_2+UV$
UV-based processes	$UV+O_3$, $UV+H_2O_2$, $UV+O_3+H_2O_2$
Photocatalytic redox processes	Semiconductor (TiO_2, ZnO)/UV
Supercritical water oxidation	High temperature and pressure
Sonolysis	Ultrasound
Electron beam and γ-rays	Beam generator

γ-ray irradiation [2]. In conventional oxidation technologies, the role of common oxidants such as chlorine and permanganate is well-known to directly oxidize and thus decompose water contaminants. Meanwhile, AOTs are based on further activation of oxidants, such as ozone, hydrogen peroxide, peroxymonosulfate (PMS), and persulfate, to generate other transient species such as hydroxyl radicals ($^{\bullet}OH$) and sulfate radicals ($SO_4^{\bullet -}$) that demonstrate much higher oxidation capability than the oxidant sources. There are many alternative ways of generating such oxidizing species, including TiO_2 photocatalysis and sonolysis. The various combinations of these technologies such as $UV/H_2O_2/O_3$ and Fenton-like reaction are preferred for their high reactivity and efficiency and they are already in use for large-sale industrial water treatment. For example, the well-known system, Fenton reagent leads to the generation of hydroxyl radicals in the presence of iron (Reaction 1):

$$Fe^{2+} + H_2O_2 \rightarrow Fe^{3+} + {}^{\bullet}OH + OH^- \quad (1)$$

This technology is currently applied for the treatment of industrial discharges, soil, and groundwater remediation. However, the majority of the other technologies are still in the developmental stage.

AOTs are characterized by a specific oxidation pathway. Extremely reactive hydroxyl radicals are commonly formed during the reaction initiated by the chemicals or equipment used for AOTs. In this chapter, we mostly focus on more common hydroxyl radical-based AOTs while sulfate radicals also play a role as a strong oxidizing species in certain types of AOTs (sulfate radicals will be briefly mentioned in Section 7.2). The hydroxyl radicals readily attack organic contaminants. In sequential reactions, the organic contaminants are transformed to simpler organic molecules that are eventually mineralized to CO_2, H_2O, and inorganic species (i.e., Cl^-, NO_3^-, SO_4^{2-}). The general scheme and concept of AOTs are demonstrated in Fig. 1. The mechanism of organic decomposition using AOTs involves a sequential reaction pathway that includes the following steps: (a) generation of hydroxyl radicals, (b) formation of carbon center radicals in organic compounds by hydroxyl radicals attack, (c) transformation of carbon center radicals to peroxyl radicals by the addition of oxygen, (d) degradation of peroxyl radicals to form simpler organic molecules, and (e) repetition of the cycle until complete organic mineralization [2]. Due to rapid oxidation reactions, AOTs are characterized by high reaction rates and short treatment times, which makes them promising in water and wastewater treatment.

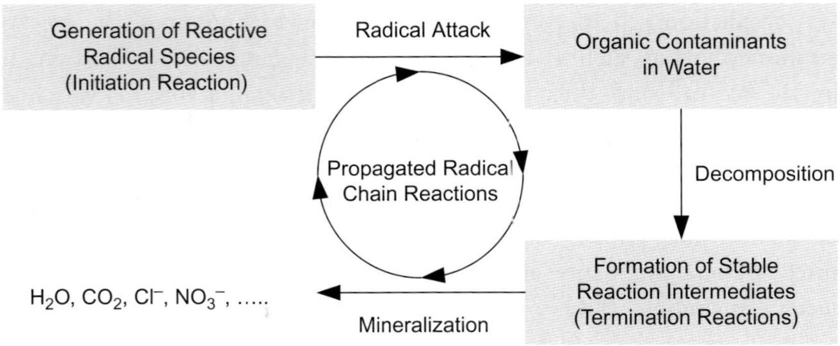

Figure 1 General scheme and concept of AOTs. Reactive radical species ($^{\bullet}OH$, $O_2^{\bullet -}$) generated from various sources (TiO_2/UV, UV/H_2O_2, O_3, UV/O_3) attack organic contaminants in water to break down and transform their molecular structure to simple intermediates via chain reactions, and then finally mineralize them to water, carbon dioxide, and inorganic species.

2. TiO_2 PHOTOCATALYTIC PROCESSES: GREEN CHEMISTRY AND ENGINEERING

Research efforts in photocatalysis have dramatically expanded since the discovery of the photocatalytic properties of TiO_2 and the demonstration of its effectiveness to generate hydroxyl radicals in the presence of UV [3]. TiO_2 photocatalysis is of particular interest because of its environmentally friendly features. The process can completely oxidize virtually all organic contaminants (nonselective) without addition of any other chemicals for the reaction, and thus produce no harmful end products in most cases. Especially, TiO_2 photocatalysis forms no disinfection by-products unlike other chemical oxidation processes when sufficient time is allowed for organic mineralization. In general, the photocatalytic process has features of a green engineering process. Although various materials (oxides: TiO_2, ZnO, ZrO_2, CeO_2, SnO_2, Fe_2O_3, $SbrO_4$; sulfides: CdS, ZnS) have been used for photocatalysis, generally TiO_2 is the most promising photocatalyst, considering its energy efficiency, durability, photostability, water insolubility, and nontoxicity. In this chapter, we overview TiO_2-based AOTs for water treatment and reuse, and we are particularly interested in their current challenges and advances for their potential in full-scale practical applications.

3. MECHANISM OF TiO_2 PHOTOCATALYSIS

The general mechanism of heterogeneous photocatalysis on TiO_2 for the destruction of water contaminants was proposed by Ollis et al. [4],

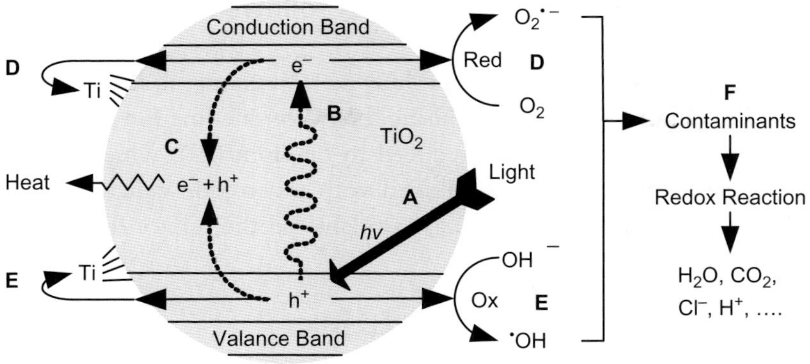

Figure 2 Schematic demonstration of the photochemical steps in semiconducting TiO_2. A: Light irradiation to TiO_2 surface, B: generation of electrons and holes as charge carriers, C: recombination of the electrons and holes, D and E: reductive and oxidative pathways of substances absorbed on TiO_2 sites, and F: redox reaction to decompose organic contaminants in water.

summarized in a book by Ollis and Al-Ekabi [3], discussed by Hoffmann and his coworkers [5,6], and overviewed in several review articles [7–11]. The primary photochemical events in TiO_2 semiconductors are simplified in Fig. 2. In its electronic structure, TiO_2, like other semiconductors, exhibits a void energy region between the top of the filled valance band and the bottom of the vacant conduction band, which is called the band gap (BG).

3.1 Generation of charge carrier species and their recombination

If a photon energy, $h\nu$, irradiating TiO_2 exceeds its BG, absorption of the photon by TiO_2 excites an e^- from the valance band to the conduction band while an electron vacancy (i.e., electron hole or simply hole), h^+ is left in the valance band (Reaction 2) [3,9]. High energy from short wavelength light (UV or near-UV photons) is required for the excitation due to the high BG energy (E_G) of TiO_2 (below UV region at 387 nm corresponding to 3.2 eV for the anatase crystal phase of TiO_2). Most of electron and hole pairs generated either simultaneously recombine in the bulk or migrate to the TiO_2 surface and then recombine, releasing useless thermal energy (Reaction 3) [3,9]. This recombination reaction is detrimental to the process and limits the activity of the catalyst [3].

$$TiO_2 + h\nu \rightarrow e^- + h^+ \quad (2)$$

$$e^- + h^+ \rightarrow \text{Heat} \quad (3)$$

3.2 Adsorption of chemicals to TiO_2 followed by their redox pathways

Eventually, some of the electron and hole pairs migrating to the surface get involved in redox reactions even during their short lifetime (on the order of nanoseconds). Titanium (IV) is reduced to titanium (III), which is finally transformed to titanium (IV) combined with superoxide radical anions, if electron acceptors such as oxygen are available on the surface (Reactions 4 and 5) [3,9].

$$Ti^{IV} + e^- \rightarrow Ti^{III} \tag{4}$$

$$Ti^{III} + O_2 \rightarrow Ti^{IV} - O_2^{\bullet -} \tag{5}$$

Meanwhile, the generated holes are utilized for the generation of hydroxyl radicals and direct oxidation of organics, R (Reactions 6–8) [3,9] or they can be combined with the electron from a donor species, depending on the reaction mechanism:

$$Ti^{IV} - OH^- + h^+ \rightarrow Ti^{IV} - OH^\bullet \tag{6}$$

$$Ti^{IV} - H_2O + h^+ \rightarrow Ti^{IV} - OH^\bullet + H^+ \tag{7}$$

$$R_{ads} + h^+ \rightarrow R^+_{ads} \tag{8}$$

Because of the short lifetime of photo carriers, the prerequisite for Reactions 6–8 is the adsorption of substances such as water and organic molecules on the TiO_2 surface and lattice oxygen (O_L^{2-}) (Reactions 9–11) [3,9]. This facilitates the redox reaction at the interface of TiO_2 solid and the water:

$$O_L^{2-} + Ti^{IV} + H_2O \rightarrow O_LH^- + Ti^{IV} - OH^- \tag{9}$$

$$Ti^{IV} + H_2O \rightarrow Ti^{IV} - H_2O \tag{10}$$

$$TiO_2 + R \rightarrow R_{ads} \tag{11}$$

3.3 Radical attack on organics

The hydroxyl radicals, the primary oxidizing species in the photocatalytic system, initiate chain reactions, leading to the generation of other radicals and subsequently oxidation of organics [7,12]. Even though it is not necessary for the reaction that hydroxyl radicals and organics are adsorbed at the TiO_2 sites, the adsorbed forms are much more helpful to increase the overall reactivity, compared to those free from TiO_2 sites (Reactions 12–15) [3,9]:

$$Ti^{IV} - OH^{\bullet} + R_{ads} \rightarrow Ti^{IV} + R'_{ads} \quad (12)$$

$$Ti^{IV} - OH^{\bullet} + R \rightarrow Ti^{IV} + R' \quad (13)$$

$$^{\bullet}OH + R_{ads} \rightarrow R'_{ads} \quad (14)$$

$$^{\bullet}OH + R \rightarrow R' \quad (15)$$

Other radicals and oxidants ($HO_2^{\bullet}$, H_2O_2) are also generated and are involved in redox reactions to decompose organic contaminants in water (Reactions 16–18) [3,9]:

$$e^- + Ti^{IV} - O_2^{\bullet -} + 2H^+ \rightarrow Ti^{IV} + H_2O_2 \quad (16)$$

$$Ti^{IV} - O_2^{\bullet -} + H^+ \rightarrow Ti^{IV} + HO_2^{\bullet} \quad (17)$$

$$H_2O_2 + {}^{\bullet}OH \rightarrow H_2O + HO_2^{\bullet} \quad (18)$$

After the photoexcitation process and the generation of reactive species, a series of reactions lead to complete mineralization of the parent compound. These reactions include hydroxyl radical attack, hydroxylation (e.g., $^{\bullet}OH$ addition, reaction with O_2, and elimination of $HOO^{\bullet}$), dihydroxylation, hydration, hydrogen abstraction, deprotonation, decarboxylation, and one-electron transfer reactions [3].

3.4 Photocatalysis and photolysis

Photolysis is a chemical reaction in which a chemical compound is broken down by the high photon energy provided by a short wavelength light (typically UV) after absorbing light of sufficient energy. Many compounds absorb light in the UV-C range of 200–280 nm, which has sufficient energy to break the chemical bonds in the molecule. The photolytic reaction performed in the absence of a catalyst (photolysis) typically exhibits very slow kinetics, compared to that in the presence of a catalyst (photocatalysis). For chlorophenol decomposition, TiO_2 photocatalysis at 340 nm is more efficient than direct photolysis at 290 nm with higher photon energy because the photocatalytic reaction occurs through the formation of hydroxyl radicals [13]. It was also reported that the mineralization of humic acid by TiO_2 photocatalysis was two times faster than that by photolysis under the same UV irradiation [14]. Direct photolysis even under short wavelength UV irradiation is not significant for many organic compounds because of low absorption of the radiation.

3.5 Parameters affecting reaction kinetics

Factors affecting heterogeneous TiO_2 photocatalysis, including TiO_2 loading, initial concentration of reactant, UV wavelength, radiant flux, quantum yield, oxygen, and solution pH, have been well-established [8]. The initial decomposition rates of organic compounds are proportional to the surface area of TiO_2 since the reaction is in a true heterogeneous regime (Fig. 3a). However, at too high loading of TiO_2 mass, its screening effect prevents part of TiO_2 surface from being irradiated with UV and the TiO_2 particles scatter the UV light. The critical or optimum amount of TiO_2 (typically ranging from 0.1 to 3.0 g/L particles in slurry systems) is experimentally determined, considering the geometry of TiO_2 (e.g., particle size, degree of agglomeration), reactor configuration, UV wavelength and irradiance, and many other process parameters. The kinetics of heterogeneous TiO_2 photocatalysis generally follow a Langmuir–Hinshelwood equation although recent studies by Serpone and his coworkers have pointed out some dogmas and misconceptions in heterogeneous photocatalysis [15–17]. For diluted solution (contaminant concentration, $C_o < 1$ mM) in most cases, the reaction follows the apparent first order whereas for solutions with C_o higher than 5 mM, the reaction rate is at maximum (zero order) (Fig. 3b). An UV source emitting light at a

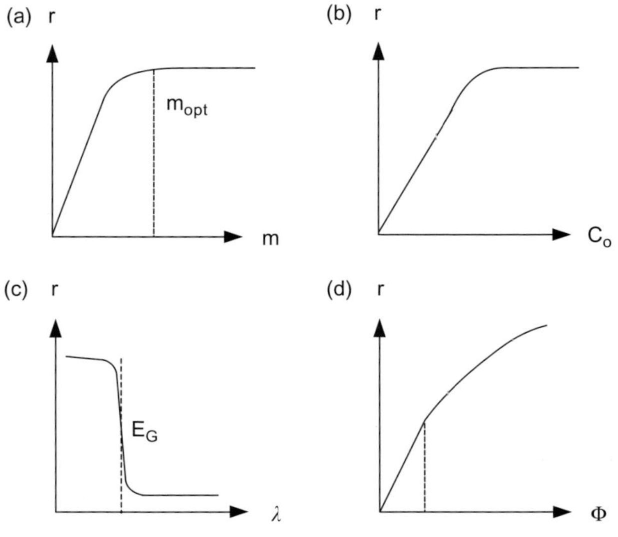

Figure 3 Parameters affecting decomposition reaction rate of organic compounds in TiO_2 photocatalysis: (a) TiO_2 loading *m*, (b) initial concentration of reactant C_o, (c) UV wavelength λ, and (d) radiant flux Φ.

wavelength λ with a certain photon energy equal to or above the BG of TiO_2 should be used for the activation of TiO_2. Anatase TiO_2 ($E_G = 3.2$ eV) requires $\lambda < 387$ nm, while rutile TiO_2 ($E_G = 3.0$ eV) needs $\lambda < 400$ nm (Fig. 3c). The reaction rate r is proportional to the radiant flux Φ (energy per unit time that is radiated from a source) at low values. However, above a certain value (~ 25 mW/cm^2), r becomes proportional to $\Phi^{1/2}$ (Fig. 3d).

Quantum yield is defined as the number of molecules of the contaminants undergoing transformation divided by the number of photons absorbed by the catalyst. Although the quantum yield depends on various conditions such as the nature of a catalyst, contaminant, and water-matrix characteristics (solution properties), it is fundamentally important since the activities of different catalysts for the same reaction can be compared in terms of their quantum yields. Molecular oxygen plays a crucial role, either inhibiting or facilitating the reaction depending on the degradation pathway and mechanism of contaminants. Its primary role is to act as an electron acceptor to produce superoxide radical anions, as discussed in Section 3.2. TiO_2 is an amphoteric material with point of zero charge around 6.0–6.4. Consequently, solution pH dramatically affects the adsorption of ionic species on the catalyst surface as well as their concentration in the electrical double layer around the catalyst. In addition, the oxidation/reduction potential of organic compounds, inorganic ions, and oxidizing species is a function of solution pH. As a result, changes in surface chemistry of TiO_2 and chemicals in water under different pH conditions affect the band edge of TiO_2 and adsorption of the contaminants, resulting in different reaction rates and reaction intermediates formation for many compounds. Other parameters affecting the reaction include the presence of coexisting chemicals and ions, and temperature.

4. PHOTOCATALYTIC WATER AND WASTEWATER TREATMENT AND PURIFICATION

Due to the nonselective attack of hydroxyl radicals, TiO_2 photocatalysis can decompose virtually all organic contaminants and oxidize various inorganic anions. The quality of polished water depends on the properties of water and wastewater and the treatment parameters, as discussed in Section 3.5. The extent of the treatment should be decided according to the purpose of treatment and regulations.

Table 2 summarizes the photocatalytic oxidation and degradation of some compounds [7–9,18]. A wide range of anions such as nitrate, sulfide

Table 2 Photocatalytic oxidation, reduction, and degradation of various compounds

Group	Substrate
Inorganic anions	NO_2^-, (NH_4^+), H_2S, SH^-, S^{2-}, SO_3^{2-}, $S_2O_3^{2-}$, H_3PO_4, CN^-
Heavy metals	Dichromate, Cr(VI), Pb^{2+}, Mn^{2+}, Ti^+, Co^{2+}, Hg^{2+}, Hg(II), $HgCl_2$, CH_3HgCl, Ag(I), Cu(II), Cd(II), Au(III), Pt(VI), Pd(II), Ni(II)
Alkanes	Isobutene, pentane, heptane, cyclohexane, paraffins, chloromethane, bromomethane, chloroethane
Aliphatics	Methanol, ethanol, propanol, glucose, acids (formic, ethanoic, propanoic, oxalic, butyric, malic)
Alkenes	Cyclohexene, chloroethylene, chloroethene
Aromatics	Bezene, toluene, xylene, naphthalene, chlorobenzene, chloronitrobenzene, biphenyls, polychlorinated biphenyls, acids (benzoic, aminobenzoic, phthalic, salicylic, hydroxybenzoic, chlorobenzoic)
Phenolic compounds	Phenol, chlorophenol, fluorophenol, hydroquinone, catechol, methylcatechol, cresol, nitrophenols
Surfactants	Sodium dodecyl sulfate, polyethylene glycol, sodium dodecyl benzene sulfonate, trimethyl phosphate
Herbicides and pesticides	Atrazine, prometron, propetryne, bentazon, monuron, DDT, parathion, lindane
Dyes	Methylene blue, rhodamine B, methyl orange, fluoroscein, Congo red
Activated sludge	Mixture of various organic compounds
Microorganisms	*Escherichia coli, Bacillus pumilus,* phage
Biological toxins	microcystin-LR (RR, YR, LA)

Source: Adapted and modified from Refs. [7–9,18].

and sulfite, and cyanide can be easily oxidized to harmless or less toxic compounds [19–22]. The elements, (N, S, and P) are typically transformed to their maximum oxidation state. For example, nitrite is oxidized to nitrate and sulfide and sulfite are converted to sulfate. Metals found in water resources (especially industrial wastewater, groundwater, and surface water in mining areas) impose a serious concern. Their toxicity depends on their valence states. It is practical for TiO_2 photocatalysis to convert such toxic metals to their less toxic or nontoxic forms, or remove them from the water streams. Metal ions (M), if present, are reduced by the electrons generated in the conduction band of TiO_2 (Equation 19) [23]:

$$M^{n+} + ne^- \rightarrow M^0 \tag{19}$$

The treatment feasibility counts on the standard reduction potential of the metals. Ag(I), Cr(VI), Hg(II), and Pt(II) were easily treated by TiO_2 photocatalytic redox reaction, while Cd(II), Cu(II), and Ni(II) were not removed effectively [24]. Another metal removal approach is the photodeposition of metal ions on Pt-loaded TiO_2 [25]. For example, Pb^{2+} concentration in Pt–TiO_2 suspension was reported to decrease over time while the color of TiO_2 became brown, implying deposition of Pd to the surface of Pt–TiO_2. Similar results were observed with other metal ions such as Mn^{2+}, Ti^+, Hg^{2+}, and Co^{2+}.

Most of the common organic compounds found in water and wastewater, including phenolic compounds, chlorinated chemicals, surfactants, dyes, and pesticides, can be decomposed regardless of their molecular structure due to the nonselective attack of hydroxyl radicals. Many of these organic chemicals are classified as priority pollutants by the U.S. Environmental Protection Agency (EPA) or are in the drinking water contaminant candidate list [26–31]. Their structure typically affects reaction kinetics and intermediates formation. Many research studies have focused on the photocatalytic degradability and reaction mechanisms of organic compounds, and identification of reaction by-products and their toxicity. In many cases, the reaction pathways are too complicated to elucidate the detailed steps.

Degradation of target compounds is fast. Typically, the double bonds in organic chemicals are susceptible to hydroxyl radical attack. The dearomatization of aromatic compounds (e.g., toluene, benzene, and phenol) is quick even in the presence of substituents such as Cl, NO_2, and OCH_3 on the aromatic ring [13,32,33]. An aliphatic chain bound to the aromatic ring is easy to fragment but its mineralization to CO_2 is significantly slow since intermediates formed during the reaction such as formate and acetate ions are relatively stable [34,35]. So far, complete mineralization of almost all organic compounds was observed. Due to the high stability of the triazine aromatic ring, *s*-triazine herbicides were reported to be transformed to nontoxic cyanuric acid as a final product [36]. For persistent chlorinated compounds, such as chlorophenol and chlorobenzoic acid, the chloride ions are easily removed and the dechlorinated compounds are then available for biological treatment system [13,37]. Nitrogen-containing compounds are transformed to NH_4^+ and NO_3^- [38]. Sulfur-containing chemicals are mineralized to sulfate ions [39]. Organophosphorus pesticides produce phosphate ions [35,40].

Recent advances in analytical instruments such as high-performance liquid chromatography, gas chromatography, and mass spectrometry make it possible to detect and identify reaction intermediates during the photocatalytic degradation event of organic compounds. One of the most extensive studies so far on intermediate identification and degradation pathways in TiO_2 photocatalysis was conducted by Jenks and his coworkers [41,42]. Using 4-chlorophenol (4-CP) as a model compound, they experimentally proved that after the photoexcitation process and the generation of reacting species, a series of cascade reactions (demonstrated in Section 3.3) could lead to complete mineralization of 4-CP. Initially, 4-CP undergoes bifurcation forming either hydroquinone by substitution [41] or 4-chlorocatechol by hydroxylation [42] at ratios that depend on reaction conditions. Hydroquinone undergoes further reactions to from hydroxybenzoquinone, 1,2,4-benzenetriol, 1,2,4,5-benzenetetraol (traces), and oxidative cleavage of the benzene ring at either the C1–C2 or C3–C4 bonds to form acyclic derivatives. Oxidative cleavage of the ring of 1,2,4-benzenetriol occurs via electron transfer to form a radical cation of this compound followed by trapping of the radical cation by superoxide. This mechanism results in the formation of dioxetanes, which subsequently collapse to form open-ring six-carbon compounds of either acid-aldehyde (break of C3–C4 bond) or diacid (break of C1–C2 bond). Major acyclic intermediates are (*E*)- or (*Z*)-4-oxohex-2-enedioic acids, oxobutanedioic acid, propanedioic acid, and ethanedioic acid. Other smaller intermediates, prior to the formation of acetic acid, formic acid, and formaldehyde, include 1,2-ethanediol and hydroxyacetic acid. While degradation of 4-chlorocatechol undergoes a degradation pathway that includes formation of 5-chloro-1,2,4-benzenetriol by hydroxylation, 1,2,4-benzenetriol by substitution, and small quantities of 4-chloropyrogallol by hydroxylation. Ring cleavage forms compounds as previously discussed and chlorine-bearing compounds such as (*E*)-3-chloro-4-oxohex-2-enedioic acid, 3-chloro-4,5-dioxopent-2-enoic acid, and chlorofumaric acid. Many other smaller intermediates in the degradations were identified by Jenks and his coworkers [41,42].

In addition, TiO_2 photocatalysis has strong disinfection function toward microorganisms. The widespread use of antibiotics and the emergence of more resistant strains of microorganisms in water induce an immediate need to develop alternative disinfection systems. The TiO_2 photocatalytic process is practically useful for killing pathogenic microorganisms. The hydroxyl radicals are highly toxic and reactive to microorganisms like other organic substances. Photocatalytic inactivation of bacteria such as *Escherichia coli* and *Bacillus pumilus*, as well as several phages has been investigated [43–45]. TiO_2 photocatalysis initially promotes peroxidation of the polyunsaturated phospholipid component of the lipid membrane and thus induces a major disorder in the cell and damages essential functions, leading to death of microorganisms [46]. In addition, TiO_2 photocatalysis has demonstrated high decomposition and detoxification efficiency toward biological toxins, especially cyanobacterial toxins in drinking water resources [47–49].

5. APPLICATIONS OF TiO_2 PHOTOCATALYSIS

Conventional water and wastewater treatment and reuse systems are composed of several unit operations in series, including coagulation, sedimentation, filtration, disinfection, and advanced treatments (e.g., activated carbon adsorption). In spite of the combination of the processes, the quality of water, especially potable water is hard to ensure. Moreover, these systems in combination are difficult to operate and manage, and thus require many technicians and engineers. Considering space limitations and types of water and wastewater that need to be recycled under certain conditions, these compact and reliable TiO_2 photocatalytic reactors are good candidates for systems that can serve as stand-alone or complementary and supplementary to the existing or future treatment systems. TiO_2 photocatalysis is also useful for treating water and wastewater and reusing them in areas with restricted space (target-specific small size systems). For instance, during long-term missions of a shuttle in space exploration by the National Aeronautics and Space Administration, there is a necessity to recycle water in order to assure conditions of self-sufficiency [50]. Recently, an emerging issue in drinking water industry is the presence of biological toxins in drinking water sources. Eutrophication of water resulting from human activities causes hazardous algal blooms (HABs) of cyanobacteria. The HABs have been the cause of great concern in water treatment industries and for authorities since HABs contain and release biological toxins [51,52]. The cyanobacterial toxins such as microcystin-LR cause skin irritations and liver damage or affect the nervous system [53]. TiO_2-based

AOTs have shown promising results in the degradation and detoxification of such biological toxins [47,49].

6. CHALLENGES AND ISSUES IN TiO_2 PHOTOCATALYSIS FOR WATER TREATMENT

6.1 Enhancement of photocatalytic activity

Since the activity of TiO_2 is primarily controlled by the composition and availability of its near-surface regime irradiated by UV light, small size and high surface area of TiO_2 make it attractive for the photocatalytic reaction. In addition to the surface area and morphology of TiO_2, its crystallographic properties (crystal phase and crystallite size), defect structures on its surface and in bulk, and electronic structure largely affect its catalytic activity [54]. For example, the anatase crystal phase has been generally known for having a higher activity than the other phases (brookite and rutile) and amorphous TiO_2 particles have no considerable photocatalytic activity due to many defects in the bulk [55–58]. So, controlling the morphological, crystallographic, and electronic properties of TiO_2 material via alternative synthesis procedures could be one approach to develop a catalyst with high enough activity to make the process attractive for large-scale applications [59–64].

6.2 Reactor design and immobilization of TiO_2

There are two types of photocatalytic reactors: (a) reactors utilizing TiO_2 as a suspension of ultrafine particles and (b) reactors incorporating TiO_2 nanoparticles immobilized on a support material, as shown in Fig. 4. Suspension-type reactors are often used to study degradation kinetics since they are characterized by large catalytic surface area (i.e., high TiO_2 loading) and low mass transfer limitations. However, the systems impose the requirement for filtering the effluent to remove TiO_2 particles with nanoscale size (typically below 100 nm) before discharging of the polished water containing TiO_2 [65]. Recent studies show that possible toxicity of TiO_2 nanoparticles damages brain microglia and human lung epithelial cells [66,67]. In addition, public perception and practical (i.e., catalyst reuse) and esthetic reasons require the complete separation of TiO_2 particles from the effluent. The postfiltration process is tedious and costly and unfortunately it does not guarantee complete removal of the ultrafine particles. In most cases, TiO_2 photocatalysis is considered as a refining process at the end of the treatment train and the use of suspension-type reactors is not a choice. Searching for answers and solutions to the concerns and questions on the

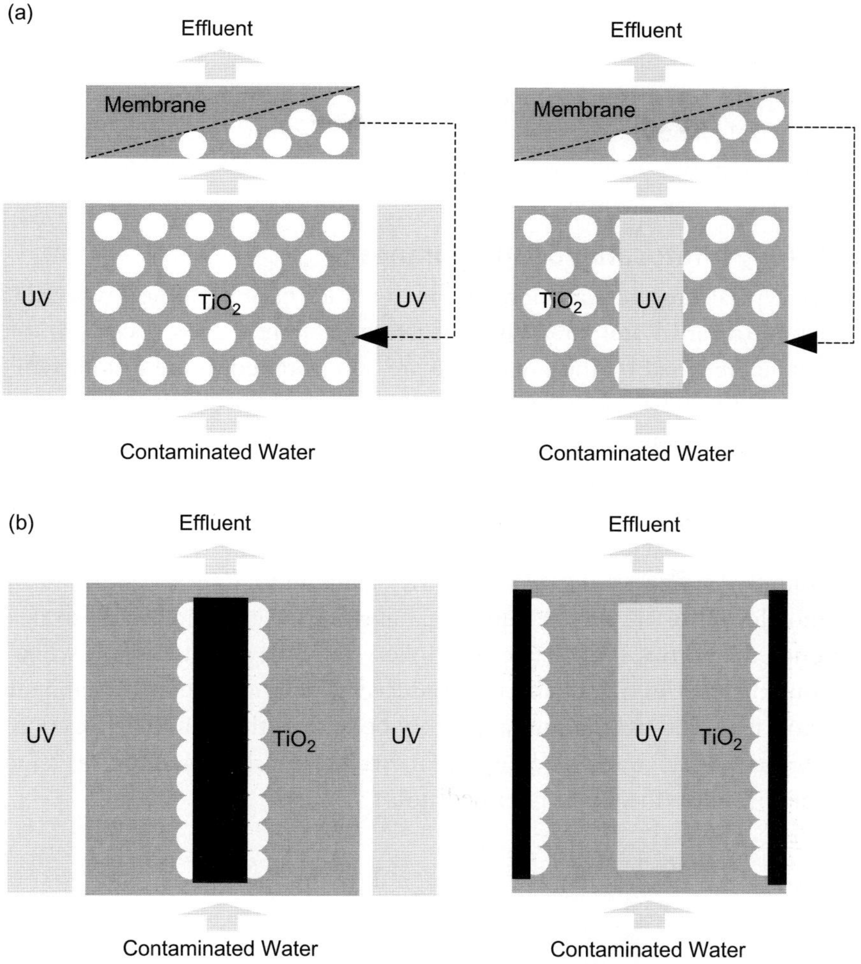

Figure 4 Photocatalytic reactors employing (a) TiO_2 particles in suspension and (b) immobilized TiO_2 coatings on a support. UV light can be installed outside or inside of the reactors, depending on the application of the reactors.

potential impact of nanosize TiO_2 to human health and the environment mandates immobilization of TiO_2 onto various substrates for use in a variety of applications [68–70]. This will be more sustainable and a key feature for practical applications. However, as expected, immobilized-type TiO_2 reactors exhibit a low catalytic activity due to limitations in catalyst loading (surface area) on a support and catalyst activation at the near surface, partial loss of catalyst by attrition, and possible mass transfer limitations. The first two problems can be solved by novel preparation routes that aim at the precise fabrication of immobilized nanoporous TiO_2 catalyst with enhanced surface area and finely tuned nanoscale dimensions for better adhesion to the support, as discussed in Section 6.1 and will be mentioned in detail in Section 7.1. The third problem can be addressed by the development of innovative photocatalytic reactors that significantly reduce or eliminate the influence of mass transfer.

6.3 TiO_2 photocatalytic membranes

Since typical organic and inorganic membranes have a physical separation function toward water contaminants (i.e., no chemical decomposition), the pollutants retained by the membrane are concentrated in the system while the treated water is clean enough for its specific usage. The retentate with high organic concentration should be treated appropriately before its discharge into the environment. As a result, special attention is given to the fabrication of TiO_2 photocatalytic membranes, which interestingly combine the physical separation of organic contaminants and their simultaneous chemical decomposition [71–74]. As illustrated in Fig. 5, TiO_2 is immobilized on a porous substrate, allowing water molecules to pass through the porous TiO_2 active layer followed by the porous support when a pressure across the membrane is introduced. The photocatalytic membrane reactors may gain tremendous popularity because of their multiple functions: photocatalysis of organic compounds and physical separation of target contaminants and reaction intermediates.

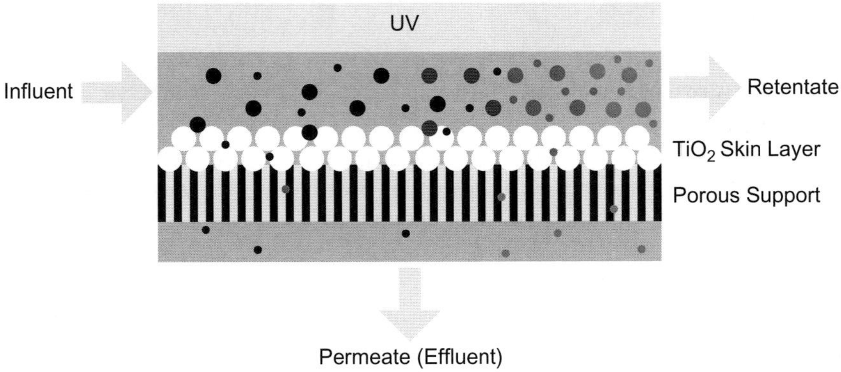

Figure 5 Concept of a TiO_2 photocatalytic membrane, exhibiting multifunction of photocatalysis, separation, and antibiofouling. Photocatalytic decomposition of organic chemicals results in the formation of simpler intermediates. Note cascaded changes in the color of contaminants with different sizes represent their photocatalytic decomposition and increase in the number of contaminants expresses their retention and accumulation when passing the membrane in parallel.

Choi et al. demonstrated such a multifunction of TiO_2 membranes in terms of the decomposition of methylene blue dye and creatinine, destruction of biological toxins (microcystin-LR), and inactivation of pathogenic microorganisms (*E. coli*) [75]. They also observed high and reliable organic rejection efficiency of the TiO_2 membrane when it was used to treat dissolved organic carbon solution obtained from an activated sludge plant for wastewater treatment. Interestingly, the inherent antibiofouling properties of TiO_2 membrane, which are highly important in membrane research and industry, were investigated [76]. TiO_2 membranes irradiated by UV exhibited less flux decline over time. The antibiofouling property can be explained by the fact that the decomposition of the organic foulants accumulated on the TiO_2 membrane surface simultaneously occurs due to hydroxyl radical attack during their filtration.

6.4 Sustainable solar energy-based TiO_2 photocatalysis

For the excitation of the TiO_2 to generate electrons and holes and subsequently hydroxyl radicals, the required energy must be higher than its BG energy. This corresponds to radiation with wavelength (λ) less than 387 nm for anatase TiO_2. Unfortunately, these wavelength cutoffs are in the near-UV range ($\lambda = 300$–400 nm). This inhibits the utilization of solar light for TiO_2 activation because only 4–5% of the incoming solar energy on the earth's surface is in the UV range. In order to utilize visible light for TiO_2 excitation, dye-sensitized or metal ion-doped TiO_2 has been developed and showed promising results for the degradation of water contaminants under visible light irradiation [77,78]. Introduction of anionic dopants, especially nitrogen, to TiO_2 also makes it possible to achieve TiO_2 BG narrowing [79–81]. The activation of TiO_2 under visible light can facilitate the development of promising processes for the remediation of contaminated water resources using solar light as a sustainable energy source without complicated facilities for generating and introducing UV light.

6.5 TiO_2 catalyst deactivation and fouling

Partial catalyst deactivation and fouling were observed in solution containing certain inorganic species and metals by either catalyst active site blockage (i.e., Cl^-, PO_4^{3-}, Cr(VI), Na^+, Cu^{2+}, Fe^{2+}, Al^{3+}) or attachment of inorganic precipitates formed during the process (i.e., MnO_2, MnOOH, $Fe(OH)_3$) [82–85]. Other inorganic species (i.e., HCO_3^-, CO_3^{2-}) can cause organic destruction inhibition by absorbing UV radiation or by scavenging hydroxyl radicals [84]. Catalyst deactivation might be caused by strong interaction and chemisorption of intermediate reaction products on catalyst active sites, and certain cleaning procedures using heat treatment and UV could result in complete and partial reactivation depending on the chemicals adsorbed on the catalyst surface and the extent of the catalyst deactivation. Fortunately, it was also observed that fouling was at least partially reversible using water wash and it was completely eliminated with proper feed solution pretreatment (i.e., ion exchange) [85].

7. CURRENT ADVANCES IN TiO_2 PHOTOCATALYSIS

7.1 Environmental nanotechnology/advanced oxidation nanotechnologies

As discussed in Section 6.1, the crystallographic, electronic, and structural properties of TiO_2 are of importance with respect to UV light utilization, catalytic sites for the hydroxyl radical generation, and reactant accessibility to/from TiO_2. Recently, in order to control the physicochemical properties of TiO_2 at the nano level and thus achieve maximum photocatalytic performance in versatile environmental applications, nanotechnology has been introduced in this research area, especially in the field of synthesis of new catalytic TiO_2 materials. This results in more efficient TiO_2-based advanced oxidation nanotechnologies (AONs) for water treatment, purification, recycle, and reuse. Among the synthetic routes, modified sol–gel methods employing pore templating agents including block copolymers and surfactant molecules have been attractive during the last decade as promising approaches for the tailor-designing of TiO_2 structure [49,54,74,76,86–91].

Sol–gel preparation methods refer to room temperature wet chemistry-based formation of solid inorganic materials from molecular precursors [91,92]. The technology is applied for the versatile preparation of powders, catalytic films, inorganic membranes, monoliths, fibers, reactive coatings, sensors, and optics [92–94]. Fig. 6 shows the surfactant template-based sol–gel synthesis of TiO_2. The approach utilizes self-assembled surfactants as a pore template or particle growth template. Finally, TiO_2 inorganic materials with highly porous network or well-defined TiO_2 nanoparticles are formed.

Based on the synthesis approach, by changing the type, chain length, and concentration of surfactant, it is possible to control the physicochemical properties of TiO_2 (i.e., morphological structure, crystal phase, defect structure, impurities, BG energy, hydrophilicity) and thus improve its organic adsorption capacity and catalytic activity [49,54,74,76,88]. Especially, this technique is very useful in fabricating TiO_2 films and

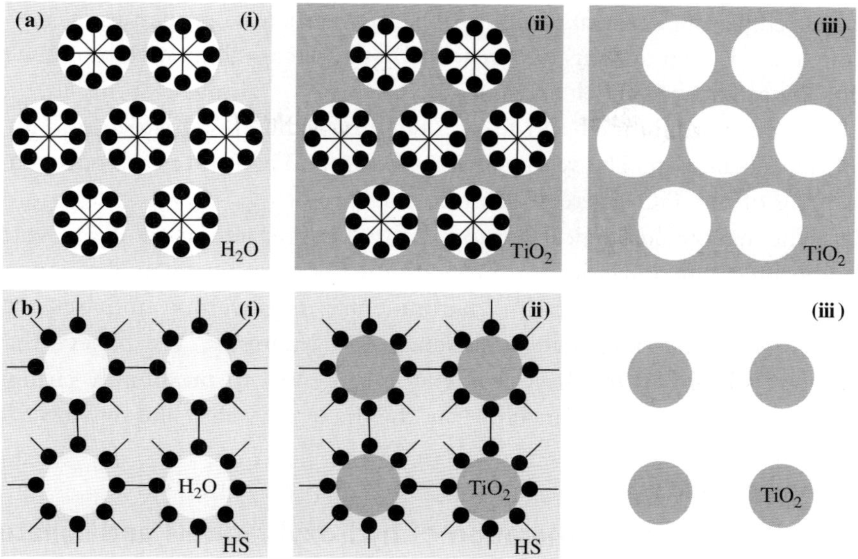

Figure 6 Synthesis approaches of engineered TiO_2 via sol-gel method employing surfactant self-assembly as (a) a pore template and (b) particle growth template. (a) Synthesis of TiO_2 with mesoporous inorganic network: (i) surfactant molecules are self-organized in water-rich environment, forming surfactant head group outside towards water molecules and its tail group inside free from water, (ii) titanium alkoxide precursor is hydrolyzed and condensed to form TiO_2 inorganic network around the self-assembled surfactant, forming a surfactant organic template-embedded TiO_2 inorganic matrix, and (iii) porous TiO_2 inorganic network is formed after removal of the organic template by thermal treatment or organic extraction. (b) Synthesis of TiO_2 nanoparticles: (i) surfactant molecules are self-organized in water-poor environment (bulk hydrophobic solvent (HS) with small portion of water), forming surfactant head group inside towards water molecules and its tail group outside towards HS, (ii) titanium alkoxide precursor is hydrolyzed and condensed to form TiO_2 inorganic network in the water phase, inside of self-assembled surfactant, forming TiO_2 inorganic core/surfactant organic shell structure, and (iii) well-defined TiO_2 nanoparticles are formed after removal of the organic template.

membranes with engineered properties to improve their performance. In order to minimize the hydraulic resistance through a membrane, an asymmetric mesoporous TiO_2 membrane was fabricated by changing the concentration of a surfactant in a sol–gel synthesis of TiO_2 [76]. The membrane showed a hierarchical change in pore diameter and porosity from 2–6 nm and 46.2%, 3–8 nm and 56.7% to 5–11 nm and 69.3% from the top to the bottom layer, and exhibited improved water permeability without sacrificing organic retention and photocatalytic activity. Interestingly, a nitrogen-containing surfactant (dodecylammonium chloride) as a pore templating material to tailor-design the structural properties of TiO_2 and as a nitrogen dopant to narrow its BG (as discussed in Section 6.4) was introduced in a sol–gel synthesis of TiO_2 [49]. Nitrogen atoms in the surfactant were diffused and incorporated into the crystal lattice of TiO_2 during calcination. The synthesis of mesoporous TiO_2 and *in situ* nitrogen-doping of the TiO_2 were concurrently achieved.

The nanotechnological approach for the preparation of TiO_2 with engineered functionalities and properties for environmental applications is interdisciplinary, integrating environmental engineering and science, chemical engineering, materials science, and chemistry. We believe that the TiO_2-based AONs have tremendous potential to profoundly change current science and engineering in the field of water and wastewater treatment.

7.2 Simultaneous generation of hydroxyl radicals and sulfate radicals

Sulfate radicals have been proven to be strong oxidizing species like hydroxyl radicals [95]. Among various methods, sulfate radicals are generated by catalytic decomposition of PMS in a homogeneous pathway with the aid of transition metals (M: Fe^{2+}, Co^{2+}, Ag^{+}, Mn^{2+}, Ni^{2+}) and/or UV radiation (Equation 20):

$$M^{2+} + HSO_5^- \rightarrow M^{3+} + SO_4^{\bullet -} + OH^- \tag{20}$$

Oxidants used in sulfate radicals-based AOTs are currently used for pools and spa disinfection. Of the transition metals investigated for sulfate radical generation, cobalt ions (Co^{2+}) showed the best performance [96]. Although this system is promising for degrading environmental pollutants in water, health concerns associated with the adverse effect of dissolved cobalt (i.e., cobalt salts are used) in water still need to be addressed. Anipsitakis and his coworkers responded to the request by using commercially available cobalt oxide particles (Co_3O_4) and first demonstrated the heterogeneous PMS activation [97]. More recently, Yang and her coworkers immobilized and distributed well-defined 10–15 nm nanocrystalline Co_3O_4 particles on the surface of 30–40 nm TiO_2 nanoparticles as a support material for the Co_3O_4 particles to disperse [98]. The Co_3O_4/TiO_2 nanocomposite showed enhanced decomposition of 2-chlorophenol by sulfate radical attack. In the PMS/Co_3O_4/TiO_2 system, a feasibility of simultaneous generation of sulfate radicals and hydroxyl radicals was first attempted by introducing UV irradiation to the system and some promising preliminary results were obtained [99]. The simultaneous generation of two strong oxidizing species, sulfate radicals and hydroxyl radicals, are scientifically and practically important, although the mechanism of simultaneous generation of both radicals and their effect on the degradation of organic compounds are still being investigated.

Sulfate radicals were reported to be efficient for the decomposition of 2,4-dichlorophenol, atrazine, and naphthalene [95]. Since sulfate radicals (2.5–3.1 V) possess similar reduction potential to hydroxyl radicals (1.9–2.0 V) at neutral pH, they can also be effective in degrading certain organic compounds. However, the organic attack mechanisms of the two radicals have differences and thus different reaction intermediates are formed. The stability of the intermediates determines the overall mineralization of organic substrates using hydroxyl radicals and sulfate radicals. At acidic condition, they both have similar reduction potentials, but sulfate radicals are more selective for oxidation than hydroxyl radicals, which react rapidly with organic molecules by initial hydroxylation or hydrogen abstraction [100].

8. ECONOMIC ASPECT OF TiO_2 PHOTOCATALYSIS

Cost is comprised of capital costs, and operation and maintenance costs. The capital costs can be reduced by designing more efficient systems. The operation costs greatly depend on many factors such as the type and concentration of pollutants, level of treatment, and catalyst dose and loading method (fixed and slurry), in addition to the pretreatment (e.g., removal of particles causing high turbidity, which inhibit UV light penetration) and posttreatment (e.g., membrane separation in case of suspension-type TiO_2 reactors) costs. Since more efficient systems have been developed, aiming at increase in efficiency and decrease in cost, the estimated costs of such systems have been roughly compared with those of the conventional treatment technologies used. Recent studies have been devoted to scientifically and technically improve TiO_2 photocatalysis, particularly solar energy-based detoxification process. The costs for solar photocatalytic process were compared with those of activated carbon and UV/H_2O_2 systems [101]. The analysis showed that the cost of photocatalysis rapidly decreases while the cost of the other conventional systems more or less remains the same between current and projected period. A field test of a solar photocatalytic process to detoxify BTEX pollutants (benzene, toluene, ethyl benzene, xylene) demonstrated that the treatment cost is competitive with those of conventional treatment technologies [102]. Many similar results supporting the competitiveness and effectiveness of TiO_2 photocatalysis were reported with other compounds such as trichloroethylene, pesticides, and polychlorinated biphenyls [103–105]. In most studies so far, the cost for typical TiO_2-based AOTs seems higher at this moment than those of conventional technologies mainly due to the UV energy requirement, whose cost fortunately is now decreasing. However, potentially greater cost reductions are expected particularly in case of solar-activated TiO_2 systems and AONs.

9. CONCLUDING REMARK: ADVANCED OXIDATION NANOTECHNOLOGIES AND SUSTAINABILITY

The potential applications of the TiO_2 photocatalytic process strongly depend on future development in photocatalytic engineering. As we discussed earlier, many scientific hurdles including visible light activation and enhancement of catalytic activity, and technical and practical challenges including immobilization of TiO_2 and fabrication of TiO_2 membrane reactors have been currently solved by introducing nanotechnological material synthesis approaches and new reactor design and concepts. The TiO_2-based AONs seem promising for water and wastewater treatment due to the following aspects:

- Chemical stability of TiO_2 (insoluble in water) in all pH range
- Relatively low cost of TiO_2
- No chemical additives required
- System applicable at low and high pollutant concentration
- Absence of inhibition or low inhibition by ions present in water
- Nonselective radical attack by the generated oxidizing species
- Complete mineralization for almost all organic pollutants
- Removal of toxic anions, harmful metals, and nonbiodegradable organics
- Disinfection and detoxification of drinking water
- Antibiofouling properties
- Possibility of using sustainable solar energy
- Practical for combination and integration with other treatment methods
- Decreasing overall costs

However, because of low quantum yield, TiO_2-based AONs are used in limited devices with optimum treatment capacity. In order for TiO_2 photocatalysis to become a competitive process for full-scale applications, many efforts and research studies should be focused on easy scale-up, effective reactor design, UV fouling reduction, and TiO_2 fouling prevention. It is believed that the catalyst activity should be increased by at least one order of magnitude [2,3,7–9]. The synthesis of more efficient TiO_2, which can be activated under visible light irradiation, has tremendous impacts on the development of sustainable solar energy-based water treatment systems. Meanwhile, it should also be emphasized that the unique properties of nanostructured TiO_2 with high catalytic activity, which has recently developed with the help of nanoscience and nanotechnology, could potentially lead to unexpected threat to the environment and thus research studies on its environmental impacts and risks should be pursed in parallel.

ACKNOWLEDGMENTS

D.D.D. acknowledges financial support for studies on TiO_2-based AONs from the National Science Foundation through a CAREER award (BES 0448117) and the Center of Sustainable Urban Engineering (SUE) at the University of Cincinnati. Elias Stathatos also acknowledges financial support from the Greek General Secretariat of Research and Technology through the E&T bilateral GR-USA cooperation program (05NON-EU-521). Since most of the work demonstrated in this chapter was conducted during his Ph.D. research at the University of Cincinnati and the writing of this chapter was initiated during his postdoctoral research at the National Risk Management Research Laboratory of US Environmental Protection Agency, Hyeok Choi greatly appreciates financial and research support from the two institutions. This chapter has not been subjected to internal policy review of the U.S. Environmental Protection Agency. Therefore, the research results do not necessarily reflect the views of the agency or its policy. Mention of trade names and commercial products does not constitute endorsement or recommendation for use.

REFERENCES

[1] B.E. Reed, M.R. Matsumoto, R. Viadero, R.L. Segar, R. Vaughan, D. Masciola, Water Environ. Res. 71 (1999) 584.
[2] H. Al-Ekabi (Ed.), J. Adv. Oxidation Technol. 1 (1996) 1.
[3] D.F. Ollis, H. Al-Ekabi (Eds.), Photocatalytic Purification and Treatment of Water and Air, Elsevier Science, Amsterdam, The Netherlands, 1993.
[4] D.F. Ollis, E. Pelizzetti, N. Serpone, Environ. Sci. Technol. 25 (1991) 1523.
[5] S.T. Martin, H. Herrmann, W. Choi, M.R. Hoffmann, J. Chem. Soc., Faraday Trans. 90 (1994) 3315.
[6] S.T. Martin, H. Herrmann, M.R. Hoffmann, J. Chem. Soc., Faraday Trans. 90 (1994) 3323.
[7] M.R. Hoffmann, S.T. Martin, W. Choi, D.W. Bahnemann, Chem. Rev. 95 (1995) 69.
[8] J.-M. Herrmann, Topics Catal. 34 (2005) 49.
[9] K. Kabra, R. Chaudhary, R.L. Sawhney, Ind. Eng. Chem. Res. 43 (2004) 7683.

[10] U. Diebold, Surf. Sci. Reports 48 (2003) 53.
[11] A.L. Linsebigler, G. Lu, J.T. Yates, Chem. Rev. 95 (1995) 735.
[12] C.S. Turchi, D.F. Ollis, J. Catal. 119 (1990) 483.
[13] J.C. D'Oliveira, G. Al-Sayyed, P. Pichat, Environ. Sci. Technol. 24 (1990) 990.
[14] D. Gonenc, M. Bekbolet, Water Sci. Technol. 44 (2001) 205.
[15] D.F. Ollis, J. Phys. Chem. B 109 (2005) 2439.
[16] N. Serpone, J. Adv. Oxid. Technol. 10 (2007) 111.
[17] A.V. Emeline, V.K. Ryabchuk, N. Serpone, Catal. Today 122 (2007) 91.
[18] D.M. Blake, Bibliography of Work on Photocatalytic Removal of Hazardous Compounds From Water and Air, National Renewable Energy Laboratory, NREL/TP-430-22197, Gorden, 1999 and 2001.
[19] A. Zafra, J. Gacia, A. Milis, J. Domennech, J. Mol. Catal. 70 (1991) 343.
[20] S.N. Frank, A.J. Bard, J. Phys. Chem. 81 (1977) 1484.
[21] H. Hidaka, T. Nakamura, A. Ishizaha, M. Tsuchiya, J. Zhao, J. Photochem. Photobiol. A 66 (1992) 367.
[22] C.H. Pollema, J.L. Hendrix, E.B. Milosavljevic, L. Solujic, J.H. Nelson, J. Photochem. Photobiol. A 66 (1992) 235.
[23] J.M. Herrmann, J. Disdier, P. Pichat, J. Catal. 113 (1988) 72.
[24] M.R. Prairie., L.R. Evans, S.L. Martinez, in: W.W. Eckemfelder, A.R. Bowers, J.A. Roth (Eds.), Chemical Oxidation: Technology for the Nineties, Second International Symposium, Technomic Publishing Company, Lancast, PA, 1994, p. 428
[25] K. Tanaka, K. Harada, S. Murata, Solar Energy 36 (1986) 159.
[26] U.S. EPA, http://www.epa.gov/ost/, accessed July 2009.
[27] U.S. EPA, http://www.epa.gov/oppt/
[28] U.S. EPA, http://www.epa.gov/scipoly/oscpendo/index.htm, 2002, accessed July 2009.
[29] U.S. EPA, http://www.epa.gov/ogwdw/
[30] AWWA., Water Quality and Treatment: A Handbook of Community Water Supplies, McGraw Hill, New York, 1999.
[31] U.S. EPA, Office of Ground Water and Drinking Water, MTBE in Drinking Water, http://www.epa.gov/safewater/mtbe.html, 2001, accesed July 2009.
[32] C. Maillard-Dupuy, C. Guillard, P. Pichat, New J. Chem. 18 (1994) 941.
[33] L. Amalric, C. Guillard, N. Serpone, P. Pichat, J. Environ. Sci. Health A 28 (1993) 1393.
[34] J.M. Herrmann, J. Disdier, P. Pichat, Appl. Catal. B 17 (1998) 15.
[35] M. Kerzhentsev, C. Guillard, J.-M. Herrmann, P. Pichat, Catal. Today 27 (1996) 215.
[36] E. Pelizzeti, V. Maurino, C. Minero, O. Zerbinati, E. Borgarello, Chemosphere 18 (1989) 1437.
[37] H. Tahiri, Y.A. Ichou, J.M. Hermann, J. Photochem. Photobiol. A 114 (1998) 219.
[38] C. Maillard, C. Guillard, P. Pichat, New J. Chem. 16 (1992) 821.
[39] G.K.C. Low, S.R. McEvoy, R.W. Matthews, Environ. Sci. Technol. 25 (1991) 460.
[40] K. Harada, T. Hisanaga, K. Tanaka, Water Res. 24 (1990) 1415.
[41] X. Li, J.W. Cubbage, T.A. Tezlaff, W.S. Jenks, J. Org. Chem. 64 (1999) 8509.
[42] X. Li, J.W. Cubbage, T.A. Tezlaff, W.S. Jenks, J. Org. Chem. 64 (1999) 8525.
[43] C. Wei, W.Y. Lin, Z. Zainal, N.E. Williams, K. Zhu, A.P. Kruzic, R.L. Smith, K. Rajeshwar, Environ. Sci. Technol. 28 (1994) 934.
[44] S.S. Block, D.Y. Goswami, J. Sol. Energy Eng. 119 (1997) 85.
[45] T. Matsunaga, M. Okochi, Environ. Sci. Technol. 29 (1999) 501.
[46] P.-C. Maness, S. Smolinski, D.M. Blake, Z. Huang, E.J. Wolfrum, W.A. Jacoby, Appl. Environ. Microbiol. 65 (1999) 4094.
[47] L.A. Lawton, P.K.J. Robertson, B.J.P.A. Cornish, M. Jaspers, Environ. Sci. Technol. 33 (1999) 771.

[48] A.J. Feitz, D.T. Waite, G.J. Jones, B.H. Boyfen, P.T. Orr, Environ. Sci. Technol. 33 (1999) 243.
[49] H. Choi, M.G. Antoniou, M. Pelaez, A.A. de la Cruz, J.A. Shoemaker, D.D. Dionysiou, Environ. Sci. Technol. 41 (2007) 7530–7535.
[50] NASA, Advanced Life Support Project Plan, Report CTSD-ADV-348 Rev C, JSC 39168, 2002.
[51] W.W. Carmichael, Toxicon 34 (1996) 299.
[52] M.G. Antoniou, A.A. de la Cruz, D.D. Dionysiou, J. Environ. Eng. 131 (2005) 1239.
[53] K. Sivonen, Phycologia 53 (1996) 12.
[54] H. Choi, E. Stathatos, D.D. Dionysiou, Topics Catal. 44 (2007) 513.
[55] Y.H. Hsien, C.F. Chang, Y.H. Chen, S. Cheng, Appl. Catal. B 31 (2001) 241.
[56] K.M. Schindler, M. Kunst, J. Phys. Chem. B 94 (1990) 8222.
[57] M.A. Fox, M.T. Dulay, Chem. Rev. 93 (1993) 54.
[58] G.P. Fotou, S. Pratsinis, Chem. Eng. Commun. 151 (1996) 251.
[59] L.-Q. Wu, P. Huang, N. Xu, J. Shi, J. Membr. Sci. 173 (2000) 263.
[60] B. Ohtani, Y. Ogawa, S.-I. Nishimoto, J. Phys. Chem. B 101 (1997) 3746.
[61] K.Y. Jung, S.B. Park, J. Photochem. Photobiol. A 127 (1999) 117.
[62] J. Yang, J.M.F. Ferreira, Mater. Lett. 36 (1998) 320.
[63] B. Tryba, A.W. Morawski, M. Inagaki, Appl. Catal. B 46 (2003) 203.
[64] S. Vemury, J.S.E. Pratsinis, J. Am. Ceram. Soc. 78 (1995) 2984.
[65] W. Xi, S.V. Geisen, Water Res. 35 (2001) 1256.
[66] T.C. Long, N. Saleh, R.D. Tilon, G.V. Lowry, B. Veronesi, Environ. Sci. Technol. 40 (2006) 4346.
[67] L.K. Limbach, Y. Li, R.N. Grass, M.A. Hintermann, M. Muller, D. Gunther, W.J. Stark, Environ. Sci. Technol. 39 (2005) 9370.
[68] F. Bosc, A. Ayral, P.-A. Albouy, L. Datas, C. Guizard, Chem. Mater. 16 (2004) 2208.
[69] G. Balasubramanian, D.D. Dionysiou, M.T. Suidan, V. Subramanian, I. Baudin, J.M. Laîné, J. Mater. Sci. 38 (2003) 823.
[70] P.C.A. Alberius, K.L. Frindell, R.C. Hayward, E.J. Kramer, G.D. Stuchy, B.F. Chmelka, Chem. Mater. 14 (2002) 3284.
[71] M.D. Moosemiller, C.G. Hill Jr., M.A. Anderson, Sep. Sci. Technol. 24 (1989) 641.
[72] M.A. Anderson, M.J. Gieselmann, Q. Xu, J. Membr. Sci. 39 (1988) 243.
[73] S.-Y. Kwak, S.H. Kim, S.S. Kim, Environ. Sci. Technol. 35 (2001) 2388.
[74] H. Choi, E. Stathatos, D.D. Dionysiou, Appl. Catal. B 63 (2006) 60.
[75] H. Choi, M.G. Antoniou, A.A. de la Cruz, E. Stathatos, D.D. Dionysiou, Desalination 202 (2007) 199.
[76] H. Choi, A.C. Sofranko, D.D. Dionysiou, Adv. Funct. Mater. 16 (2006) 1067.
[77] E. Bae, W. Choi, Environ. Sci. Technol. 37 (2003) 147.
[78] Y.-M. Lin, Y.-H. Tseng, J.-H. Huang, C.C. Chao, C.-C. Chen, I. Wang, Environ. Sci. Technol. 40 (2006) 1616.
[79] R. Asahi, T. Morikawa, T. Ohwaki, K. Aoki, Y. Taga, Science 293 (2001) 269.
[80] H. Irie, Y. Wanatabe, K. Hashimoto, J. Phys. Chem. B 107 (2003) 5483.
[81] C. Burda, Y. Lou, X. Chen, A.C.S. Samia, J. Stout, J.L. Gole, Nano Lett. 3 (2003) 1049.
[82] J. Gimenez, M.A. Aguado, S. Cervera-March, J. Mol. Catal. A 105 (1996) 67.
[83] G. Colon, M.C. Hidalgo, J.A. Navio, J. Photochem. Photobiol. A 138 (2001) 79.
[84] R.A. Burns, J.C. Crittenden, D.W. Hand, V.H. Selzer, L.L. Sutter, S.R. Salman, J. Environ. Eng. 125 (1999) 77.
[85] J.C. Crittenden, Y. Zhang, D.W. Hand, D.L. Perram, E.G. Marchand, Water Environ. Res. 68 (1995) 270.
[86] E. Stathatos, H. Choi, D.D. Dionysiou, Environ. Eng. Sci. 24 (2007) 13.
[87] H. Choi, Y.-J. Kim, R.S. Varma, D.D. Dionysiou, Chem. Mater. 18 (2006) 5377.
[88] H. Choi, E. Stathatos, D.D. Dionysiou, Thin Solid Films 510 (2006) 107.

[89] F. Bosc, A. Ayral, P.-A. Albouy, C. Guizard, Chem. Mater. 15 (2003) 2463.

[90] P. Yang, D. Zhao, D.I. Margolese, B.F. Chmelka, G.D. Stucky, Chem. Mater. 11 (1999) 2813.

[91] D.M. Antonelli, J.Y. Ying, Angew. Chem. Int. Ed. Engl. 34 (1995) 2014.

[92] C.J. Brinker, G.W. Scherer, Sol-Gel Science: The Physics and Chemistry of Sol-Gel Processing, Academic Press, San Diego, CA, 1990.

[93] L.L. Hench, J.K. West (Eds.), Chemical Processing of Advanced Materials, Wiley, New York, 1992.

[94] J.H. Fendler, Nanoparticles and Nanostructured Films, Wiley-VCH, Weinheim, Germany, 1998.

[95] G.P. Anipsitakis, D.D. Dionysiou, Environ. Sci. Technol. 37 (2003) 4790.

[96] G.P. Anipsitakis, D.D. Dionysiou, Environ. Sci. Technol. 38 (2004) 3705.

[97] G.P. Anipsitakis, E. Stathatos, D.D. Dionysiou, J. Phys. Chem. B 109 (2005) 13052.

[98] Q. Yang, H. Choi, D.D. Dionysiou, Appl. Catal. B 74 (2007) 170.

[99] Q. Yang, H. Choi, Y. Chen, D.D. Dionysiou, Appl. Catal. B 77 (2008) 300.

[100] P. Neta, R.E. Huie, A.B. Ross, J. Phys. Chem. Ref. Data 17 (1988) 1027.

[101] D. Blake, H. Link, K. Eber, Solar photocatalytic detoxification of water, in: K.W. Boer (Ed.), Advances in Solar Energy: An Annual Review of Research and Development, Vol. 7, Kluwer Academic Publishers, Dordrecht, The Netherlands, 1992, Chapter 5, p. 202

[102] J.C. Crittenden, Y. Zhang, D.W. Hand, D.L. Perram, E.G. Marchand, Water Environ. Res. 68 (1996) 270.

[103] C.S. Turchi, M.S. Mehos, H.F. Link, Design and Cost of Solar Photocatalytic Systems for Groundwater Remediation: National Renewable Energy Laboratory Technical Paper-432-4865, 1992.

[104] A. Vidal, A.I. D'yaz, A. El Hraiki, M. Romero, I. Muguruza, F. Senhaji, J. Gonzalez, Catal. Today 54 (1999) 283.

[105] D.F. Ollis, N. Serpone, Heterogeneous photocatalysis in the environment: applications to water purification, in: D.F. Ollis, N. Serpone (Eds.), Photocatalysis: Fundamentals and Applications, John Wiley and Sons, New York, 1989 (Chapter 18).

CHAPTER 9

Membrane Bioreactors: Theory and Applications to Wastewater Reuse

Pierre Bérubé*
Department of Civil Engineering, University of British Columbia, Vancouver, BC V6T 1Z4, Canada

Contents

1. INTRODUCTION

Membrane bioreactors (MBRs) are increasingly used for wastewater treatment applications. The global market for MBRs is growing at a rate of over 10% annually and is expected to exceed 1.3 billion USD by 2010 [1]. The increase in popularity of MBRs is mainly due to the ability of these systems to effectively remove contaminants of concern present in wastewaters [2]. This ability also makes MBRs ideally suited for wastewater reuse applications.

The use of MBRs for the treatment of wastewater for reuse applications is presented in this chapter. The physical components of MBRs are reviewed, the parameters that affect the performance of these components are discussed, and the expected contaminant removal efficiencies are

*Corresponding author.
E-mail address: berube@civil.ubc.ca

Sustainability Science and Engineering, Volume 2 ISSN 1871-2711, DOI 10.1016/S1871-2711(09)00209-8

presented. An example, describing the use of an MBR in a full-scale wastewater reuse applications, as well as a summary of published studies that investigated the use of MBRs for wastewater reuse, are included at the end of the chapter.

2. MEMBRANE BIOREACTOR FUNDAMENTALS

MBRs combine biological and membrane treatment to effectively remove contaminants of concern from wastewaters. As illustrated in Fig. 1, MBRs are similar to conventional activated sludge systems (CASs) with the exception that the biomass (i.e., microorganisms) responsible for removing the contaminants of concern are retained within the bioreactor component of the system using membranes (Fig. 1b and c) rather than secondary clarifiers (Fig. 1a). Early designs of MBRs simply replaced the secondary clarifier of CASs with an external membrane (Fig. 1b). However, most MBRs are now designed with the membrane submerged within the bioreactor component of the system (Fig. 1c). The treatment performances of external and submerged MBRs are similar; however, the capital and operating costs for submerged membrane systems are typically much lower than those for external systems, and comparable to those for CASs [53]. Primary clarifiers are typically used prior to CASs to remove material that can easily settle by gravity, reducing the contaminant load on these systems.

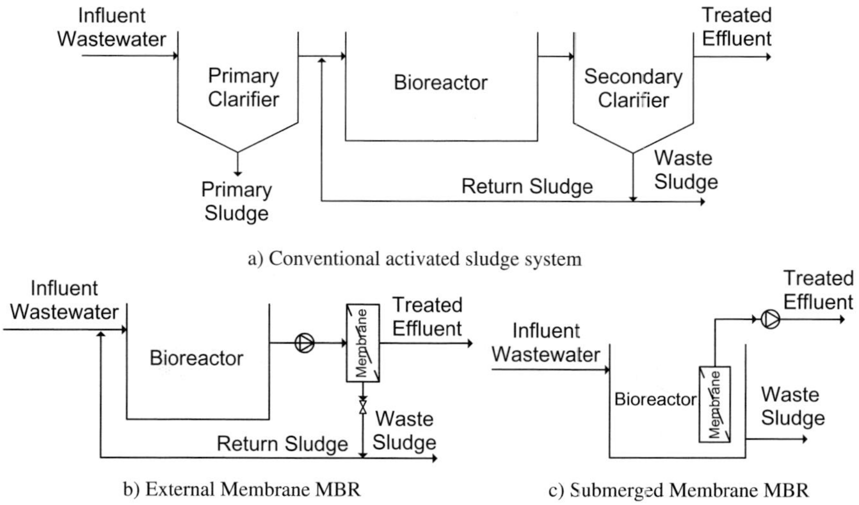

Figure 1 Bioreactor process schematics.

However, since MBRs can sustain higher loading rates than CASs, primary clarifiers are typically not required. On the other hand, fine mesh screening (0.5–3 mm) is typically used prior to MBRs to remove large debris and fine material (e.g., hair), which can negatively affect the performance of the membrane component of these systems.

Although MBRs and CASs are relatively similar, their process configurations in wastewater reuse applications differ substantially. This is because CASs are not as effective at removing the contaminants of concern present in wastewaters (see Section 3), and therefore the effluent from these systems typically must be further treated (e.g., using granular media filtration) before reuse applications [3]. Typical MBR and CAS configurations used for wastewater reuse applications are presented in Fig. 2. As illustrated in the figure, MBRs usually have fewer unit processes than CASs, leading to reduced system complexity and improved operability [4]. In addition, the total life cycle cost of MBRs is lower than that of CASs with granular media filtration [5]. It is also interesting to note that the total life cycle cost of producing reverse osmosis (RO) quality water from seawater is approximately twice that of producing RO quality water from MBR effluents [6]. The final effluent from MBRs and CASs is typically disinfected to kill and/or inactivate any pathogens that may remain in the treated effluent.

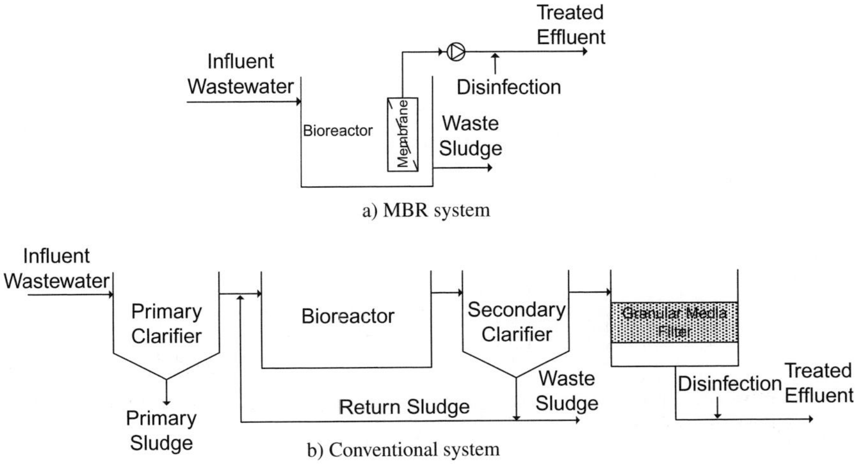

Figure 2 Typical process configuration for wastewater reuse application.

2.1 Bioreactor component of MBRs

The bioreactor component of an MBR is designed to remove biologically degradable contaminants of concern. A brief introduction to bioreactor fundamentals is presented below.

2.1.1 Bioreactor fundamentals

A number of bioreactor configurations exist and are commonly used in wastewater treatment applications [7]. All are governed by the kinetics of biomass growth and substrate (i.e., contaminant) consumption (i.e., removal). For a simple, completely mixed flow through bioreactor, the extent of substrate consumption can be estimated using a mass balance approach as depicted verbally and numerically in Eqs. (1) and (2), respectively. Under steady-state conditions, these equations can be simplified to yield Eq. (3).

$$\begin{matrix}\text{Change in mass of} \\ \text{substrate in system}\end{matrix} = \begin{matrix}\text{Mass of substrate} \\ \text{entering system}\end{matrix} - \begin{matrix}\text{Mass of substrate} \\ \text{leaving system}\end{matrix} - \begin{matrix}\text{Rate at which substrate} \\ \text{is consumed}\end{matrix} \tag{1}$$

$$\frac{\mathrm{d}S}{\mathrm{d}t} V_{\mathrm{r}} = QS_0 - QS - V_{\mathrm{r}} r_{\mathrm{S}} \tag{2}$$

where S is the concentration of substrate in the bioreactor (and bioreactor effluent) $[\mathrm{M/L^3}]$, V_{r} the volume of the bioreactor $[\mathrm{L^3}]$, Q the flow rate through the bioreactor $[\mathrm{L^3/T}]$, S_0 the concentration of substrate in the influent to the bioreactor $[\mathrm{M/L^3}]$, and r_{S} the rate at which the substrate is consumed $[\mathrm{M/L^3T}]$:

$$V_{\mathrm{r}} = \frac{(S_0 - S)Q}{r_{\mathrm{S}}} \tag{3}$$

As presented in Eq. (3), the size of a bioreactor required to achieve a given amount of substrate consumption (i.e., contaminant removal) is smaller when the rate at which substrate can be consumed is high. The rate at which substrate is consumed can be estimated using Eq. (4).

$$r_{\mathrm{S}} = \frac{kXS}{k_{\mathrm{S}} + S} \tag{4}$$

where k is the maximum rate of substrate consumption per unit mass of biomass $[1/\mathrm{T}]$, X the concentration of biomass in the bioreactor $[\mathrm{M/L^3}]$, and k_{S} the half saturation constant for the substrate being consumed by the biomass $[\mathrm{M/L^3}]$.

Easily biodegradable substrates have a high maximum rate of substrate consumption and are therefore rapidly consumed. Less biodegradable substrates can also be consumed rapidly by maintaining a high biomass concentration in the bioreactor. For this reason, bioreactors used in wastewater treatment applications are not designed as flowthough systems, but rather have some mechanism by which the biomass is retained in the bioreactor. For MBRs, the membrane component of the system retains the biomass within the bioreactor (Fig. 1b and c). Since membranes can retain virtually all of the biomass, relatively high biomass concentrations can be achieved in MBRs, resulting in relatively high substrate consumption rates, and therefore, relatively small bioreactor volumes. Typical biomass concentrations, measured as mixed liquor suspended solids (MLSS), in MBRs range from 8 to 12 g/L [8]. By comparison, the clarifier component of CAS is not as effective at retaining biomass and therefore typical biomass concentrations that can be maintained in these systems typically range from 2.5 to 4 g/L. As a result, the hydraulic retention time in MBRs, which is the ratio of the volume of the bioreactor to the flow rate through the system, can be as low as 4 h [8]. By comparison, the size of the bioreactor component of CASs is typically approximately two times larger than that of MBRs [9,10]. It should be noted that there is a practical upper limit to the biomass concentration that can be maintained in MBRs. At biomass concentrations greater than approximately 12 g/L, oxygen transfer in the bioreactor component of the system is limiting and inhibits the growth of the aerobic biomass [11]. High biomass concentrations can also negatively affect the permeate flux through the membrane component of MBRs as discussed in the next section.

The ability to retain virtually all of the biomass in the bioreactor also enables higher mean cell residence times, also commonly referred to as sludge retention times (SRTs), to be maintained in MBRs. The amount of waste biomass produced during treatment, which can be estimated using Eq. (5), typically decreases as the SRT increases [12–14]. The SRT of CASs typically ranges from 3 to 5 days, while that of MBRs is usually greater than 10 days. As a result, the amount of waste biomass produced by MBRs is typically approximately 15–50% less than that produced by CASs [10,12,15].

$$M_{\mathrm{EB}} = \frac{Y}{1 + k_{\mathrm{d}}\theta}(S_0 - S) \tag{5}$$

where M_{EB} is the mass of excess biomass produced [M/L^3], Y the ratio of mass of biomass formed to the mass of substrate consumed [–], K_{d} the endogenous biomass decay coefficient [1/T], θ the SRT, which is the ratio of the volume of the bioreactor to the waste sludge flow rate from the bioreactor [T], and (S_0-S) the amount of substrate consumed $[M/L^3]$.

The ability to rapidly and effectively remove contaminants using relatively small bioreactor volumes and the limited extent of waste sludge production are some of the principal advantages that MBRs have over CASs [16]. It should be noted that the overall amount of waste sludge produced by MBRs can be relatively similar to that produced by CASs since MBRs typically do not have primary clarifiers, which remove the easily settlable contaminants from the waste stream prior to biological treatment (i.e., primary sludge), and because MBRs remove more contaminants than CASs [17].

Although the mass balance equations, rate kinetics, and sludge yields associated with bioreactor configurations used in wastewater treatment and reuse applications are more complex than those presented above for a simple flow through systems, the overall conclusions from the above discussion apply to all biological processes and bioreactor configurations. A more detailed discussion on bioreactor kinetics, beyond the scope of the present discussion, can be found in Metcalf and Eddy [7] and Bailey and Ollis [16]. Most bioreactors are designed to promote the growth of aerobic biomass capable of degrading organic contaminants (Fig. 1). However, bioreactors can also be designed to promote the growth of biomass that is capable of removing nutrients such as nitrogen and/or phosphorus (Fig. 3a and b), or can be designed as anerobic systems. A detailed discussion on these different bioreactor configurations can be found in Metcalf and Eddy [7] and Bailey and Ollis [18].

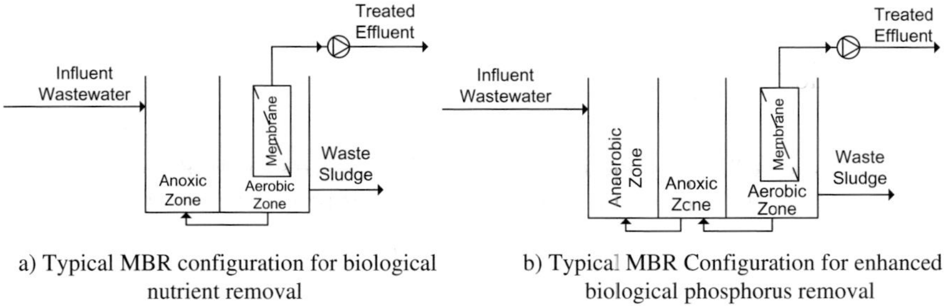

Figure 3 Bioreactor system configurations for enhanced nutrient removal.

2.2 Membrane component of MBRs

The membrane component of MBRs is selected to retain the biomass within the system [8]. Details of the types of membranes typically used in MBRs and the parameters that affect their performance are discussed below.

2.2.1 Categories of membranes

Pressure-driven membranes are typically classified based on the size of the material that they can retain. Microfiltration and ultrafiltration membranes, also commonly referred to as low-pressure membranes, are predominantly used to remove particulate material. On the other hand, nanofiltration and RO membranes, also commonly referred to as high-pressure membranes, are used to remove soluble material. Considering that the primary role of the membrane component of MBRs is to separate the biomass particles from the treated wastewater, low-pressure membrane systems are typically used for MBR applications.

Low-pressure membranes, hereafter simply referred to as membranes, essentially function as sieves, retaining particulate material that is larger than the pore sizes of the membranes. Particles in MBRs typically range in size from 1 to 200 μm [2], suggesting that membranes with pore sizes slightly smaller than this would effectively separate the biomass. However, experience has shown that the performance of membranes with pore sizes of approximately 0.5–1 order of magnitude smaller than the size of the biomass is better than that of membranes with larger pore sizes. Membranes with larger pores tend to get clogged internally by biomass and other solids in the mixed liquor. As discussed below, the resistance to the permeate flow increases as membrane pore size decreases, and therefore membranes with pore sizes smaller than 0.04 μm are also seldom used in MBR applications. As a result, membranes with pore sizes ranging from 0.04 to 0.4 μm (i.e., microfiltration and ultrafiltration) are typically used in MBR applications. Table 1 lists some of the characteristics a few different types of commercially available membranes used in MBR applications.

As previously discussed, both external and submerged membrane configurations are used in MBR applications. External membrane configurations are stand-alone systems to which the solution to be filtered (i.e., mixed liquor) is pumped in a recirculating loop (Fig. 1b). A flow restriction on the return line provides the pressure necessary to drive the permeate through the membrane, while the relatively high cross-flow velocity scours the membrane surface, preventing excessive accumulation (i.e., fouling) of retained material. The solution being filtered is typically confined to the inside of the membrane and flows from the inside to the

Table 1 Characteristics of membranes commonly used in MBR applications

	Ge-Zenon	Norit	Siemens/ Memcor	Kubota
Type	UF	UF	MF	MF
Pore size	0.04	0.03	0.02	0.4
Configuration	Hollow fiber	Tubular	Hollow fiber	Flat sheet
Material	PVDF	PVDF	PVDF	Polysulphone
Operation	Submerged air sparged	External air lift	Submerged air sparged	Submerged air sparged

Note: PVDF, polyvinylidene fluoride.

outside of the membrane (i.e., inside-out flow). External membrane systems are usually operated with a constant pressure and variable permeate flux (i.e., permeate flux decreases as membrane fouls). Submerged membrane systems are integrated into the bioreactor component of MBRs (Fig. 1c). A permeate pump provides the vacuum necessary to drive the permeate through the membrane and air sparging is used to scour the membrane surface. The solution being filtered is not confined by the membrane and flows from the outside to the inside of the membrane (i.e., outside-in flow). Submerged membranes are typically operated with a constant flow and variable transmembrane pressure (i.e., transmembrane pressure increases as membrane fouls).

Tubular membranes are typically used in external membrane MBR configurations where the permeate flows from the inside to the outside of the membrane. These membranes have internal diameters typically greater than 5 mm [8] and are grouped into modules containing multiple tubes (Fig. 4a). Tubular membranes with smaller internal diameters are not typically used in MBR applications since they can be plugged by the particulate material present in the mixed liquor. Hollow fiber and flat sheet membranes are usually used in submerged MBRs configurations. In these systems, the solution to be filtered is not confined within the membrane (i.e., outside-in flow) and therefore plugging is typically not a concern. Hollow fiber membranes usually have external diameters ranging from 1 to 3 mm and are potted into modules consisting of a dozens to thousands of

a) System of tubular membranes[1] b) System of hollow fiber membranes[2] c) System of flat sheer membranes[3]

Figure 4 Membrane configurations. *Source*: 1, Norit Americas Inc.; 2, GE Water and Process Technologies; 3, Sanitherm Engineering.

fibers (Fig. 4b). Flat sheet membranes are typically mounted onto panels, which are stacked into assemblies (Fig. 4c).

A number of polymeric materials are commonly used in membrane applications (Table 1). Membranes made from hydrophobic polymers, such as polyvinylidene fluoride (PVDF), can withstand repeated exposure to the relatively harsh cleaning agents commonly used for membrane cleaning (e.g., hypochlorous and citric acids) and therefore are commonly used in commercial membrane applications. However, membranes made from hydrophobic polymers tend to foul more readily than those made from hydrophilic polymers. For this reason, most membranes used in MBRs consist of proprietary blends of hydrophobic and hydrophilic polymers to maximize stability and minimize fouling. Inorganic membranes, made from materials such as ceramics, are increasingly being considered for wastewater treatment applications because they are much more durable than polymeric membranes. Although extensively used in many industrial applications, the current cost of these membranes makes them prohibitively expensive for municipal wastewater treatment applications. However, this may change in the near future as innovative inorganic membrane designs and manufacturing approaches are developed.

2.2.2 Flux and fouling in membrane systems

Flow through clean, low-pressure membranes can be modeled as Poiseuille flow through a network of capillary tubes (i.e., pores) using Eqs. (6) and (7) for constant-pressure, variable-flux systems, and for constant-flux, variable-pressure systems, respectively:

$$J = \frac{fr^2 \Delta P}{8\mu\tau\delta} = \frac{\Delta P}{\mu R_{\mathrm{m}}} \tag{6}$$

$$\Delta P = \frac{J 8\mu\tau\delta}{fr^2} = J\mu R_{\mathrm{m}} \tag{7}$$

where J is the permeate flux $[\mathrm{L}^3/\mathrm{L}^2\mathrm{T}]$; f the fraction of open pore area on the membrane surface [–]; r the radius of the membrane pores [L]; ΔP the pressure drop across the membrane, also commonly referred to as the transmembrane pressure $[\mathrm{F}/\mathrm{L}^2]$; μ the viscosity $[\mathrm{FT}/\mathrm{L}^2]$; τ the pore tortuosity factor [–]; δ the effective thickness of the membrane [L]; and R_{m} the membrane resistance to the permeate flow [1/L].

As presented in Eqs. (6) and (7), the resistance to the permeate flow is proportional to the thickness of the membrane and inversely proportional to the square of the size of the membrane pores. The permeate flux, or transmembrane pressure, that is achievable is also a function of the viscosity of the liquid being filtered [19]. As illustrated in Fig. 5, the biomass viscosity

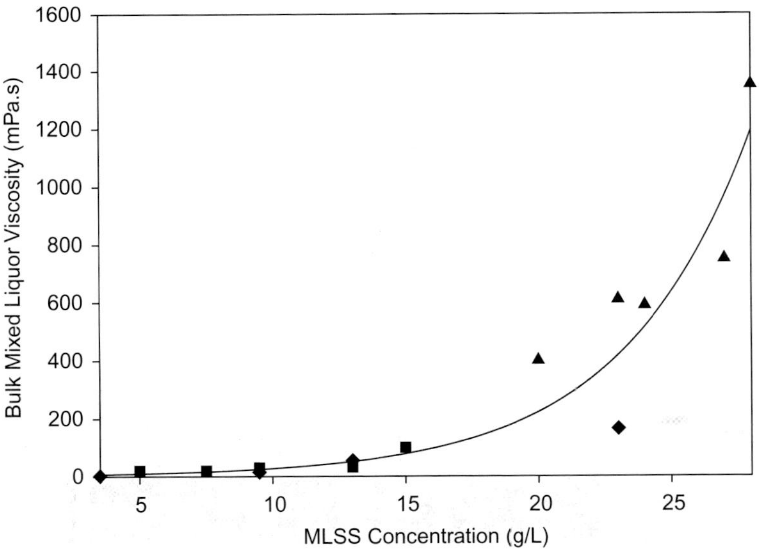

Figure 5 Relationship between MLSS concentration and bulk mixed liquor viscosity in MBRs treating wastewater. Adapted from Takemura et al. [62], Nagaoka et al. [58], and Sato and Ishii [61]. Solid line: Exponential relationship fitted to reported data as per Krauth and Staab [59].

tends to increase very rapidly at concentrations exceeding approximately 10,000 mg/L (measured as MLSS). For this reason, the biomass concentration in an MBR is typically less than approximately 12 g/L. The viscosity also affects the back-transport of foulants away from a membrane surface, as discussed below.

During operation, the material that is retained by the membrane can accumulate, plugging membrane pores, reducing the diameter of membrane pores, and/or forming a cake layer on the membrane surface (Fig. 6), resulting in an increase in the overall resistance that the permeating flow must overcome to pass through the membrane.

A number of mechanisms and relatively complex models have been developed to mechanistically describe fouling processes [2,8,20]. Although these models can provide insight into the different fouling mechanisms, no comprehensive model currently exists that can effectively describe fouling in MBRs. However, the overall effect of fouling on the permeate flux through a membrane can be estimated using a simple empirical resistance-in-series relationship as presented in Eqs. (8) and (9) for constant-pressure, variable-flux systems, and for constant-flux, variable-pressure systems, respectively. Over time, as fouling progresses, the resistances due to pore plugging, pore constriction, and cake formation increase, resulting in a decrease in the permeate flux (for constant pressure systems) or an increase in the transmembrane pressure (for constant permeate flux systems) [21], as presented in Fig. 7. A scanning electron microscope image of a clean and

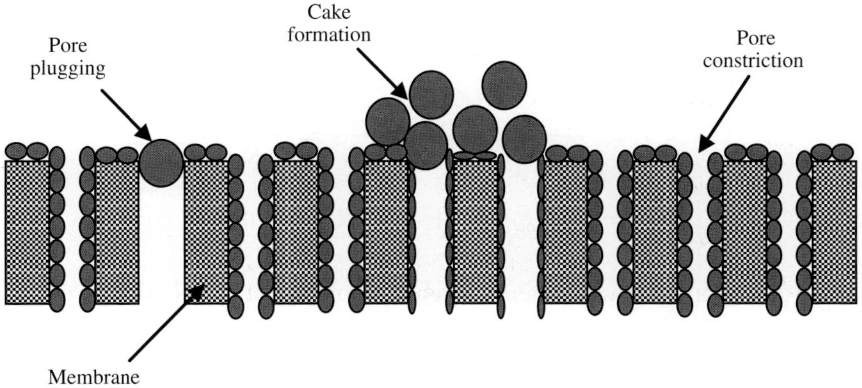

Figure 6 Pore plugging, pore constriction, and cake formation.

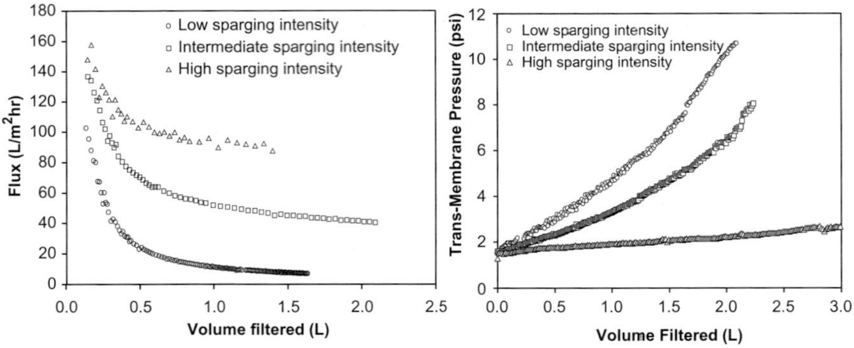

a) Constant pressure variable flux system b) Constant flux variable pressure system

Figure 7 Effect of fouling on membrane performance.

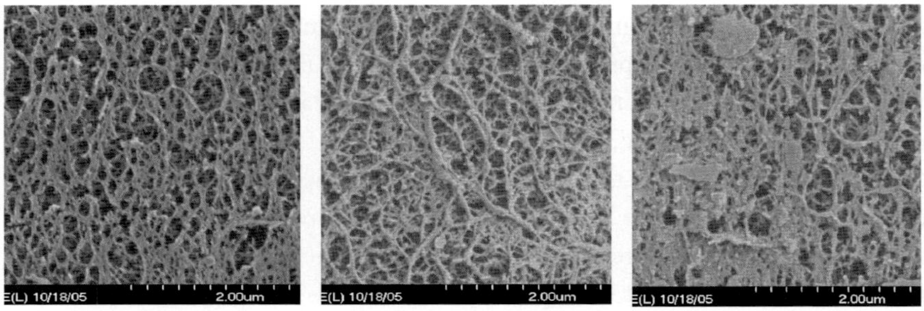

a) Clean membrane b) Partially fouled membrane c) Heavily fouled membrane

Figure 8 Fouled membrane surface (hollow fiber polymeric membrane system) [60].

fouled membrane is presented in Fig. 8.

$$J = \frac{\Delta P}{\mu(R_{\mathrm{m}} + R_{\mathrm{p}} + R_{\mathrm{r}} + R_{\mathrm{c}})} = \frac{\Delta P}{\mu(R_{\mathrm{T}})} \tag{8}$$

$$\Delta P = J\mu(R_{\mathrm{m}} + R_{\mathrm{p}} + R_{\mathrm{r}} + R_{\mathrm{c}}) = J\mu(R_{\mathrm{T}}) \tag{9}$$

where R_{m} is the resistance offered by the membrane [1/L], R_{p} the increased resistance due to pore plugging [1/L], R_{r} the increased resistance due to a reduction in the membrane pore size [1/L], R_{c} the increased resistance due to the formation of a layer (i.e., cake) of retained material on the membrane surface [1/L], and R_{T} the total resistance to the permeate flow [1/L].

It should be noted that although fouling negatively affects membrane performance by increasing the overall resistance to the permeate flow, fouling can also positively affect the performance of membranes. This is because the foulant layer can act as a more selective secondary membrane capable of removing material that is smaller than the membrane pores.

Fouling can be minimized by enhancing the rate of back-transport of retained material away from the membrane surface. When dealing with MBRs, the principal mechanisms responsible for the back-transport of retained material are expected to be inertial lift, shear-induced diffusion, and surface transport [20]. Although significantly different, each of these mechanisms can be promoted by providing high shear forces at a membrane surface. In external tubular systems, this is usually achieved by maintaining relatively high cross-flow velocities within the membranes. Typical cross-flow velocities inside tubular membranes range from 2 to 4 m/s. Higher cross-flow velocities can negatively affect biomass activity while lower velocities are not sufficient to prevent fouling. In submerged membrane systems, high surface shear forces are usually achieved by sparging air at the base of the membranes. Although the resulting bulk cross-flow velocity induced by the rising air bubbles is relatively low (0.2–0.5 m/s), surface shear forces comparable to those maintained in external tubular systems can be generated by the turbulent conditions that surround rising air bubbles [22]. The relatively high shear forces present in MBRs tend to break up the flocs in the mixed liquor, reducing the overall size of the particles in solution [2]. The diffusion resistance associated with smaller flocs can be less than that for larger ones, and as a result, the overall contaminant removal rate can be higher when the size of particles in the mixed liquor is smaller [23].

Membrane relaxation or back-flushing cycles can be used to further reduce the extent of fouling. During a relaxation cycle, the permeate flow is interrupted while the high surface shear forces are maintained. This provides an opportunity for accumulated foulants to be transported away from the membrane surface while no new foulants accumulate. During back-flushing cycles, the permeate flow is reversed while the high surface shear forces are maintained. The reversed flow can further enhance foulant back-transport by lifting away accumulated foulants from a membrane surface. The extent of fouling control can be enhanced by using a solution containing a chemical cleaning agent (e.g., chlorine or citric acid) for back-flushing. Relaxation/back-flush cycles typically last 10–60 s and typically occur every 5–60 min. The effect of back-flushing on the permeate flux in a submerged hollow fiber MBR is presented in Fig. 9.

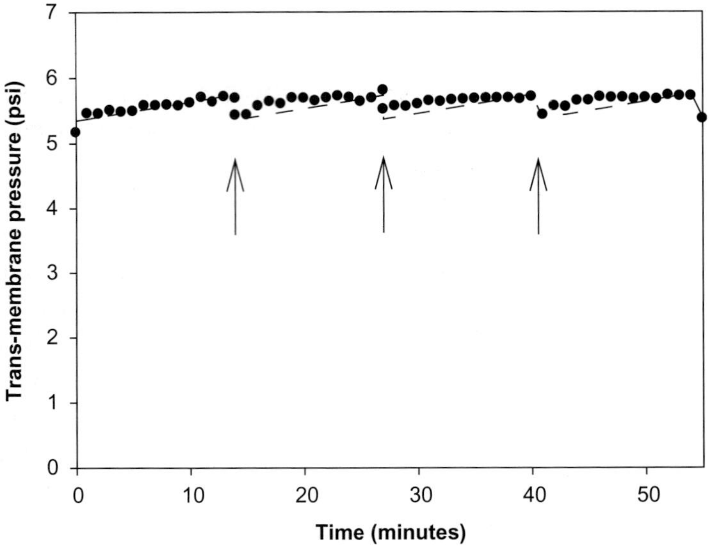

Figure 9 Transmembrane pressure in a submerged membrane MBR during successive back-flush cycles (constant-flux, variable-pressure system; unpublished operating data for University of British Columbia pilot-scale MBR; arrows indicate when a back-flush cycle was performed).

Regardless of the fouling reduction measures used, some irreversible fouling will eventually occur. When the extent of irreversible fouling becomes too large, the membrane must be chemically cleaned. Fouling in MBRs is typically due predominantly to the accumulation of organic and/or biological material on the membrane surface and/or within membrane pores. Caustic chlorine solutions are commonly used to effectively remove organic and/or biological foulants. However, chlorinated by-products of concern, such as trihalomethanes (THMs) and haloacetic acids (HAAs), can be generated during cleaning with chlorine compounds. As an alternative to chlorinated cleaning agents, citric acid is also commonly used to clean membranes in MBRs. Whichever cleaning agent is used, care must be taken to ensure that the cleaning agent is chemically compatible with the membrane material.

A number of other parameters can also affect membrane fouling and a detailed analysis of these is beyond the scope of this discussion. A summary of these parameters is presented in Table 2 and additional information on how they affect membrane fouling can be found in Chang et al. [24] and Bérubé et al. [25].

Table 2 Parameters that affect permeate flux and fouling in MBRs

Membrane material
Membrane material. Hydrophilic membranes tend to foul less extensively than hydrophobic membranes.
Polarity and charge. Negatively charged membranes tend to foul less extensively than positively charged membranes.
Pore size. Membranes with pore sizes greater than 0.2 μm tend to foul more readily than those with smaller pore sizes.
Membrane configuration. Fouling control mechanisms are different for submerged and external membrane systems.
Membrane operation
Cross-flow velocity. Relatively high cross-flow velocities (2–4 m/s) are typically required to prevent fouling in external membrane systems.
Gas sparging. Air sparging is typically used in submerged membrane systems to prevent excessive fouling.
Transmembrane pressure (TMP). At a low TMP, the permeate flux increases linearly with TMP. At a high TMP, the permeate flux is mass-transfer-limited and does not increase linearly with TMP.
Operating flux. Surface fouling can be more extensive at a high operating flux.
Membrane cleaning. Relaxation and back flushing cycles can minimize fouling. Caustic bleach solutions and acidic solutions are typically effective at removing organic/biological foulants and inorganic foulants, respectively.
Operating temperature. Higher temperatures lower the viscosity of the mixed liquor, and therefore, increase the permeate flux. Greater biodegradation of soluble organic foulants also occurs at high temperatures.
Mixed liquor characteristics
Suspended solids. Higher solids concentrations typically increase the extent of fouling.
Colloidal solids. Colloidal solids have a greater tendency to accumulate at the surface of membranes and can clog pores or form dense foulant layers.
Soluble products. Soluble microbial products are likely responsible for most of the long-term irreversible fouling in MBRs.
Inorganic precipitates. Depending on the wastewater being treated and the operating conditions, inorganic material can precipitate onto the membrane surface.
Bioreactor operation
Loading rate/hydraulic retention time.[a] Higher loading rates or lower hydraulic retention times typically lead to greater biological growth and therefore higher biomass concentrations.
Solids retention time.[a] Linked to the production of excess biomass and therefore MLSS concentration.

[a] The loading rate/hydraulic retention time and the solids retention time also affect the concentration of soluble microbial products in the MBR.

3. CONTAMINANT REMOVAL

The principal contaminants of concern in water reuse applications can be broadly classified into three groups: organic, microbial, and physical parameters. A summary of the specific concerns associated with contaminants within each group is presented in Table 3.

As presented in Table 4, MBRs are significantly more effective at removing bulk organic material than CASs, and can remove over 99% of the biodegradable organic material contained in wastewaters. The higher removal efficiency that can be achieved in MBRs is largely due to the membrane component of the system, which, as presented in Table 4, can effectively retain particulate material, and associated organic material, within the bioreactor where it can be degraded [8,31]. The organic material remaining in the effluent from MBRs consists mainly of soluble and relatively poorly biodegradable microbial products generated during treatment. The bioreactor component of MBRs is also capable of degrading some of the organic material that can foul membranes [30]. As a result, the fouling of the membrane component of MBRs is less extensive than that which occurs during direct membrane filtration of wastewater (Table 4). In addition, MBRs are more robust than CASs and can produce a consistently high-quality effluent even when the hydraulic or organic load to the system is variable [12,16,32].

The extent of nitrogen and phosphorus removal that can be achieved using MBRs is typically comparable to that achieved using CASs, both with and without enhanced nutrient removal [7,27,33]. However, some studies have reported higher nutrient removal efficiencies for MBRs, compared to CASs [15]. The higher removal efficiencies that can be achieved with MBRs are in part due to the ability of the membrane component of these systems to retain virtually all particulate material, and associated nutrients, within the bioreactor [15]. However, some studies have also suggested that the foulant layer that forms on membranes in MBRs can contribute directly to the removal of soluble nutrients [34].

Several studies have reported that MBRs can effectively remove some trace organic contaminants of concern such as endocrine-disrupting compounds (EDCs), as well as pharmaceutical products and personal care products (PPCPs) (Table 5). The extent of removal for many EDCs and PhACs is similar for MBRs and CASs, suggesting that biodegradation in the principal removal mechanism [28,38,40,44,45]. However, some studies have reported that certain EDCs and PPCPs can be removed to a greater extent using MBRs than CASs [42]. On the other hand, studies have

Table 3 Constituents of concern in raw municipal wastewaters

Constituents of concern	Typical concentrations in raw municipal wastewater
Organic parameters	
Bulk organic material. The overall concentration of organic material is typically measured as total organic carbon (TOC), chemical oxygen demand (COD), or biochemical oxygen demand (BOD). Most of the organic material remaining after biological treatment is relatively nonbiodegradable, soluble microbial products. The presence of organic material in treated wastewater is of concern for reuse applications since it can promote microbial growth, which can result in odor problems, exert a disinfectant demand, and generate disinfection by-products. Soluble microbial products are also of concern in MBR applications since these compounds have been documented to contribute to membrane fouling.	BOD: 110–350 mg/L COD: 250–800 mg/L TOC: 80–260 mg/L
Nutrients. Nitrogen and phosphorus are of concern since they can also promote microbial growth. In addition, when treated wastewater that contains high concentrations of nitrogen is applied to land, nitrogen can leach through the soil and may cause groundwater concentrations of nitrates to exceed those in drinking water standards.	Total nitrogen: 20–70 mg/L Total phosphorous: 4–12 mg/L
Emerging trace contaminants. The potential health risks to humans associated with endocrine-disrupting compounds (EDCs), as well as pharmaceutical products, and personal care products (PPCPs), are not well known. However, these compounds have been documented to have adverse effects on aquatic organisms.	See Table 5 for compounds reported to be present in raw wastewaters

Table 3 *(Continued)*

Constituents of concern	Typical concentrations in raw municipal wastewater
Disinfection by-products. Organic material can react with oxidants used for disinfection to generate disinfection by-products of concern. The principal disinfection by-products of concern are trihalomethanes (THMs) and haloacetic acids (HAAs). These compounds have been linked to cancer and developmental problems in humans. In addition, the oxidants that react with the organic material are consumed in the process, reducing the extent of disinfection that can be achieved.[a]	THMs: 0 (μg/L) HAAs: 0 (μg/L)
Microbial parameters	
Microbial pathogens. Bacteria, protozoa and viruses, can be highly infectious, significantly affecting human health, and in some cases, can cause death.	Coliforms: 10^6 to 10^9/100 mL Viruses: 10^1 to 10^4 PFU/100 mL
Physical parameters	
Suspended solids. Not directly of concern but can shield microbial pathogens from disinfectants. Suspended solids are also of concern since they can make reuse water aesthetically unappealing. In addition, suspended solids can cause operational problems in reuse applications such as clogging of nozzles used in irrigation applications.	Total suspended solids 120–400 mg/L
Turbidity. Also not of direct concern, and is often used as a surrogate measurement for suspended solids.[b]	

Source: Adapted from Asano et al. [26].

[a] THMs and HAAs are typically not present in raw wastewaters since they are formed during disinfection.

[b] The turbidity of raw municipal wastewater is usually too high to be reliably reported.

Table 4 Effluent quality from CASs, membranes, and MBRs

Parameter	CASs	Membranes only	MBRs
COD (mg/L)	17–110	7–24	1–40
BOD (mg/L)	1.5–43	0.5–7	0.5–2.5
TSS (mg/L)	2–12	0–2.2	<1
Without enhanced nutrient removal			
Total nitrogen (mg/L-N)	13–21	11–27	3–12
Total phosphorous (mg/L-P)	0.3–1.1		0.4–2
With enhanced nutrient removal			
Total nitrogen (mg/L-N)	7.5–16		7.5–15
Total phosphorous (mg/L-P)	0.7–1.1		0.1–0.25
Coliforms (log removal)	2–3		3.5–7
Virus/coliphage[a] (log removal)	1–3	0.5	1.3 to 5
Fouled membrane resistance[b] (1/m)		40×10^{-9}	4×10^{-9}

Source: Adapted from side-by-side studies comparing the performance of MBRs, and CASs or membrane systems [3,14,27–30].

[a] Lower range corresponds to removal immediately after membrane cleaning, while upper range corresponds to removal after a few days of operation.

[b] Resistance after 100 days of operation;

demonstrated that membranes on their own are not capable of effectively removing trace soluble organic contaminants such as EDCs and PPCPs [37,46]. The removal of hydrophobic EDCs and PPCPs is believed to occur predominantly via the adsorption of these compounds onto biomass, and the subsequent retention of these compounds within the bioreactor component of the system for a long enough period of time to be degraded. Some studies also suggest that the foulant layer that forms on membrane surfaces also contributes to retaining EDCs and PCPPS in the bioreactor component of MBRs [43]. As presented in Table 5, only some EDCs and PCPPS can be removed using MBRs, and of the compounds that are removed, some are not consistently removed by all MBRs. It is likely that operating parameters, such as biomass concentration and fouling control measures, affect the ability of MBRs to remove EDCs and PhACs [16,28,37].

Table 5 – Removal of trace organic contaminants of concern using MBRs

Effectively removed		Not effectively removed
Personal care products	*EDCs*	*Personal care products*
Cafeine[1,3,8]	Androstenedione[1,8]	DEET[3]
DEET[1,8]	Cholesterol[2]	TCEP[1,8]
Oxybenzone[1,8]	Coprostanol[2]	*Pharmaceutical compounds*
Pharmaceutical products	Estradiol[5]	Carbamazepine[1,5,8]
Acetaminophen[1,8]	Estriol[7,8]	Diazepam[8]
Carbamazepine[3]	Estrone[1,5,8]	Diclofenac[1,3,8]
Erythromycin-H_2O[3,8]	Ethinylestradiol[1,5]	Dilatin[1,3,8]
Estradiol valerate[7]	Progesterone[8]	Erythromycin-H_2O[1]
Fluoxetine[3]	Stigmastanol[2]	Fluoxetine[1,8]
Gemfibrozil[3,8]	Testosterone[8]	Gemfibrozil[1]
Hydrocodone[3]	Total estrogenic activity[6,8]	Hydrocodone[1,8]
Ibuprofen[1,8]	17-α-Estradiol[7]	Ibuprofen[3]
Medrogestone[7]	17-β-Estradiol[7,8]	Lopromide[1,8]
Naproxen[1,3,8]	*Others*	Meprobamate[1,3,8]
Norgestrel[7]	NDMA[8]	Phenazone[5]
Pentoxifylline[8]		Propyphenazone[5]
Sulfamethoxazole[3]		Sulfamethoxazole[1,8]
Triclosan[1,3,8]		Venlafaxine[7]
Trimegestone[7]		*EDCs*
Trimethoprim[3,8]		Estrone[7]
17-α-Dihydroequilin[7]		Thinly estradiol[7]

Sources: 1, [35]; 2, [36]; 3, [37]; 4, [38]; 5, [39]; 6, [40]; 7, [41]; 8, [42]; 9, [43].

Although disinfection by-products (DBPs) are typically not present at significant concentration in wastewaters, they can be present in the effluent from MBRs. The DBP formation potential of the effluent from MBRs has been reported to be relatively low, ranging from 9 to 13 μg/L for trihalomethanes (THMs) [40]. However, some studies have reported relatively high concentrations of THMs in MBR effluents, ranging from 182 to 689 ug/L [47]. These high THM concentrations were attributed to the chlorine used in the periodic membrane back-wash cycles.

MBRs are also effective at removing microbial contaminants (Table 4), and therefore provide treatment that is equivalent to disinfection. Log removal values[1] in the order of 7 have been reported for bacteria, such as *Escherichia coli,* and other large microbial pathogens, such a *Giardia* and *Cryptosporidium*, using MBRs. These microbial pathogens are effectively removed using MBRs since they are larger than the pore sizes of the membrane component of the system. Viruses on the other hand are smaller than the pore size of membranes used in MBRs [29,48]. Nonetheless, MBRs have been reported to be capable of achieving log removal values as high as 5 for viruses [28,40,48–50]. Although only approximately 0.5 log removal can be attributed to retention of viruses by the membrane itself, approximately one log removal can typically be attributed to adsorption of viruses onto biomass and predation within the bioreactor [3,28,29]. The remaining removal has been attributed to retention by the foulant layer that forms on the membrane surface during operation [28,29,40,48]. For this reason, virus removal efficiencies are typically lower immediately following membrane cleaning and increase over time as the membrane surface becomes fouled [28,29,40,48]. The physical integrity of the membrane component of MBRs has also been reported to significantly affect the ability of these systems to effectively remove microbial pathogens [51]. Although membrane integrity testing is typically performed in drinking water treatment applications, it is seldom performed on MBRs used for wastewater treatment [16]. However, integrity testing protocols, similar to those developed for drinking water applications, could be developed to ensure that maximal contaminant removal efficiencies are being achieved in MBR applications for wastewater reuse.

A summary of the advantages and disadvantages of MBRs for wastewater reuse applications are listed in Table 6.

4. APPLICATION OF MEMBRANE BIOREACTORS FOR WASTEWATER REUSE

An example, describing the use of an MBR in a full-scale wastewater reuse application, as well as a summary of published studies that investigated the use of MBRs for wastewater reuse, are presented in the sections that follow.

[1]The removal of microbial pathogens is typically expressed based on a log removal value, which is the log of the ratio of the concentration of microbial pathogens in the raw wastewater to the concentration of microbial pathogens in the treated effluent.

Table 6 Summary of advantages of MBRs for reuse applications

Advantages	Disadvantages
• Produces high-quality effluent with reuse potential • System can be configured for enhanced nutrient removal • Process performance not affected by variations in influent load and quality • Process performance not affected by settling characteristics of biomass • Relatively small footprint • Relatively low sludge production • Relatively easy to automate	• Long-term history of operation is not available • Membrane configurations are not standardized and most are proprietary • Membranes typically must be replaced every 7 to 10 years • Membrane replacement is relatively expensive • Pilot testing required to design full-scale system

4.1 Example of MBR application for wastewater reuse: City of Key Colony Beach (Florida): MBR used as pretreatment for reverse osmosis for wastewater reuse[2]

By 1998, the City of Key Colony Beach's (Fig. 10) municipal wastewater treatment plant had reached its rated capacity, limiting development opportunities. The city anticipated that stringent effluent requirements of 5 mg/L biochemical oxygen demand (BOD) and total suspended solids (TSS), 3 mg/L total nitrogen (TN), and 1 mg/L total phosphorous (TP) would likely be imposed in the future because of the coastal discharge. The existing treatment facility would be unable to achieve this effluent quality without significant capital upgrade. In addition, the community wanted to irrigate the local golf course using recycled wastewater, but substantial salinity removal would be required because of seawater intrusion into the municipal sewer network.

The wastewater treatment technology selected was the ZenoGem® MBR, provided by GE Water and Process Technologies. The MBR process was selected for several reasons, among them the ability to achieve a high quality, particulate free effluent on a very compact footprint and the ability to generate an effluent suitable for direct RO treatment without pretreatment [i.e., MBR effluent silt density index (SDI) <3]. The ZenoGem® process was put in operation in June 1999 and the RO process was put in operation in December 1999. The ZenoGem® process had treated all flows to the wastewater treatment facility since

[2]Information information provided by GE water Water & process Process technologiesTechnologies.

Figure 10 Aerial view of City of Key Colony.

commissioning, while the RO unit has been operated for only intermittent periods because the water distribution system has not yet been constructed.

ZeeWeed® are proprietary hollow fiber membranes that are immersed within the bioreactor in direct contact with the mixed liquor. The ZeeWeed® hollow fiber membranes are contained in bundles called modules, which are assembled into cassettes of 8–12 modules. The membrane modules are directly immersed in the aeration tank, in direct contact with the mixed liquor. Through the use of a centrifugal pump, a vacuum varying between 13.8 and 62 kPa (2–9 psi) is applied to a header connecting the membrane modules. The vacuum draws the treated water through the hollow fiber membranes. The treated water passes through the hollow fibers and is pumped out by the permeate pump (see Fig. 11, "ZenoGem® Conceptual Process"). All particulate matter and the mixed liquor solids are rejected at the surface of the membrane. The ZeeWeed® membranes are automatically back-pulsed on a regular basis using collected permeate. A coarse bubble air diffuser is located at the base of each membrane module. The airflow provided by the diffuser scours the external surface of the membrane, transferring the rejected solids away from the membrane surface. This airflow also provides a portion of the biological oxygen requirements. Supplemental coarse or fine bubble diffuser grids may be used to supply the remainder of the biological oxygen

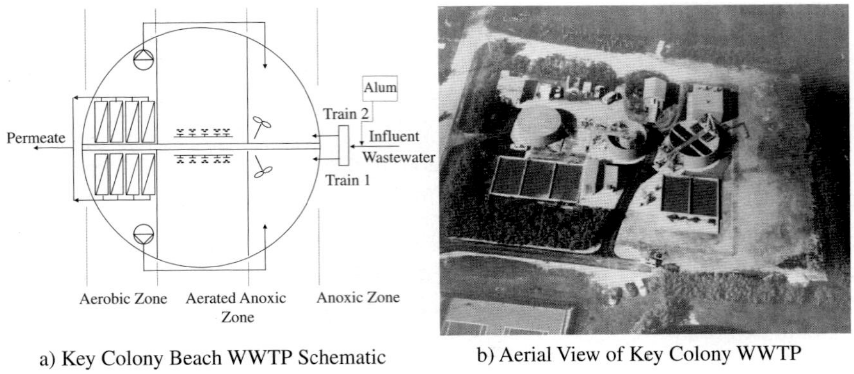

Figure 11 Key Colony MBR.

requirements. Sludge is wasted directly from the aeration tank at the operating MLSS concentration between 10,000 and 15,000 mg/L. The high biomass concentration allows the ZenoGem® process to be operated at reduced organic loading rates (i.e., low food/microorganism ratio) and elevated solids retention times (>15 days). Year-round nitrification is ensured because the operating SRT greatly exceeds the minimum SRT required for nitrification, which is typically 5–7 days under winter operating conditions. ZenoGem® bioreactors are ideally suited for denitrification as well. Since the ZeeWeed® membranes eliminate the need for secondary clarification, it is not necessary for the operators to concern themselves with the settling properties of the mixed liquor. The anoxic zone can be sized for optimal nitrogen removal and with the high MLSS concentrations, a total nitrogen removal efficiency of over 90% is readily achieved.

The aeration tank at Key Colony is separated into two trains, with each train divided into three distinct zones separated by concrete baffles (Zones 1, 2, and 3) as presented in Fig. 11. The raw wastewater, after passing through the rotating drum screens, is fed into Zone 1, which has a combined (Train 1+Train 2) operating volume of 210,000 L. A small amount of air is used for mixing in this zone, but the dissolved oxygen (DO) concentration is maintained less than 0.2 mg/L. The majority of denitrification occurs in this zone. The mixed liquor flows by gravity (through a submerged cutout) from Zone 1 to Zone 2. Zone 2 has a combined operating volume of 280,000 L and is aerated at a limited rate to achieve a DO concentration in the range of 0.2–0.8 mg/L. The DO in Zone 2 is maintained at a low enough concentration to allow both nitrification and denitrification to occur in Zone 2, minimizing the ammonia and nitrate concentration entering Zone 3. The mixed liquor flows by gravity (through a submerged cutout) from Zone 2 to Zone 3. Zone 3 has a combined operating volume of 260,050 L. This zone also contains the ZeeWeed® membranes, and is fully aerobic, being aerated by a grid of coarse bubble diffusers to achieve a DO concentration greater than 2 mg/L. Any of the remaining ammonia and soluble carbon (measured as BOD) will be oxidized in this zone. Mixed liquor from Zone 3 is recirculated back to Zone 1 at a flow rate of 6250 L/min, approximately four to eight times the influent flow rate.

Since startup, the effluent BOD, TSS, and total nitrogen concentrations have remained below 5, 5, and 3 mg/L, respectively, and with alum addition, it is possible to achieve total phosphorus concentrations less than 1 mg/L. ZenoGem® effluent turbidity has consistently measured <0.2 NTU. The RO unit has been operated for only intermittent periods because the water distribution system has not yet been constructed. There has been no evidence of RO membrane fouling, indicating that the ZenoGem® effluent was entirely suitable for direct feed to the RO. For the period of time when the RO unit has been operated, the reduction in conductivity has averaged greater than 98%. The total dissolved solids (TDS) of the RO permeate has been measured at less than 100 mg/L, resulting in a high-quality water entirely suitable for irrigation purposes.

Since the commissioning of the Key Colony plant, there have been a number of improvements to the membrane cassette configuration. Some of these improvements have already been implemented at Key Colony. For example, with the addition of automatic valves, the membrane air can now be cycled between the two trains, resulting in a close to 50% reduction in aeration requirements. As the membrane aeration makes up a substantial portion of the total operating costs of the ZeeWeed®/ZenoGem® system, this reduction in net aeration requirements will equate to significant operating cost savings over the life of the plant. Other improvements to the design include more efficient spacing of the individual membrane elements in the cassettes, resulting in the ability to increase membrane surface area per cassette by over 20%, while at the same time improving the efficiency of the membrane aeration system, resulting in a reduced membrane cleaning frequency. This new cassette configuration has been in operation at Key Colony since August 2000, and the performance of the new membrane cassettes has equaled or exceeded the performance of the original cassette configuration.

Table 7 Summary of results from MBR studies on wastewater reuse

Source and objective	MBR system details	MBR effluent quality	Comments
Ahn et al. [30]: Compared the performance of direct membrane filtration and MBR systems (municipal wastewater)	*Bioreactor* MLSS: 4000–7000 mg/L *Membrane* Submerged hollow fibers Polyethylene Pore size: 0.1 μm	COD: <9 mg/L BOD: <0.8 mg/L TOC: <3.8 mg/L Total suspended solids: <0.3 mg/L Turbidity: <0.1 NTU Total nitrogen: <10 mg/L Ammonia-N: <2.2 mg/L Total phosphorous: <1.6 mg/L HPC assay: $0.02–4.2 \times 10^4$ CFU/100 mL Coliphage: 0 PFU/100 mL	Fouling resistance increased much faster in direct membrane filtration than in the MBR. Also, filtration resistance in the MBR was one order of magnitude lower than in the direct filtration system
Al-Malack [12]: Investigated effect of MLSS and organic loading rate on system performance (synthetic wastewater)	*Bioreactor* Volume: 20 L MLSS: 3000–15,000 mg/L OLR: 0.1–1.2 kg COD per kg MLSS per day *Membrane* Submerged air sparged Tubular: 1.27 cm ID Material: polyester Pore size: 20–40 μm	COD: 18–224 mg/L Total suspended solids: 0 mg/L Total coliform: 1.7×10^7 to 9×10^7 MPN/100 mL Sludge production: 0.26 mg VSS/mg COD	Poor COD removal occurred immediately following sudden increases in organic loading rate. Increase in MLSS concentration increased the COD removal efficiency

Chae et al. [34]: Characterized effectiveness of MBR at organic and nitrogen removal at different HRTs (municipal wastewater)	*Bioreactor* Vertical anoxic(an)–aerobic(ar) system MLSS: 8700 mg/L(an), 4200 mg/L(ar) Volume: 1.3 m^3 HRT: 4–10 h *Membrane* Polytetrafluoroethylene Pore size: 0.45 μm	Total suspended solids: 100% removal COD: 94–97% removal Total nitrogen: 62–76% removal Total phosphorous: 42–77% removal Total coliform: 11 counts/100 mL *E.coli*: 2 counts/100 mL	Nitrification efficiency decreased when hydraulic retention time (HRT) reduced to 4 h. Higher range of removal efficiencies corresponded to an HRT of 10 h and lower range corresponded to an HRT of 4 h
Côté P. et al. [50]: Demonstrated effects of high biomass concentrations, sludge age, and bioreactor configuration (municipal wastewater)	*Bioreactor* Two systems: one aerobic, one anoxic–aerobic MLSS: 5000–15,000 mg/L SRT: 10–50 days *Membrane* Submerged air sparged hollow fibers MWCO: 200,000 Da	Total suspended solids: 100% removal COD: > 96% removal Ammonia-N: 80–99% removal Total nitrogen: 36–80% removal Total phosphorous: 15% removal Total coliform: > 6 log removal Viruses (bacteriophage): 4 log removal	Sludge production was 0.25 kg TSS/kg COD removed, and was approximately 50% lower than for typical conventional system. Lower range of nitrogen removal for aerobic system and higher range for anoxic–aerobic system
Fatone et al. [38]: Demonstrated feasibility of MBRs in wastewater reuse applications (municipal wastewater)	*Bioreactor* Anoxic–aerobic system MLSS: 4800–9000 mg/L HRT: 6–8 h *Membrane* Submerged hollow fibers Pore size: 0.04 μm	Total suspended solids: 0 mg/L COD: 4–11 mg/L Ammonia-N: 0.1 mg/L NO_3^--N: 0.9–5.2 mg/L Total nitrogen: 1.1–5.4 mg/L Total phosphorous: 31–57% removal	Approximately 80% removal of trace aromatic hydrocarbons in wastewater (i.e., acenaphthylene, acenaphthene, fluorene, anthracene, chrysene)

Table 7 *(Continued)*

Source and objective	MBR system details	MBR effluent quality	Comments
Guglielmi et al. [27]: Compared performances of MBRs and conventional tertiary systems in wastewater reclamation for irrigation (municipal wastewater)	*Bioreactor* Anoxic–aerobic system Volume: 7.9 m^3 MLSS: 12,000 mg/L SRT: 12–15 days *Membrane* Submerged hollow fiber Pore size: 0.04 μm	COD: 12.9 ± 3 BOD: 2.0 ± 0.5 TKN: 1.9 ± 1.5 Ammonia-N: 0.6 ± 0.5 Total nitrogen: 8.1 ± 4.5 Total phosphorous: 1.2 ± 0.8 TSS: < 1 NTU *E. coli*: 1 CFU/100 mL	MBR proved to be more reliable than conventional activated sludge system with tertiary filtration, especially for microbial contaminants
Helmig et al. [41]: Investigated ability of MBRs to removal pharmaceuticals (pharmaceutical manufacturing wastewater)	*Bioreactor* Volume: 2.7 m^3 HRT: 14.8 h SRT: 20 days *Membrane* Submerged hollow fiber Pore size: 0.04 μm	COD: 94% removal BOD: 99% removal Total suspended solids: non-detectable Total nitrogen: 60% removal Total phosphorous: 38% removal	Some hormones and oral contraceptives were removed to near or below detection limits. However, several key pharmaceuticals were resistant to MBR treatment
Innocenti et al. [13]: Determined effect of SRT and MLSS on performance of an MBR (municipal wastewater)	*Bioreactor* Batch anoxic/aerobic MLSS: 4000–17,000 mg/L SRT: 10 to > 200 days *Membrane* Pore size: 0.02 μm Submerged hollow fiber	COD: 19–40 mg/L Total suspended solids: 0 mg/L Ammonia-N: 0.2–0.5 mg/L Total nitrogen: 6.2–13.3 mg/L Total phosphorous: 0.9–1.1 mg/L	Sludge production at 10, 190, and > 200 days was 0.56, 0.08, and 0.02 g MLVSS/g COD, respectively. Total nitrogen removal optimal at intermediate MLSS concentration

Jefferson et al. [52]: Compared performance of external (E) and submerged (S) MBRs (municipal wastewater)	*Bioreactor* Volume: 0.035 m^3 (S), 0.38 m^3 (E) MLSS: 200 mg/L (S), 430 mg/L (E) *Membrane* Flat sheet (S), tubular (E) Pore size: 0.4 μm (S), 4000 kDa (E)	COD: 2.5–31 mg/L Turbidity: 0.2 NTU Total coliform: 3–5 log removal	Slightly better removal for external MBR. However, the MBRs were operated under considerably different conditions
Jefferson et al. [53]: Compared performance of MBRs, biological aerated filters (BAF) and membrane aeration bioreactor (MABR) (synthetic gray water)	*Bioreactor* Volume: 0.035 m^3 HRT: 12 h *Membrane* Submerged plate and frame Pore size: 0.4 μm	COD: <10 mg/L Total coliform: 7 log removal	Performance independent of MLSS concentration in the range of 400–8000 mg/L. MBR met reuse requirements 100% of the time, while BAF and MABR systems did not
Kumar et al. [54]: Investigated the use of an MBR prior to RO treatment (municipal wastewater)	*Bioreactor* Anoxic–aerobic configuration *Membrane* Flat sheet submerged Pore size: 0.4 μm	BOD: <2 mg/L TOC: <10 mg/L Turbidity: <0.13 NTU Total coliform: <2 MPN/100 mL Coliphage: <10 PFU/100 mL (80% of the time)	Only results from MBR presented

Table 7 *(Continued)*

Source and objective	MBR system details	MBR effluent quality	Comments
Li et al. [55]: Compared the performance of single-stage bioreactor configuration and powdered activated carbon (PAC) addition on nitrogen removal in MBRs (synthetic wastewater)	*Bioreactor* Volume: 17.2 L MLSS: 4100–13,500 mg/L *Membrane* Submerged hollow fibers Pore size: 0.2 μm Polyvinylidene	NH_4^+-N: 92–98% removal Total nitrogen: 30–65% removal	Higher MLSS concentrations (in excess of 12,000 mg/L) led to improved nitrogen removal by enhancing anoxic microenvironments. PAC addition did not affect nitrogen removal
Liu et al. [56]: Investigated the effect of PAC addition on the performance of MBRs (municipal wastewater)	*Bioreactor* Attached growth MBR Volume: 21 L *Membrane* Submerged hollow fibers Polysulfone Pore size: 0.05 μm	COD: <25 mg/L BOD: <1.5 mg/L Ammonia-N: <0.8 mg/L Total nitrogen: <15 mg/L	PAC addition decreased the effluent DOC to <15 mg/L and reduced the extent of fouling
Lozier and Fernandez [57]: Characterized performance of MBR as pretreatment for RO (municipal wastewater)	*Bioreactor* MLSS: 10,000–14,000 mg/L HRT: 5.6–7.6 h SRT: 16–22 days *Membrane* Submerged hollow fibers Pore size: 0.04 μm	Total suspended solids: 0.3–0.4 mg/L Turbidity: 0.16–0.27 NTU COD: 13.8–15 mg/L BOD: 0.57–1.05 mg/L Total phosphorous: 0.18–3.38 mg/L Ammonia-N: 0.11–5.77 mg/L Total nitrogen: 16–27 mg/L	MBR effluent exceeded RO feed water quality criteria

Monti et al. [15]: Compared performance of MBR and conventional systems configured for enhanced nutrient removal (municipal wastewater)	*Bioreactor* Volume: 2.2 m^3 HRT: 7–10 h SRT: 10 days MLSS: 6000–8000 mg/L Anaerobic–anoxic–aerobic *Membrane* Submerged hollow fiber Pore size: 0.04 μm	COD: 90% removal (average) Sludge yield: 0.23–0.28 g VSS/g COD Ammonia-N: nondetectable Total phosphorous: 1.85 ± 0.065 mg/L	Sludge yield of MBR approximately 15% lower than for conventional system. Greater denitrification was observed for a conventional system, possibly due to anoxic conditions in the secondary clarifier. Substantially lower effluent phosphorus concentrations were achieved in MBR
Oota et al. [28]: Determined the ability of MBRs to remove viruses and select endocrine disrupting compounds (municipal wastewater)	*Bioreactor* Anoxic–aerobic system HRT: 6 h *Membrane* Submerged flat plate Pore size: 0.4 μm	BOD: <5 mg/L TOC: <3.8 mg/L *E.coli*: 100% removed Viruses (coliphage): 5 log removal Nonylphenol: 0.1 ng/L Bisphenol A: 0.02–0.03 ng/L DEPH: <0.2 ng/L Benzophenone: 0.01 ng/L 17β-Estradiol: nondetectable	Hypothesized that viruses are removed by adsorption onto biomass and retention by foulant layer that forms on membrane surface
Soriano et al. [32]: Compared performance of MBR and conventional systems at low SRT and HRT (municipal wastewater)	*Bioreactor* Volume: 225 L HRT: 5–15 h SRT: 2–7 days	COD: >90% removal Total nitrogen: >65% removal (when SRT >3 days)	Consistently better performance of MBR. Overall nitrogen removal declined substantially when the SRT was less than 3 days

Table 7 *(Continued)*

Source and objective	MBR system details	MBR effluent quality	Comments
Spring et al. [36]: Compared ability of MBRs and conventional activated sludge systems at removing EDCs and other trace contaminants (municipal wastewater)	*Bioreactor* MLSS: 10,000–15,000 mg/L SRT: >40 days *Membrane* Submerged hollow fibers Pore size: 0.04 μm	Cholesterol: 96% removal (average) Coprostanol: 100% removal (nondetectable) Stigmastanol: 100% removal (nondetectable) Bisphenol A: <12.6 ng/L Estrone: <1.2 ng/L 17β-Estradiol: <1.1 ng/L 17α-Estradiol: <1.6 ng/L	MBRs could remove approximately 10% more cholesterol, coprostanol and stigmastanol from municipal wastewaters than conventional treatment
Tam et al. [40]: Compared performance of MBRs and conventional activated sludge+MF as pre-treatment for RO (municipal wastewater)	*Bioreactor* Anoxic–aerobic configuration *Membrane* Pore size: 0.4 μm Submerged hollow fiber	BOD: <2 mg/L COD: 17.5 mg/L (average) Total suspended solids: <2 mg/L Ammonia-N: 0.4 mg/L (average) NO_3^--N: 1.9 mg/L (average) *E. coli*: 7 log removal (average) Viruses: 4.7 log removal (average) THMFP: 9–13.5 μg/L Total estrogens: 80% removal	Removal of viruses decreased significantly following membrane cleaning, and increased over time thereafter
Tao et al. [31]: Investigated effect of anoxic-aerobic zone configurations on performance of MBR (municipal wastewater)	*Bioreactor* Various anoxic–aerobic configurations MLSS: 4000–13,000 mg/L HRT: 4.5–12 h SRT: 14–28 days	Turbidity: <0.2 NTU TOC: <5 mg/L Ammonia-N: <1 mg/L Total nitrogen: <12 mg/L	Membrane pore size did not affect permeate quality

	Membrane Submerged hollow fiber and flat sheet Pore size: 0.035–0.4 μm		
Wintgens et al. [45]: Investigated ability of full-scale MBRs to remove EDCs (landfill leachate)	*Bioreactor* Volume: 180 m^3 Anoxic–aerobic configuration MLSS: 25,000 mg/L *Membrane* External tubular Cross-flow: 5 m/s	Bisphenol A: over 99% removal	Removal attributed to adsorption onto biomass and subsequent biodegradation
Wong et al. [29]: Determined contribution of membrane, biomass and foulant layer to virus removal (synthetic wastewater)	*Bioreactor* Volume: 19 L MLSS: 6000 mg/L *Membrane* Pore size: 0.4 μm Submerged hollow fiber	Virus (MS-2): 1.3–3 log removal	Removal increased over time as membrane became fouled. Approximately 0.5 and 0.6 log removal was attributed to retention by the membrane and removal by the biomass, respectively. Approximately 1.3–1.8 log removal attributed to foulant layer
Yoon et al. [33]: Compared performance of MBRs configured for enhanced nutrient removal and	*Bioreactor* Anoxic–anaerobic–aerobic Volume: 15.9 m^3 SRT: 20–60 days	BOD: 0.3–2.1 mg/L Ammonia-N: 91.9% removal (average) Total nitrogen: 3.8–17.1 mg/L Total phosphorous: 0.06–1.4 mg/L	Relatively similar BOD, TN and TP removal efficiencies with MBR and conventional

Table 7 *(Continued)*

Source and objective	MBR system details	MBR effluent quality	Comments
conventional systems with chemical phosphorous removal (municipal wastewater)	*Membrane* Polyethylene Submerged air sparged hollow fiber Pore size: 0.4 μm		activated sludge system followed by coagulation/ flocculation and gravity settling
Zhang and Farahbakhsh [3]: Compared microbial removal efficiency in MBRs can conventional activated sludge systems (municipal wastewaters)	*Bioreactor* MLSS: 10,000 mg/L HRT: 6 h SRT: 18 days *Membrane* Submerged hollow fibers Pore size: 0.04 μm	Coliforms: 5.7 log removal Coliphage: 3.1–5.5 log removal	MBRs can achieve better microbial removal in fewer steps than conventional activated sludge. Effluent from MBRs and conventional activated sludge with tertiary filtration were relatively similar in quality

4.2 Summary of literature published on MBRs for wastewater reuse

The use of MBRs for wastewater reuse applications is still in its infancy and research teams worldwide are focusing their attention on characterizing the performance of MBRs for wastewater reuse and developing approaches to optimize the treatment efficacy. Table 7 summarizes the results from some recent studies on the use of MBRs for wastewater reuse applications.

REFERENCES

[1] Business Communication Company, Membrane Bioreactors in the Changing World Water Market, Wellesley, MA, 2006.
[2] J. Mallevialle,, P.E. Odendaal, M.R. Wiesner (Eds.), Water Treatment Membrane Processes, McGraw Hill, New York, 1996.
[3] K. Zhang, K. Farahbakhsh, Removal of native coliphages and coliform bacteria from municipal wastewaters by various wastewater treatment processes: implications to water reuse, Water Res. 41 (2007) 2816–2824.
[4] G.T. Daigger, J.C. Lozier, G.V. Crawford, Water reuse applications using membrane technology, in: Proc. WEFTEC, Dallas, TX, CD-ROM, 2006.
[5] P. Côté, M. Liu, MBR beats tertiary filtration for indirect water reuse, Desal. Water Reuse 13 and 14 (2004) 32–40.
[6] P. Côté, S. Siverns, S. Monti, Comparison of membrane based solutions for water reclamation and desalination, Desalination 182 (2005) 251–257.
[7] Metcalf and Eddy., Wastewater Engineering, Treatment and Reuse, 4th ed., McGraw Hill, New York, 2003.
[8] T. Stephenson, S. Judd, B. Jefferson, K. Brindle, Membrane Bioreactors for Wastewater Treatment, IWA Publishing, London, 2000.
[9] P. Côté, M. Masini, D. Mourato, Comparison of membrane options for wastewater reuse and reclamation, Desalination 167 (2004) 1–11.
[10] M.A. Gander, B. Jefferson, S.J. Judd, Aerobic MBRs for domestic wastewater treatment: a review with cost considerations, Sep. Purif. Technol. 18 (2000) 119–130.
[11] P. Cornel, M. Wagner, S. Krause, Investigation of oxygen transfer rate in full-scale membrane bioreactors, Water Sci. Technol. 47(11) (2003) 313–319.
[12] M.H. Al-Malack, Performance of an immersed membrane bioreactor (IMBR), Desalination 214 (2007) 112–127.
[13] L. Innocenti, D. Bolzonella, P. Pavan, F. Cecchi, Effect of sludge age on the performance of a membrane bioreactor: influence on nutrient and metals removal, Desalination 146 (2002) 467–474.
[14] J. Wagner, K.H. Rosenwinkel, Sludge production in membrane bioreactors under different conditions, Water Sci. Technol. 41(10 and 11) (2000) 251–258.
[15] A. Monti, E.R. Hall, R.N. Dawson, H. Husain, H. Kelly, Comparative study of biological nutrient removal (BNR) processes with sedimentation and membrane based separation, Biotechnol. Bioeng. 96(4) (2006) 740–752.
[16] T. Melin, B. Jefferson, D. Bixio, C. Thoeye, W. De Wilde, J. De Koning, J. van der Graaf, T. Wintgens, Membrane bioreactor technology for wastewater treatment and reuse, Desalination 187 (2006) 271–282.
[17] B. Lejean, F. Luck (n.d.), Assessment of the membrane bioreactor technology and European market outlook, IDS Water White Paper, http://www.idswater.com/water/europe/Whitepaper_Membrane_Bioreactor/155/paper_information.html

[18] J.E. Bailey, D.F. Ollis, Biochemical Engineering Fundamentals, 2nd ed., McGraw Hill, New York, 1986.
[19] F. Meng, B. Shi, F. Yang, H. Zhang, New insights into membrane fouling in submerged membrane bioreactor based on rheology and hydrodynamics concepts, J. Membr. Sci. 302 (2007) 87–94.
[20] G. Belfort, R.H. Davis, A.L. Zydney, The behaviour of suspensions and macromolecular structures in cross-flow microfiltration, J. Membr. Sci. 96 (1994) 1–58.
[21] M. Cheryan, Ultrafiltration and Microfiltration Handbook, Technomic Publishing, Lancaster, PA, 1986.
[22] C.C.V. Chan, P.R. Bérubé, E.R. Hall, Shear profiles inside gas sparged submerged hollow fiber membrane modules, J. Membr. Sci. 297 (2007) 104–120.
[23] Y.Q. Liu, Y. Liu, J.H. Tay, Relationship between size and mass transfer resistance in aerobic granules, Lett. Appl. Microbiol. 40(5) (2005) 312–315.
[24] I.S. Chang, P. LeClech, B. Jefferson, S. Judd, Membrane fouling in membrane bioreactors for wastewater treatment, J. Environ. Eng. 128 (2002) 1018–1029.
[25] P.R. Bérubé, E.R. Hall, P.M. Sutton, Parameters governing the permeate flux in an anaerobic membrane bioreactor treating low strength municipal wastewater: a literature review, Water Environ. Res. 78 (2006) 887–896.
[26] T. Asano, F.L. Burton, H.L. Leverenz, R. Tsuchihashi, G. Tchobanogous, Water Reuse: Issues, Technologies and Applications, McGraw Hill, New York, 2007.
[27] G. Guglielmi, D. Chiarani, G. Andreottola, G. Ziglio, Membrane bioreactor technology for agricultural reuse of municipal wastewater: a comparative study, in: Proceedings of 2nd IWA National Young Water Professionals Conference, Berlin Germany, CD-ROM, 2007.
[28] S. Oota, T. Murakami, K. Takemura, K. Noto, Evaluation of MBR effluent characteristics for reuse purposes, Water Sci. Technol. 51(6 and 7) (2005) 441–446.
[29] H.M. Wong, C. Shang, G.H. Chen, Membrane bioreactor for wastewater treatment and reuse: virus removal, in: Proceedings of European Symposium on Environmental Biotechnology, Oostende, Belgium, 2004, pp. 133–136.
[30] K.H. Ahn, K.G. Song, I.T. Yeom, K.Y. Park, Performance comparison of direct membrane separation and membrane bioreactor for domestic wastewater treatment and reuse, Water Sci. Technol.: Water Supply 1(5–6) (2001) 315–323.
[31] G. Tao, K. Kekre, Z. Wei, T.C. Lee, B. Viswanath, H. Seah, Membrane bioreactors for water reclamation, Water Sci. Technol. 51(6 and 7) (2005) 431–440.
[32] G.A. Soriano, M. Erb, C. Garel, J.M. Audic, A comparative pilot-scale study of the performance of conventional activated sludge and membrane bioreactor under limiting operating conditions, Water Environ. Res. 75(3) (2003) 225–231.
[33] T.I. Yoon, H.S. Lee, C.G. Kim, Comparison of pilot scale performances between membrane bioreactor and hybrid conventional wastewater treatment systems, J. Membr. Sci. 242 (2004) 5–12.
[34] S.R. Chae, S.T. Kang, S.M. Lee, E.S. Lee, S.E. Oh, Y. Watanabe, H.S. Shin, High reuse potential of effluent from an innovative vertical submerged membrane bioreactor treating municipal wastewater, Desalination 2002 (2007) 83–89.
[35] J. deCarolis, S. Adham, J. Oppenheimer, K. Kinser, L. Webb, Evaluation of membrane bioreactor technology and desalting membranes for wastewater reuse, in: Proc. WEFTEC, Washington, DC, 2005, pp. 4134–4146.
[36] A.J. Spring, D.M. Bagley, R.C. Andrews, S. Lemanik, P. Yang, Removal of endocrine disrupting compounds using a membrane bioreactor and disinfection, J. Environ. Eng. Sci. 6 (2007) 131–137.
[37] S.A. Snyder, S. Adhan, A.M. Redding, F.S. Cannon, J. DeCarolis, J. Oppenheimer, E.C. Wert, Y. Yoon, Role of membranes and activated carbon in the removal of endocrine disruptors and pharmaceuticals, Desalination 202 (2007) 156–181.

[38] F. Fatone, P. Battistoni, F. Cecchi, Application of a membrane bioreactor for the treatment of low loaded domestic wastewater for water re-use, Water Sci. Technol. 53(9) (2006) 111–121.
[39] B. Lesjean, R. Gnirss, H. Buisson, S. Keller, A. Tazi-Pain, F. Luck. Outcomes of a 2 year investigation on enhanced biological nutrient removal and trace organics elimination in membrane bioreactors (MBR), Water Sci. Technol. 52(10–11) (2005) 453–460.
[40] L.S. Tam, T.W. Tang, G.N. Lau, K.R. Sharma, G.H. Chen, A pilot study for wastewater reclamation and reuse with MBR/RO and MF/RO systems, Desalination 202 (2007) 106–113.
[41] E.G. Helmig, J.D. Fettig, L. Cordone, API removal for pharmaceutical manufacturing wastewater – results from process development, pilot testing and scale-up, in: Proc. WEFTEC, Washington, DC, 2005, pp. 207–226.
[42] B. Mansell, J. Kuo, C.C. Tang, S.J. Huitric, W.H. Horvath, J.F. Stahl, Evaluation of removals of NDMA, hormones, pharmaceuticals, and personal care products in a membrane bioreactor, in: Proc. WEFTEC, Washington, DC, 2005, pp. 7411–7427.
[43] J. Oppenheimer, J. DeCarolis, S. Adham, S. Lagos, N. Tran, S. Snyder, G. Lorraine, Evaluation of membrane bioreactor and reverse osmosis for removal of endocrine disrupting compounds, pharmaceuticals and personal care products, in: Proc. WEFTEC, New Orleans, LA, CD-ROM, 2004.
[44] J.E. Drews, J. Hemming, S. Ladenburger, J. Schauer, W. Sonzogni, Endocrine disrupting activity changes in water reclamation systems, in: Proc. WEFTEC, Washington, DC, 2005, pp. 3196–3212.
[45] T. Wintgens, M. Gallenkemper, T. Melin, Removal of endocrine disrupting compounds with membrane processes in wastewater treatment and reuse, Water Sci. Technol. 50(5) (2004) 1–8.
[46] N. Jang, X. Ren, G. Kim, C. Ahn, J. Cho, I.S. Kim, Characteristics of soluble microbial products and extracellular polymeric substances in the membrane bioreactor for water reuse, Desalination 202 (2007) 90–98.
[47] A.M. Comerton, B.C. Andrews, D.M. Bagley, Evaluation of an MBR-RO system to produce high quality reuse water: microbial control, DBP formation and nitrate, Water Res. 39 (2005) 3982–3990.
[48] T. Ueda, N.J. Horan, Fate of indigenous bacteriophage in a membrane bioreactor, Water Res. 34(7) (2000) 2151–2159.
[49] S. Dukes, A. von Gottburg, Membrane bioreactors for RO pretreatment, in: Proc. WEFTEC, Dallas, TX, 2006, pp. 6121–6132.
[50] P. Côté, H. Buisson, C. Pound, G. Arakaki, Immersed membrane activated sludge for the reuse of municipal wastewater, Desalination 113 (1997) 189–196.
[51] A.E. Childress, P. LeClech, J.L. Daugherty, C. Chen, G.L. Leslie, Mechanical analysis of hollow fiber membrane integrity in water reuse applications, Desalination 180 (2005) 5–14.
[52] B. Jefferson, P. LeClech, S. Smith, A. Laine, S. Judd, The influence of membrane configuration on the efficacy of membrane bioreactors for domestic wastewater recycling, in: Proc. WEFTEC, Anaheim, CA, CD-ROM, 2000a.
[53] B. Jefferson, A.L. Laine, S.J. Judd, T. Stephenson, Membrane bioreactors and their role in wastewater reuse, Water Sci. Technol. 41(1) (2000) 197–205.
[54] M. Kumar, J. DeCarolis, S. Adham, Integrated membrane systems for wastewater reuse, in: Proc. WEFTEC, New Orleans, LA, 2004, pp. 1561–1585.
[55] Y.Z. Li, Y.L. He, D.G. Ohandja, J. Ji, J.F. Li, T. Zhou, Simultaneous nitrification-denitrification achieved by an innovative internal-loop airlift MBR: comparative study. Biores. Technol. 99 (2008) 5867–5872.

[56] Y. Liu, L. Wang, B. Wang, H. Cui, J. Zhang, Performance improvement of hybrid membrane bioreactor with PAC for water reuse, Water Sci. Technol. 53(10 and 11) (2005) 338–391.
[57] J. Lozier, A. Fernandez, Using a membrane bioreactor/reverse osmosis system for indirect potable reuse, Water Sci. Technol.: Water Supply 1(5) (2001) 303–313.
[58] H. Nagaoka, S. Ueda, A. Miya, Influence of bacterial extracellular polymers on the membrane separation activated sludge process, Water Sci. Technol. 34(9) (1996) 165–172.
[59] K.H. Krauth, K.F. Staab, Pressureized biomembrane reactor for wastewater treatment, Hydrotop 94 (1994) 555–562.
[60] H. Lin, Modeling Fouling in Membrane Systems, M.A.Sc. Thesis, Department of Civil Engineering, The University of British Columbia, 2007.
[61] T. Sato, Y. Ishii, Effects of activated sludge properties on water flux of ultrafiltration membrane used for human excrement treatment, Water Sci. Technol. 23 (1991) 1601–1608.
[62] K. Takemura, N. Ohkuma, M. Ohnishi, H. Andoh, Y. Okuno, Membrane separation activated sludge process for sewage treatment, Proc. Membrane Technology in Environmental Management, Tokyo, Japan, 1999, pp. 440–443.

PART *Four*

Concentrate Disposal Options

CHAPTER 10

Concentrate Treatment for Inland Desalting

Christopher J. Gabelich[1,*], **Pei Xu**[2] *and* **Yoram Cohen**[3]
[1]Palos Verdes, CA 90275, USA
[2]Colorado School of Mines, Golden, CO 80401-1887, USA
[3]University of California, Los Angeles, CA 90095-1592, USA

Contents

1. INTRODUCTION

Inland desalination of brackish water can provide an important source of potable water in many parts of the world. Reverse osmosis (RO) and nanofiltration (NF) have become the technology of choice for many of these applications [1]. However, large-scale deployment of RO/NF in inland locations would necessitate operation at relatively high product water recovery to maximize water resource utilization and minimize treatment costs and environmental challenges associated with disposal of the concentrate. Concentrate disposal is generally limited to one or two options for any given area and is directly related to land cost, energy costs, regulations, and the type and quantity of salts in the concentrate stream [2,3]. Inland concentrate disposal options include reuse of concentrates,

*Corresponding author.
E-mail address: c2gabelich@cox.net

Sustainability Science and Engineering, Volume 2 ISSN 1871-2711, DOI 10.1016/S1871-2711(09)00210-4

surface water discharge, sewer disposal, deep well injection, land applications, and evaporation ponds (followed by land filling of solids) [2]. This chapter provides an overview of post-RO/NF concentrate minimization technologies with the goal of minimizing the concentrate volume such that ultimate disposal options are more feasible. Examples of several of these technologies are also provided.

2. BACKGROUND

Inland brackish water sources often contain precursors of sparingly soluble mineral salts, which upon concentration of the reverse osmosis (RO) retentate as a consequence of the desalting process can result in mineral scaling of RO membranes. Membrane scaling impairs plant productivity (i.e., permeate flux) and product water quality. Therefore, the development of cost-effective scale-control strategies is imperative, including the optimization of conventional techniques of membrane scaling mitigation (e.g., feed pH adjustment and antiscalant dosing).

As water is passed through successive RO modules, mineral salt scale precursor ions (e.g., Ba^{2+}, Ca^{2+}, SO_4^{2-}, CO_3^{2-}, etc.) are concentrated in the membrane retentate side (i.e., feed-side) to levels that can exceed the solubility limit of various sparingly water-soluble mineral scalants such as calcium sulfate dihydrate (gypsum), calcium carbonate (e.g., calcite), and barium sulfate (barite), in addition to metal hydroxides and silica [4]. Under such conditions, sparingly soluble mineral salts may crystallize directly onto the membrane surface or form in the bulk and subsequently deposit on the membrane [5,6] forming mineral surface scales. This leads to water-permeate flux decline and potential damage to the membrane, thereby shortening its useful lifetime.

One simple solution to alleviating the mineral scaling problem is to operate at low water recoveries (i.e., lower the fraction of product water produced per volume of feedwater) in order to reduce concentration of mineral salts at the membrane surface (i.e., also known as concentration polarization). However, this approach that limits the recovery leads to increased volume of generated concentrate and introduces a serious concentrate management dilemma, which can result in significant increased process cost or even elimination of membrane desalting as a water desalination option.

In order to maximize water recovery, a majority of RO facilities use pH adjustment and antiscalants to control for these rate-limiting salts [7]. Both of these treatment techniques essentially shift the salt solubility such that scaling does not occur within the RO unit. However, even with the above processes, water recovery is limited (~50–85%) with a significant concentration volume that requires further costly processing or disposal. For many inland locations, cost-effective disposal methods are limited or unavailable. Therefore, there is a need to minimize the volume of the concentrate stream in order to reduce the concentrate management challenge.

Concentrate minimization often requires integration of one or more treatment units in combination with RO or other desalination processes. The technologies fall into four basic categories: (1) cation control, (2) anion control, (3) physical separation of anions and cations, and (4) thermal processes. Cation and anion control includes intermediate chemical demineralization (ICD) [8,9] and biological sulfate control [10], respectively. Electrodialysis (ED) and electrodialysis reversal (EDR) [11], along with capacitive deionization (CDI) [12,13], have been proposed as charge separation technologies. Examples of thermal processes are vapor compression (VC) [14], membrane distillation (MD) [15], and freeze desalination [16]. Several newer technologies such as forward osmosis (FO) [17,18], dewvaporation [19], and vibratory shear–enhanced process (VSEP) [20] have been proposed as methods to treat RO concentrate due to their potentially lower fouling propensity and/or energy usage. However, each of the aforementioned technologies is in various stages of development. While ICD has undergone pilot and demonstration testing [5,21], most of the others [e.g., biological sulfate reduction (BSR), FO, MD, and CDI] have not progressed past the bench level. These technologies could be the first step toward achieving a zero-liquid discharge (ZLD) or near-ZLD facility. Each of these processes will be discussed briefly in the following sections.

3. CONCENTRATE PRODUCTION

The limitation to high water recovery RO desalting (and thus RO concentrate minimization) can be illustrated by considering the factor (CF) by which the reject (i.e., retentate) stream is concentrated, relative to the feed [6,22,23]:

$$\mathrm{CF} = \frac{C_\mathrm{C}}{C_\mathrm{F}} = \left(\frac{1}{1 - R_\mathrm{W}}\right)[1 - R_\mathrm{W}(1 - R_\mathrm{S})] \tag{1}$$

where C_C and C_F are the retentate and feed concentrations, respectively, R_S is the nominal salt rejection ($R_\mathrm{S} = 1 - C_\mathrm{P}/C_\mathrm{F}$, where C_P is the permeate concentration), and R_W is the fractional product water recovery ($R_W = Q_P/Q_F$, where Q_P and Q_F are the permeate and feed volumetric flow rates, respectively). CF rises rapidly as product water recovery increases above about 80%, as illustrated in Fig. 1 for a salt rejection level of 97%. For example, for desalting at recovery levels of 90% and 98% (i.e., concentrate stream is 10% and 2% of the feed-flow rate, respectively) the retentate stream will be concentrated by about a factor of 10 and 48, respectively.

Evaluating the solubility limits of scalants of concern is vitally important when selecting a concentrate minimization technology. For example, for Colorado River water (CRW) desalting at 85% water recovery, the primary scalants of concern are $BaSO_4$ and $CaCO_3$ (Fig. 2). As the recovery is increased, $CaSO_4$, and silica (SiO_2) play increasing important roles in limiting product water recovery. Thermodynamic solubility calculations for the concentrate derived from 85% water recovery desalting of CRW water suggest that pH adjustment can be used to mitigate scale formation of some potential scalants (e.g., $CaCO_3$, $Mg(OH)_2$, and SiO_2), though at differing pH ranges [9]. However, it is apparent that the saturation indices of $CaSO_4$ and $BaSO_4$ are relatively independent of pH. Therefore, further desalting of

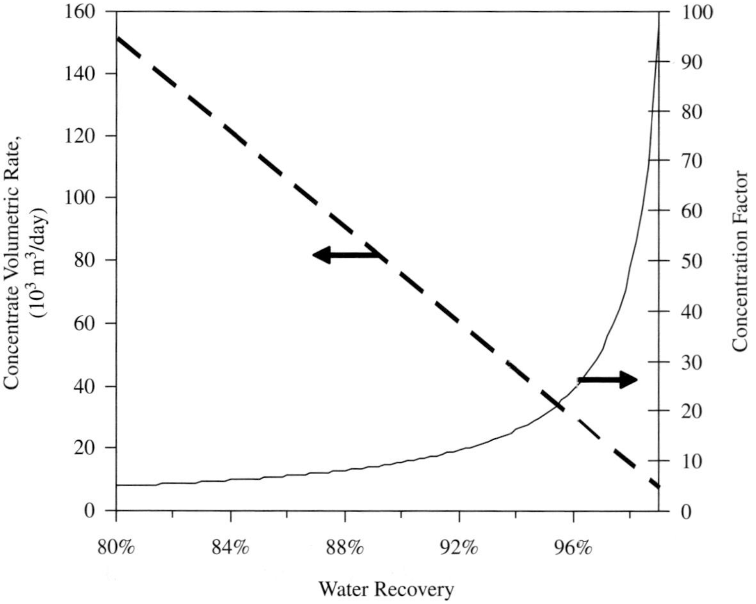

Figure 1 Variation of concentrate volumetric flow rate and concentration factor ($C_{concentrate}/C_{feed}$) with product water recovery for a 760×10^3 m^3/day membrane desalination plant with nominal salt rejection of 97%.

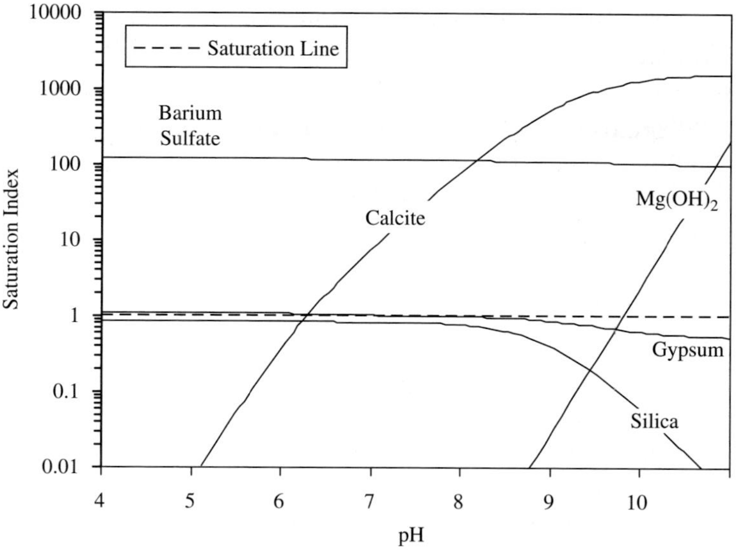

Figure 2 Saturation index of $CaCO_3$, $CaSO_4$, SiO_2, and $Mg(OH)_2$ for Colorado River water concentrate produced from RO desalting (85% water recovery at 97% salt rejection) [9].

the above CRW RO concentrate, even at low pH would be a challenge since the solution is nearly saturated with respect to $CaSO_4$ and SiO_2 and oversaturated with respect to $BaSO_4$.

To achieve the goal of 95%, total system water recovery for CRW desalting, scaling thresholds for $CaCO_3$, $BaSO_4$, and $CaSO_4$ must be overcome [9]. At the above conditions, desalting of even the relatively low-salinity CRW would result in the saturation indices, reaching values of 124, 141, 1.2, 1.2, and 1.0 at CF = 10, and 611, 1078, 5.7, 9.5, and 4.3 at CF = 48, for $BaSO_4$, $CaCO_3$, $CaSO_4$, $SrSO_4$, and SiO_2, respectively. Saturation index are defined as,

$$SI_x = \frac{IAP}{K_{sp,x}} \tag{2}$$

where IAP is the ion activity product and $K_{sp,x}$ is the solubility product for the mineral salt x. Clearly, the RO concentrate from 90% and 98% recovery desalination will either be saturated ($SI_x = 1$) or oversaturated ($SI_x > 1$) with respect to the above minerals.

It is important to recognize that in cross-flow membrane desalting, as water permeates across an RO membrane, rejected salt ions accumulate near the membrane surface, resulting in the formation of a concentration boundary layer. As the concentration and osmotic pressure at the membrane surface gradually increase (from the RO channel entrance to the exit), the effective net driving force for permeation decreases, thus, the permeate flux decreases toward the exit region (Fig. 3). The concentration of salts at the membrane surface can be approximated using the simple film model [5]:

$$\mathrm{CP} = \frac{C_\mathrm{m}}{C_\mathrm{b}} = (1 - R_\mathrm{o}) + R_\mathrm{o} \exp\left(\frac{J}{k}\right) \tag{3}$$

where C_m, C_b, and C_p are the solute concentrations at the membrane surface, in the bulk, and in the permeate, respectively, J is the permeate flux, and k is the solute feed-side mass transfer coefficient, R_o is the observed rejection ($R_\mathrm{o} = 1 - C_\mathrm{p}/C_\mathrm{b}$), and CP is the concentration polarization modulus. CP increases along the RO membrane channel, reaching its highest value at the channel exit and thus the potential for mineral scaling correspondingly increases along the membrane surface toward the exit region of the membrane flow channel. The challenge in RO desalting is thus to enable one to reach sufficiently high recovery, while reducing mineral scaling propensity.

The following sections discuss the potential of various technologies in treating RO concentrate to minimize reject volume. Depending on the water quality, treatment goals, and site-specific characteristics, one or more

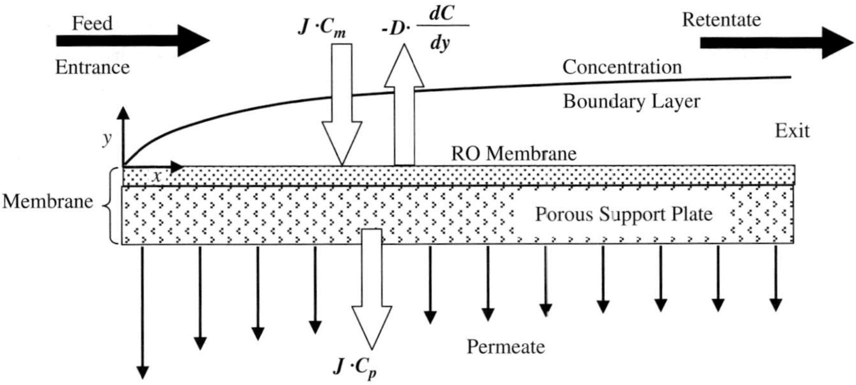

Figure 3 Schematic of cross-flow plate-and-frame RO system showing the formation of a concentration boundary layer. J is the water flux, C_m and C_p are the concentrations at the membrane surface and in the permeate, respectively. D is the solute diffusivity, and $\mathrm{d}C/\mathrm{d}y$ is the solute concentration gradient in the y-direction.

of these technologies may be applicable to any given scenario. While examples of several technologies are provided, the purpose of this section is not to provide an endorsement of any one technology.

4. RO CONCENTRATE TREATMENT TECHNOLOGIES

4.1 Cation control through intermediate chemical demineralization (ICD)

Fig. 4 provides a conceptual schematic of a two-pass RO facility with integrated ICD [21]. The term "chemical demineralizaton" is a general term for a variety of technologies that have been proposed whereby precursor scalant ions are removed from the primary RO concentrate via chemical precipitation. A number of scoping studies using a variety of conceptual process schemes to achieve high-recovery RO desalting via ICD have been conducted [24–36]. More recently, Gabelich et al. [5,21] demonstrated that upwards of 95% total system water recovery was possible for CRW RO desalting using ICD at the pilot and demonstration scales. ICD was shown to be effective in reducing the concentrations of Ca^{2+} and other scaling precursors in the primary RO concentrate below saturation or to a metastable supersaturation range (i.e., very slow precipitation kinetics) so as to allow further RO desalting of this concentrate stream.

Chemical demineralization processes rely on $CaCO_3$ precipitation (appropriate where calcium carbonate concentration is sufficiently high to enable calcite precipitation by pH adjustment) using alkaline solutions of caustic (NaOH), lime ($Ca(OH)_2$), sodium carbonate (Na_2CO_3), or through

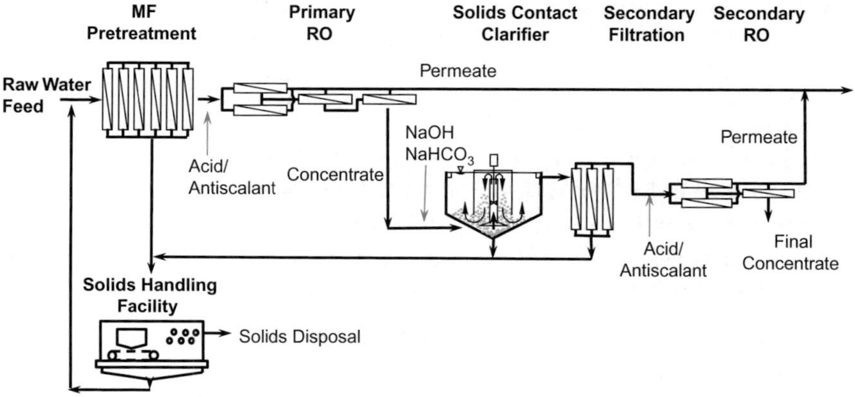

Figure 4 Conceptual schematic drawing for two-pass RO facility with integrated intermediate chemical demineralization.

combinations of each. The choice of alkaline reagent is dependent on the concentrate composition and chemical cost. For example, for CRW RO concentrate, the molar ratio of total carbonate to total calcium was approximately 1.1 [8]. Therefore, sufficient carbonate ions were available to precipitate >99% of the calcium ions as $CaCO_3$. Thus, for this particular water chemistry, NaOH dosing required the minimal dosing (on a molar basis) to induce the precipitation of $CaCO_3$ [9]. In contrast, $Ca(OH)_2$ dosing would have required the addition of Na_2CO_3 to deplete the additionally introduced calcium.

Example

Demonstration-scale studies were conducted by the Metropolitan Water District of Southern California at the U.S. Bureau of Reclamation's Water Quality Improvement Center in Yuma, Arizona [21]. These demonstration-scale studies were designed to provide process design criteria for a two-pass RO process. The demonstration studies focused on two portions of the two-stage RO process (Fig. 4), namely (1) maintaining proper pH control of the solids contact reactor (SCR) effluent and secondary RO units and (2) determining whether a 95% total system water recovery was possible at a near-production plant size. Optimizing operations of the primary RO and SCR was addressed in prior investigations [8,37] and these unit operations were run under a single, optimized operating condition for the duration of pilot testing.

For the duration of testing, the source water was CRW taken just prior to the Northern International Boundary Dam in Yuma, Arizona. The salinity of the Colorado River ranged from 700 to 900 mg/L. The primary RO concentrate stream (250 L/min) was split with approximately 190 L/min being sent to the SCR, and the remaining used for EDR testing (see Section 3.3.1). Chemical feed to the SCR included NaOH for pH control and $NaHCO_3$ to increase alkalinity. Sulfuric acid was added to the SCR effluent to adjust the pH to 7.0. SCR effluent was processed through a microfiltration (MF) skid prior to being sent to a 20–22 L/min secondary RO unit. Antiscalant and H_2SO_4 (pH 7.0) were added to prevent membrane scaling in the secondary RO units.

Operating pH and Ca^{2+}, Ba^{2+}, and Sr^{2+} removal from the SCR are shown in Fig. 5. Calcium is removed from solution as $CaCO_3$ solid and is the principal precipitate generated in the SCR. Over the range pH levels tested, Ca^{2+} removal varied from 54% to 83%. Removal of Ba^{2+} and Sr^{2+} exhibited similar pH dependence to Ca^{2+}, with Ba^{2+} removal being slightly higher than that of Ca^{2+} and Sr^{2+} slightly lesser. scanning electron microscopy (SEM) and electron dispersive spectroscopy (EDS) data of the SCR precipitate indicated that removal of Ba^{2+} and Sr^{2+} precipitated as $BaSO_4$ and $SrSO_4$, respectively, through either inclusion in the $CaCO_3$ crystalline lattice or through adsorption on the freshly precipitated $CaCO_3$ crystalline surface. Given the percentage of cation removal and MF effluent water quality, the performance of the secondary RO unit showed no signs of fouling over the 600 h of operation (Fig. 5, bottom), hence validating the ability of ICD to enhance RO recovery at practical scales.

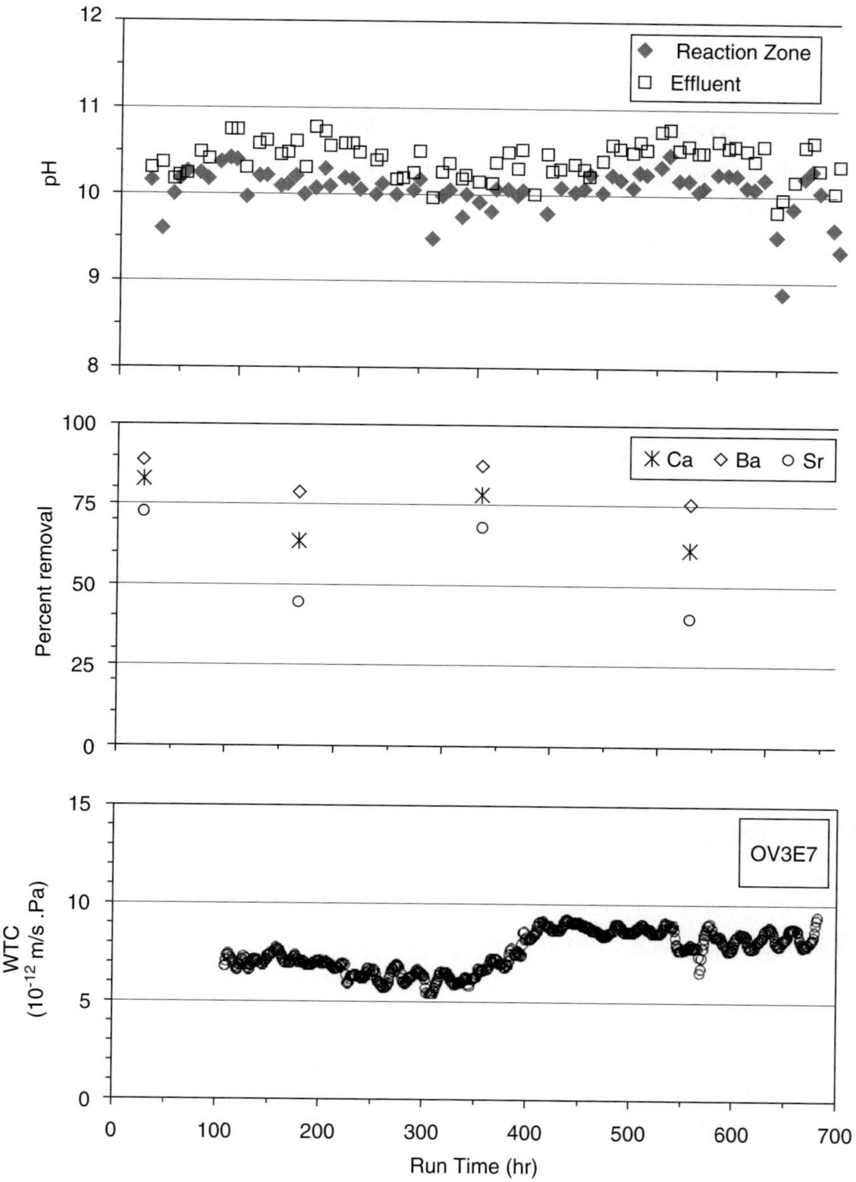

Figure 5 Intermediate chemical demineralization process data: (top) SCR pH data over time; (middle) percent cation removal; and (bottom) water transport coefficient for the terminal RO element.

4.2 Anion control through biological sulfate reduction

A unique method for controlling sulfate-based scalants is through the use of microorganisms to reduce sulfate (SO_4^{2-}) to sulfide (S^{2-}). This reduction process is known as BSR. The resulting S^{2-} can then be reoxidized to elemental sulfur (S_s) via another unit process, resulting in a solid end-product. The reduction of SO_4^{2-} via BSR would effectively lower the potential for the precipitation of $BaSO_4$, $CaSO_4$, and $SrSO_4$, thereby reducing the scaling potential of the primary RO concentrate for further treatment by another unit process. BSR has been used in the industrial setting for a number of years [38–43]. The general reactions for sulfate reduction are as follows [44]:

$$SO_4^{2-} + \text{Organic carbon} \Rightarrow S^{2-} + CO_2 + H_2O \quad (4)$$

$$SO_4^{2-} + 6H_2 + CO_2 \Rightarrow S^{2-} + \text{Biomass} + 6H_2O \quad (5)$$

From Eqs. (4) and (5), the basic process variables of the BSR process are (1) finding a suitable electron donor and carbon source, whether it be organic carbon or hydrogen (H_2) and carbon dioxide (CO_2); (2) selecting the appropriate biological community to reduce SO_4^{2-}; and (3) devising a means by which S^{2-} is rendered harmless, both in terms of inhibiting the biological reduction of sulfate, as well as to the environment in general.

4.2.1 Electron donor and carbon source

Low-molecular-weight organic compounds, such as acetate, propionate, ethanol, glucose, glycerol, malate, lactate, and sucrose, as well as H_2 and CO_2 are known electron donors and carbon sources for SO_4^{2-} reduction [37,41]. Molasses has also been shown to be effective carbon source in SO_4^{2-} reduction processes [45]. The use of appropriate carbon source is an important consideration owing to several reasons: cost, speed of assimilation, and by-product formation. For instance, lactate is assimilated more rapidly by SO_4^{2-}-reducing bacteria than acetate or ethanol, and might therefore be used initially as a carbon source to reduce the induction times for reactor operation [37,41]. However, lactate could be more expensive than ethanol or acetate, and SO_4^{2-}-reducing bacteria utilizing lactate would yield acetate as major by-product and propionate as a minor by-product in the treated effluent [46,47]. Ethanol would also yield acetate as a reaction by-product. These organic products would be poorly rejected to the RO membranes and may potentially cause organic and biological fouling. Hence, the use of lactate or ethanol may entail additional costs for acetate removal in the finished water. However, the use of H_2 and CO_2 for supporting the SO_4^{2-}-reducing bacteria might be initially expensive, but could prove advantageous in view of the fact that no organic by-products or residual will be produced requiring additional aerobic treatment [37].

4.2.2 Sulfate-reducing bacteria

The SO_4^{2-}-reducing bacteria represent a specialized group of microorganisms that use SO_4^{2-} as terminal electron acceptor for their respiration, although many species of microorganisms are capable of generating hydrogen sulfide (H_2S) metabolically, SO_4^{2-} is the primary source. In a review on SO_4^{2-}-reducing bacteria, Madigan [48] has listed 10 genera of dissimilatory organisms, divided into two broad physiological subgroups. The genera in the first group include *Desulfovibrio*, *Desulfomonas*, *Desulfotomaculum*, and *Desulfobulbus*; these bacteria utilize lactate, pyruvate, ethanol, and certain fatty acids as carbon sources, and in turn reduce SO_4^{2-} to S^{2-}. The genera in the second group include *Desulfobacter*, *Desulfococcus*, *Desulfosarcina*, and *Desulfonema*, and these bacteria specialize in the biochemical oxidation of fatty acids, especially acetate, during the concomitant reduction of SO_4^{2-} to S^{2-}. All the 10 genera are obligate anerobes, which strictly require an anerobic environment for their growth and cultivation.

4.2.3 Hydrogen sulfide control

Several researchers reported that the effects of toxicity diminished process performance due to increased levels of S^{2-} and H_2S [49–51]. It must be noted that H_2S acts as an inhibitor for the SO_4^{2-} reduction process at elevated concentrations of 16 mM or 544 ppm H_2S [50], but fluidization of the reactor bed may help in preventing the concentration from reaching inhibitory limits. These high free-H_2S concentrations caused reversible inhibition rather than acute toxicity [37]. The elimination of H_2S from BSR processes is an important consideration.

Several control methods for reducing effluent H_2S concentrations include: off-gas scrubbing, precipitation of sulfide by magnesium addition, two-phase biological processes, and enhanced partitioning of H_2S into the gas phase at thermophilic temperatures [52]. In addition to reducing aqueous phase H_2S concentrations by gas stripping, low oxygen concentrations can create selective conditions for SO_4^{2-} reduction to S_s, and less frequently to thiosulfate. Another control method would be the careful regulation of pH within the BSR process [53]. Because total sulfide exists as H_2S, HS^-, and S^{2-}, the dissolution of H_2S in water forms the following equilibrium system [45]:

$$H_2S \Rightarrow H^+ + HS^- \Rightarrow 2H^+ + S^{2-} \quad (6)$$

The chemical equilibrium of these species is pH-dependent. At pH 8, most of the total sulfides are in HS^- form, whereas at pH 6, most are in H_2S form. Low pH (<5–6) often inhibits SO_4^{2-} reduction activity and increases the solubility of metal sulfides already formed.

Example

BSR was evaluated at the bench-scale in terms of reducing SO_4^{2-} concentrations of synthetic RO concentrate when using acetate, ethanol, and H_2 and CO_2 as the electron donor and carbon sources [10]. Fig. 6 shows the influent and effluent SO_4^{2-} concentrations and percentage of SO_4^{2-} removal from fluidized bed reactors used during this study. Each of the BSR fluidized bed reactors underwent a conditioning period whereby the sulfate-reducing bacteria were exposed to ever increasing concentrations of sulfate (data not shown for acetate- and ethanol-fed reactors). In general terms, the SO_4^{2-}-reducing bacteria acclimatized faster to the high SO_4^{2-} feed when using H_2 and CO_2, followed by acetate and then ethanol. Maximum observed SO_4^{2-} removal rates were 93%, 90%, and 96% for acetate, ethanol, and H_2 and CO_2, respectively. For the acetate- and ethanol-fed reactors, periods of suboptimal performance resulted from either the operational pH falling below pH 7.5 or maintaining less than 1.0 C/S ratios. Problems associated with insufficient H_2 gas transfer and biological clogging of the media lead to an extended period of poor performance (60% SO_4^{2-} removal) for the remainder of the H_2/CO_2 test. It is important to note that when using H_2 and CO_2, mass transfer and solubility limitations of H_2 are important operational considerations. However, reactor performance was recoverable upon reverting to their respective operational set points. Lastly, acetate and ethanol consumption for each column ranged from 82% to 94% and 90% to 100%, respectively. These high levels of organic residuals would be poorly rejected by downstream RO processes and may lead to premature organic and biological fouling. However, the operational results using H_2 and CO_2 were encouraging from two standpoints: namely, achieving high process efficiency for SO_4^{2-} reduction and maintaining low organic residuals in the treated effluent.

Table 1 shows saturation indices of BSR effluent for $BaSO_4$ and $CaSO_4$ calculated using a thermodynamic solubility model (LabAnalyzer 2.0 software, OLI Systems, Morris Plains, NJ) and water quality data from the previous section's CRW primary RO concentrate (Section 3.2). The data show that given the levels of SO_4^{2-} removal achieved (greater than 90%) for any of the electron donor sources, the saturation ratios for $BaSO_4$ and $CaSO_4$ are either well within the ability of an antiscalant (less than 100) to retard $BaSO_4$ precipitation or less than unity for $CaSO_4$ scaling. Therefore, while only considering $BaSO_4$ and $CaSO_4$ scaling, BSR may be a viable technology not only to achieve high RO recoveries, but also to achieve a solid S_S end-product.

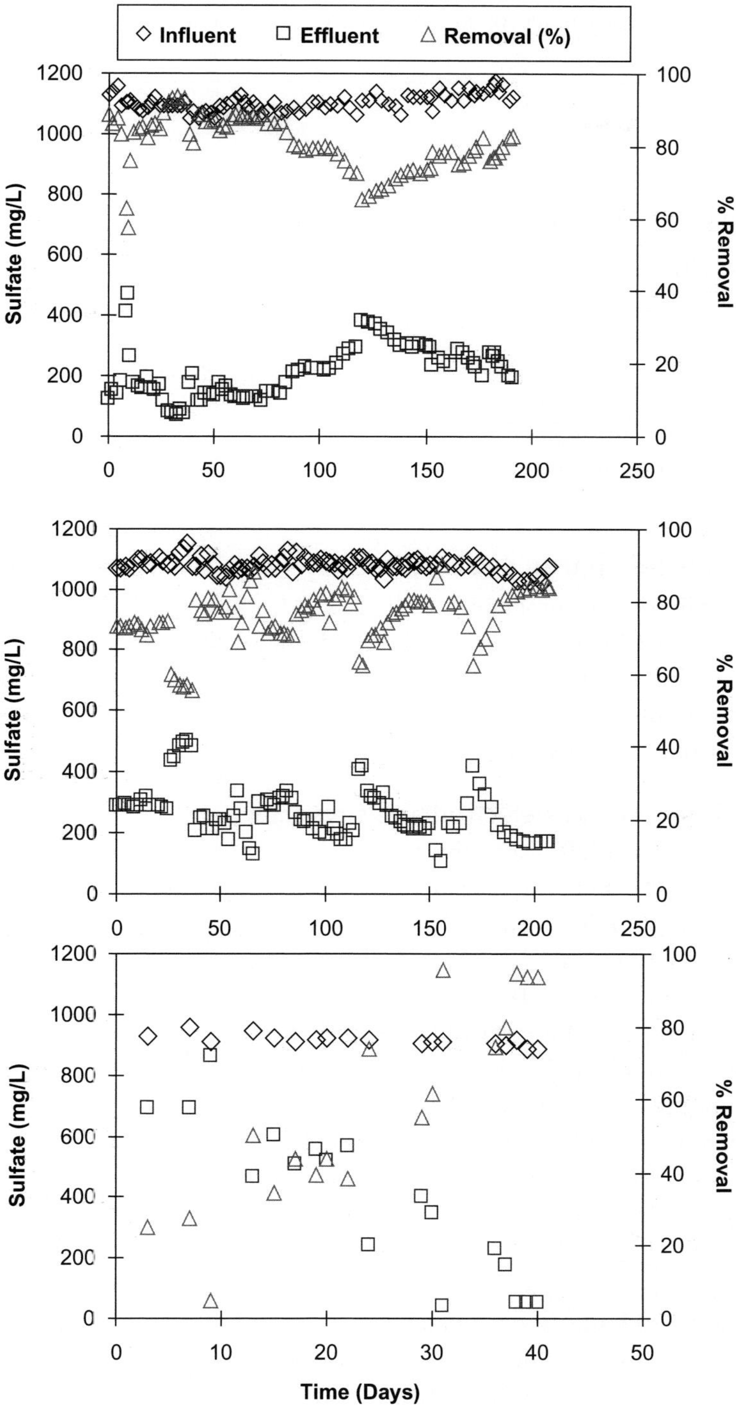

Figure 6 Influent and effluent sulfate concentration profiles and removal efficiencies for acetate (top), ethanol (middle), and H_2 and CO_2 (bottom). Periods for SO_4^{2-}-reducing bacteria to acclimatize for acetate and ethanol reactors not shown.

Table 1 Theoretical reduction in $BaSO_4$ and $CaSO_4$ saturation ratios after biological sulfate reduction

	Saturation indices	
	$BaSO_4$	$CaSO_4$
Primary RO concentrate	101	0.9
% Sulfate reduction	Secondary RO concentrate at 70% water recovery	
70	157	1.4
80	36	0.31
90	3.7	0.03

4.3 Physical charge separation

Electrically driven processes, such as ED/EDR and CDI, have been proposed to treat concentrate [11–14]. Both of these processes work by imposing an electric field that creates a physical separation between the cation and anions, thereby preventing scale-prone ion pairs from forming. Both of these processes will be discussed in general terms.

4.3.1 Electrodialysis/electrodialysis reversal

Both ED and EDR have the potential to operate at very high water recoveries. In the late 1980s, EDR was demonstrated for desalting 5000 mg/L total dissolved solids (TDS) groundwater at 94% water recovery [54]. EDR has also been applied to reclaim 8000 mg/L TDS RO concentrate to achieve RO-EDR recoveries of up to 96% [55]. However, RO is often preferable over ED and EDR, based not only on the rejection of both ionizable and nonionizable components, but also from an energy usage perspective. However, in highly sulfonated waters such as RO concentrate, ED/EDR may offer a decided advantage over RO through its ability to achieve high water recoveries without incurring scaling.

Fig. 6 shows a conventional ED configuration whereby alternating cationic and anionic transfer membranes selectively remove charged, soluble ionic constituents in the presence of an electric field [56]. Traditional ED and EDR membranes are manufactured by mixing either a cation exchange or anion exchange resin with a polymer. The membranes allow for the passage of like-charged ions, while blocking the passage of water and oppositely charged ions. Several research teams are working on improving ED/EDR membranes through development of low-resistivity nanoporous and block-polymer ED membranes that operate at much lower energy consumption than traditional ED membranes ([57]; http://www.sandia.gov/water/projects/desal2.htm). The philosophy behind these new membranes was that it would take less energy to push ions through nanoengineered materials than through a solid functionalized polymer – the standard membrane material used for conventional ED.

Moreover, by functionalizing the membranes with chemical groups having high affinities for targeted species (e.g., a nitrate binding group for nitrate), it may be possible to generate an ion-selective membrane [58,59]. Preliminary results by Lawrence Livermore National Laboratory using nanocomposite membranes indicate that significantly smaller voltage is needed to drive the same ion flux through the nanoporous membranes, indicating higher energy efficiency, although the permselectivity is not as great as with the commercial membranes [60]. A technology with lower overall energy use coupled with ion selectivity would greatly lower the costs of treating contaminated water supplies.

Example

EDR experiments were conducted using a 2.0 gpm EDR unit operating in both batch- and continuous-operation modes. The EDR unit contained 2 electrical stages, 4 hydraulic stages, and 120 cell pairs. For batch-mode tests, approximately 100 gal of primary RO concentrate and primary RO concentrate post-ICD and MF (see Section 3.1) were processed through the EDR unit with the concentrate being discarded and the product serving as the feed for the next pass. For continuous-mode testing, primary RO concentrate was run through the EDR unit at 50% water recovery. The purpose of these tests were to evaluate (1) would natural organic matter (NOM) and antiscalant carryover from the primary RO negatively impact EDR membrane performance and (2) could 95% total system water recovery be achieved using RO-EDR?

Fig. 7 provides the conductivity data over time for both the (a) continuous-mode (b) and batch-mode EDR experiments. Fig. 7a demonstrates that EDR membranes exposed to unaltered RO concentrate showed no adverse fouling after 8.5 days of operation. Therefore, in this instance, the concentrated NOM and antiscalant in the RO concentrate did not lead to immediate fouling, though longer term testing may prove otherwise. Fig. 7b and Table 2 show that upon successive treatment of the concentrate, the final salinity can be tailored for an intended end use. However, nonionizable species, such as SiO_2 and NOM, are retained in the diluate, or product water, and a polishing step may be needed to use the water for potable purposes. Research by Sethi et al. [25], using the same CRW primary RO concentrate and a bench-scale ED apparatus, confirmed that up to 80% recovery of the concentrate was possible, though further processing of the diluate was needed prior to the water being put back into production.

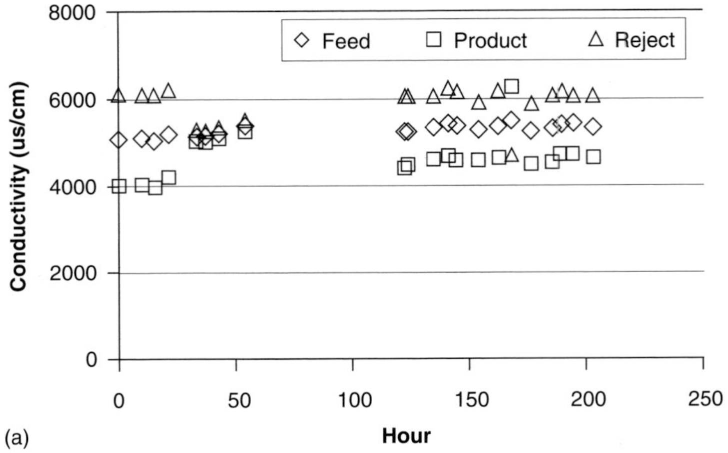

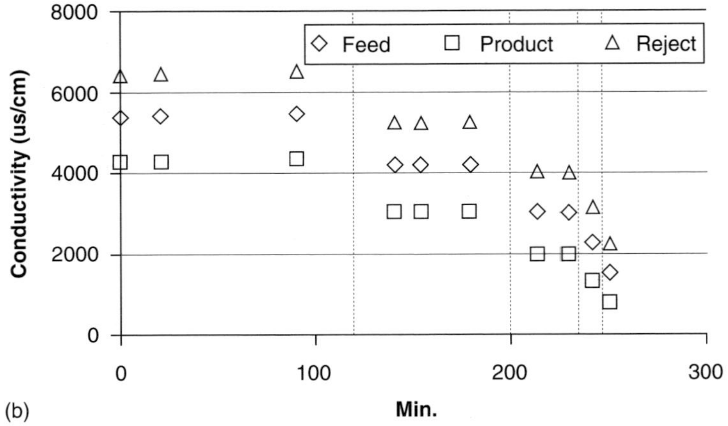

Figure 7 EDR treatment conductivity data of primary RO concentrate run at (a) steady-state (continuous-mode) and (b) in batch-mode (b). Dashed lines indicate separate passes though EDR unit.

4.3.2 Capacitive deionization

CDI is a novel technology for removing ionic species from aqueous solutions. This electrochemical process is conducted at ambient conditions and low voltages (e.g., 1 V) and requires no high-pressure pumps, membranes, distillation columns, or thermal heaters. CDI is an electrosorption process that acts as a "flowthrough" capacitor. In principle, an aqueous solution containing dissolved solids (e.g., NaCl, $CaCO_3$, and $CaSO_4$) is passed between matching pairs of carbon aerogel electrodes. Ionic species, such as sodium and chloride, are held at the charged electrode surfaces (Fig. 8) and

Table 2 Batch-mode electrodialysis reversal (EDR) water quality data

	Primary RO feed	Primary RO concentrate	Final EDR diluate
Conductivity (μs/cm)	1228	5684	759
Total hardness (mg/L as $CaCO_3$)	337	2220	104
Total alkalinity (mg/L as $CaCO_3$)	76	490	135
Ca (mg/L)	80	560	30
Ba (μg/L)	116	774	19
SO_4 (mg/L)	349	2240	85
SiO_2 (mg/L)	7.1	29.4	29.4

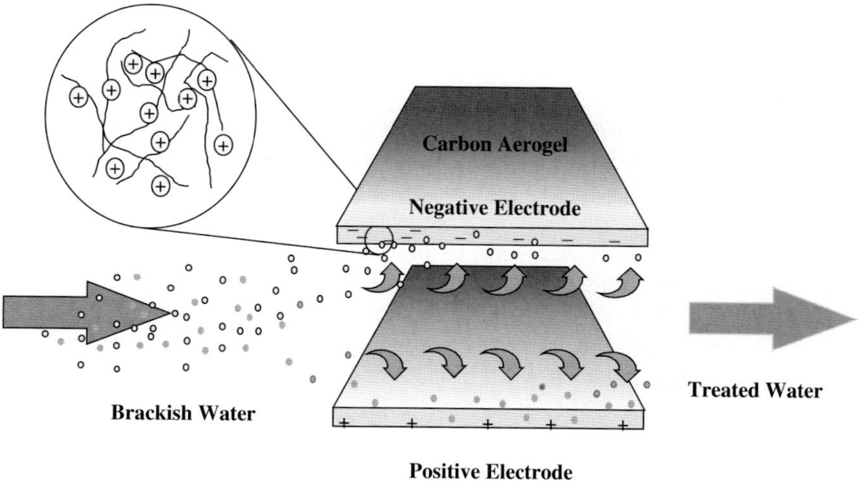

Figure 8 Graphical representation of capacitive deionization process.

are temporarily removed from the solution. The solution is continually deionized (purified) as it passes through successive electrode pairs.

Electrode materials for CDI have included porous carbon [61,62], carbon cloth [12,32,63,64], carbon nanotubes [65], and carbon aerogels [63,66,67], with carbon aerogels making up the bulk of the recent research effort. Carbon aerogels are unique, porous materials consisting of interconnected, uniform carbonaceous particles (3–30 nm) with small (< 50 nm) interstitial pores [66]. This structure leads to high density, a high specific surface area of 400–800 m^2/g low hydraulic resistance, and an exceptional electrical conductivity of ~100 S/cm. The aerogel chemical composition, microstructure, and physical properties can be controlled at the nanometer scale, giving rise to unique electrical properties.

Several separation mechanisms may be controlling ion uptake. Typically, nonreducible and nonoxidizable ions, anions, and cations are removed from solution by the imposed electric field via electrostatic attraction (without charge transfer) within the electrode/electrolyte interface. Large polyvalent oxyanions, heavy metals, and colloidal impurities may be removed by a combination of physisorption, chemisorption, electrodeposition, electrophoresis, and double-layer charging, with charge transfer possibly taking place [68]. After the electrodes become saturated with salt or impurities, the electrodes are regenerated by electrical discharge, allowing the captured salt ions to be released into a relatively small, concentrated purge stream.

Example

Limited studies have been done using CDI with natural waters [69,70]. While these studies were not conducted on high-TDS RO concentrate, they are illustrative of the problems associated with CDI in this regard. The limitations of CDI to act as a concentrate minimization technology are threefold: (1) CDI preferentially removes monovalent ions over divalent ions; (2) the limited sorption capacity of the carbon aerogel electrodes; and (3) natural organic matter easily fouls the high surface area electrodes. Fig. 9 shows the percentage removal of various ions from a blend of Colorado River water and California state project water. These data show that in a competitive environment (i.e., when multiple ions of varying valences are present), the sorption of the divalent species are limited. This finding is important in that, for RO concentrate, divalent ions control the scaling potential of the water and ultimately the water recovery of the system. Therefore, by not removing the divalent ions, the solubility of scale-prone salts within the CDI stack is decreased with the reduction of solution ionic strength, ultimately leading to scale formation.

Despite claims of sorption capacities of up to 80 mg TDS/g of aerogel, real-world applications have only achieved ~8 mg TDS/g of aerogel [71]. This finding is caused by the fact that ion selectivity is based on ionic hydrated radius [69]. As such, only pores greater than 20 nm in diameter are available as sorption sites to allow for electric double-layer formation [72].

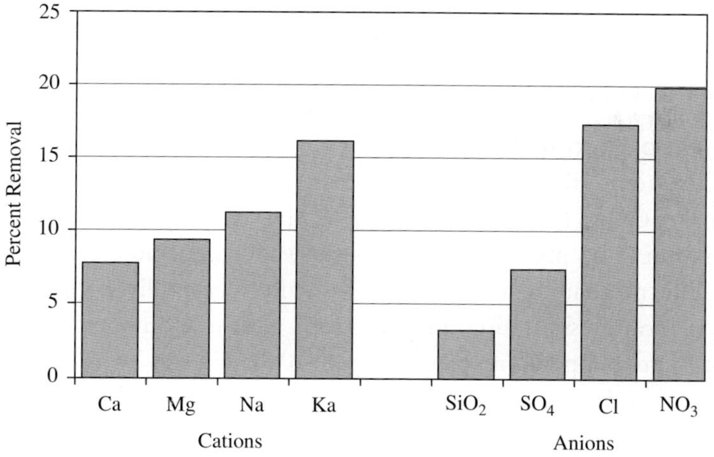

Figure 9 Carbon aerogel performance using 1 L samples of raw water that consisted of 75% Colorado River water and 25% California state project water. Mean water quality data ($n = 3$) taken using 1.4 V, 100 mL/min, at pH 8.3. Adapted from Ref. [69].

Because of the relatively small average pore size (4–9 nm) of the carbon aerogel material, only 14–42 m^2/g – less than 10% – of the aerogel surface area was available for ion sorption. Farmer et al. [66], in the original engineering study for CDI with carbon aerogels, concluded that CDI was best suited for very dilute process streams, such as contaminated groundwater, ultrapure water for semiconductor processing, and perhaps brackish water.

Lastly, in a RO system, the concentration of total organic carbon (TOC) – a mixture of natural organic material and antiscalant components – is highest in the RO concentrate. Work conducted by Gabelich et al [69,71] showed that even at moderate TOC levels, relative to RO concentrate, significant fouling issues resulted, which limited the sorption capacity of carbon aerogel electrodes. Sorption mechanisms for NOM may include electrostatic attraction as well as physical enmeshment with the carbon aerogel structure, resulting in NOM not being removed during regeneration. Electrostatic ion pumping – a simple variant of CDI – is currently being evaluated using smooth electrodes to mitigate the problems with high surface area electrodes [73].

4.4 Thermal processes

Thermal technologies are not only alternative desalting methods but also can be used to treat RO concentrate. Thermal processes include VC evaporation, MD, and freeze concentration. Because the feedwater salinity has no significant impact on the efficiency of thermal processes, thermal processes can be applied to a wide salinity range from brackish water, seawater to brine TDS greater than 300 g/L.

4.4.1 Vapor compression (VC)

In VC systems, the mechanical (mechanical VC or MVC) or thermal (thermal VC or TVC) compression of the vapor provides the heat for evaporation. The process compresses the vapor generated within the unit itself. The mechanical compressor is usually electrically or diesel driven. Thermal compression uses high-pressure steam. Compression raises the pressure and temperature of the vapor so that it can be returned to the evaporator and used as a heat source.

MVC evaporation is considered the most thermodynamically efficient process of thermal desalination processes [74]. A compressor is the driving force for this heat transfer and provides the energy required for separating the solution and overcoming dynamic pressure losses and other irreversibilities (Fig. 10) [75]. The vapor generated from the solution is pumped to the higher pressure level required on the condensing side.

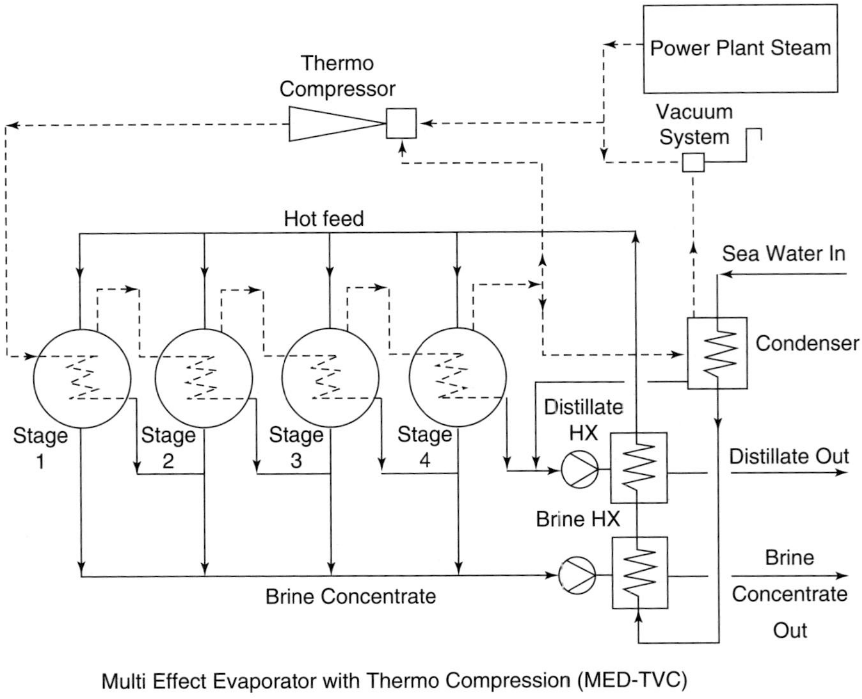

Figure 10 Schematic of multieffect evaporation with vapor compression cycle.

In this system, the distillation step is carried out at close to ambient temperatures. The advantages of MVC include simple pretreatment, and more durable and less maintenance as compared to membrane process [76]. During the last decade, intensive research and development have been carried out to improve the MVC process further by increasing the unit capacities and reducing energy. The MVC is typically limited in size to 3000 m^3/day (0.8 mgd) due to the constraints of the mechanical compressors. The MVC units with capacities up to 10,000 m^3/day (2.5 mgd) are under development [77]. TVC systems can be employed for significantly larger installations [20,000 m^3/day (5.3 mgd)].

VC process is well established and is used for seawater desalination as well as treating RO concentrate (i.e. brine concentrator application) in a near-ZLD application [14,19,78,79]. For example, brine concentrators (VC evaporators operating with seed recycle) are used in Australia to treat RO concentrate from cooling tower blowdown to achieve ZLD in power plants. Scaling is still an issue in VC process, and another disadvantage of the thermal technology is high energy consumption.

4.4.2 Membrane distillation

Membrane distillation (MD) is an emerging separation process that combines simultaneous mass and heat transfer through a hydrophobic microporous membrane [15,80,81]. The driving force for mass transfer in the process is vapor pressure difference across the membrane. A feed solution at elevated temperature is in contact with one side of the membrane and colder water is in direct contact with the opposite side of the membrane (Fig. 11); it is mainly the temperature difference between the liquids and to some extent their solute concentration, which typically

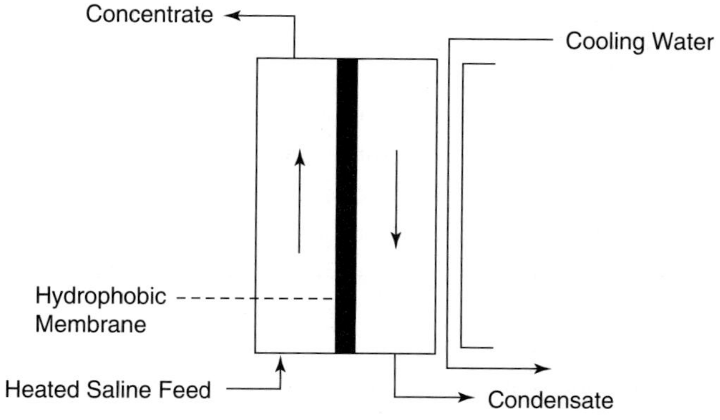

Figure 11 Schematic of air-gap membrane distillation process.

results in vapor pressure depression [78]. Significant water flux can be produced when the temperature difference is above 20 °C between the feed (e.g., 40 °C) and permeate (e.g., 20 °C) streams. By applying vacuum to the permeate side, the water flux can increase by up to 85% over traditional MD process [78]. The main difference and potential advantage of MD over RO is that the former uses the feed's vapor pressure as the driving force to penetrate the membrane instead of the high hydraulic pressures necessary for RO. This eliminates the need for the high-pressure pump and reduces the fouling and scaling problems associated with the pressure-induced concentration gradients at the membrane surface.

During the treatment of a RO concentrate with high silica concentration, MD could reduce the volume of RO concentrate by 60%, achieving an overall water recovery of 90% through RO–MD [76]. Scaling occurred on MD membrane surface at high recovery as determined by the saturation indices of mineral scalants. However, the scalants formed in treating RO concentrate did not clog membrane pores and could be removed almost completely by chemical cleaning [76].

The energy source for feed heating and/or for a vacuum system to sweep away the vapor may be low-grade thermal energy such as supplied by low-pressure steam, waste heat, as well as solar or geothermal energy. A variety of configurations can be used to induce the vapor through the membrane and to condense the penetrant gas. The common method is that the feedwater directly contacts the membrane. Condensation is typically achieved via four-process configurations: air-gap membrane distillation, direct-contact membrane distillation, sweep-gas membrane distillation, and vacuum membrane distillation. Even though MD is frequently cited as a promising desalination technology, no significant commercial operations of this technology exist at this time.

4.4.3 Freeze concentration

The freezing process is based on the natural phenomenon that occurs when ice forms in a saline solution: the resulting individual ice crystals are made up of essentially pure water [82]. Because ice crystals have great regularity and symmetry, they cannot accommodate other atoms or molecules without very severe local strain, practically every solute in the water is rejected by the advancing surface of the growing ice crystal. The "hypertonic" solution at the surface slows down the freezing part of the liquid–solid molecular exchange, by decreasing the availability of water molecules. The result is a lowering of the temperature at which the freezing and melting processes balance; that is, a depression of the freezing point [83]. For example, the freezing point of seawater is − 1.9 °C [84]. As the concentration of the hypertonic solution increases, the freezing point is continually lowered. Freeze desalination depends on the insolubility of salts in ice crystals; the crystals can be separated from the ice-brine slurry, washed, and melted to yield freshwater.

Most of the freeze desalination technologies fall into three categories: direct, indirect, or through the use of a secondary refrigerant. Indirect freeze concentration plants have been used to concentrate beers, coffees, and various juices in many countries around the world [84]. In general, freeze desalting has five basic operations: (1) precooling the feed stream; (2) partial freezing of the feed stream; (3) separation of the ice–brine mixture; (4) melting the associated ice; and (5) heat rejection. Most freezing processes pump the heat removed by crystallization to the melting ice. This is the lowest available temperature lift, and therefore the least amount of work. Heat rejection is required by all freeze desalting technologies; however, because crystallization occurs below ambient temperature, internal heat must be pumped away from the system through a secondary refrigerant. This requires additional work that other desalination processes do not encounter. If this additional work was not required, freezing would probably be the lowest energy intensive desalination process [83]. High capital and energy requirements that are only marginally competitive with other processes have forced the freeze process to be discontinued.

Example

Fig. 12 illustrates a representative change in both the ice and concentrate TDS over time [16] with freeze desalting a RO concentrate. In all, six experiments were conducted under the same conditions to provide preliminary indications of the system's flux rate. In order to create an equal measure of production, the volumetric change in concentrate over time was calculated as gallon per square foot of freezing surface area per minute – similar to that for RO. The flux of the freezing process was calculated at an average 13 gfd, comparable to that of typical RO system. However, as seen in Fig. 12, salt rejection ranged from 66% to 73%; significantly lower than that of most RO processes. In addition, while a 1.7-fold increase in concentrate TDS was observed over time, product-water TDS showed a concomitant increasing TDS. For all experiments, product-ice TDS increased from 1320 to 2150 mg/L at experiment end. This finding is significant for two reasons: (1) TDS of the ice removed from the system was never below 500 mg/L – a typical guideline for reclaimed water and (2) the removal of TDS was variable and tracked the concentrate TDS. Therefore, as recovery of the product increases, so would its salt content increase. Problems associated with freeze desalting include incomplete separation of salts from the ice slurry, fouling of the freezing surface (i.e., ice platting onto the freezing surfaces), and handling of the ice residuals.

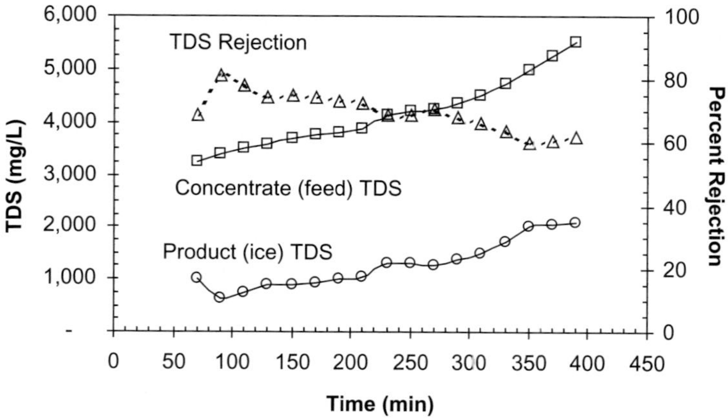

Figure 12 Mechanical freeze desalting performance data from 76 L batch experiment. Product ice double-rinsed with deionized water prior to sampling. Flux $= 6.1 \times 10^{-6}$ m/s [17].

4.5 Other technologies

4.5.1 Forward osmosis

In forward osmosis (FO), like RO, water diffuses through a semipermeable membrane that is impermeable to salt. However, unlike RO that uses hydraulic pressure as a driving force for water transport through the membrane, FO utilizes an osmotic pressure gradient across the membrane (Fig. 13) [85]. A "draw solution" having a significantly higher osmotic pressure than the saline feedwater flows on the permeate side of the membrane and water naturally diffuses across the membrane by osmosis. Osmotic driving forces in FO can be significantly greater than hydraulic driving forces in RO, potentially leading to higher water flux rates and recoveries. With the use of a suitable draw solution, very high osmotic pressure driving forces can be generated to achieve high recoveries that can lead to salt precipitation [86]. To yield potable water, the diluted draw solution is treated by another separation process that generate a stream of purified water and a stream of reconcentrated draw solution for reuse in the FO process.

The main advantages of the FO process include the relatively low fouling potential and low energy consumption as no high-pressure pump is required, provided that there is no additional energy cost associated with regeneration of the draw solution and with management of the FO concentrate. Depending upon the water composition of RO concentrate, recent studies showed that FO could further reduce RO concentrate volume by 71–81%, achieving an overall water recovery of 94–97% by combination of RO–FO processes [80,85]. However, scaling/fouling still occurs at high recovery conditions as determined by the saturation indices.

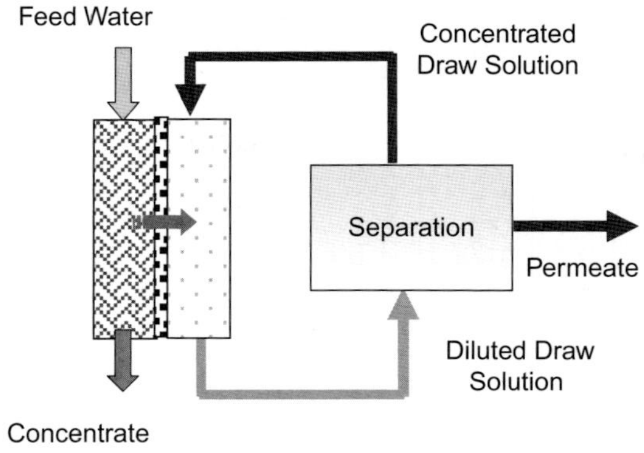

Figure 13 Schematic of forward osmosis.

Chemical cleaning was proven as being effective to remove silica and calcium sulfate scaling and 81% of the permeate flux was recovered [76].

The main challenges remain in the manufacturing of high-performance FO membranes, the need for regeneration of the draw solution, and optimization of process configuration. These are the primary obstacles to the practical applicability of FO for water production and thus large-scale commercialization of the FO process. With the suitable draw solution and appropriate semipermeable membrane, FO process can lead to salt precipitation, that is, ZLD. The technology of FO is still being developed.

4.5.2 Vibratory shear–enhanced process for membrane filtration

The vibratory shear–enhanced proces (VSEP) membrane filter pack consists of leaf elements arrayed as parallel disks and separated by gaskets. The shear waves produced by the membrane vibration cause solids and foulants to be lifted off the membrane surface and remixed with the bulk material flowing through the membrane stack. Membrane fouling and scaling is minimized through the effects of an axial, vibratory-induced shear force imposed directly on the membrane–liquid interface. A pilot scale testing on highly saline boiler blowdown water yielded a recovery of 95% using VESP RO and a solid waste (brine) with approximately 50–65% dry solids [76]. More studies are required to evaluate this technique for the purpose of concentrate minimization and recovery improvement, and its overall cost-effectiveness.

5. CONCLUSION

In summary, several technologies for concentrate treatment are emerging and some may offer the potential of enhanced water recovery and reduced concentrate. However, no one technology is appropriate for all instances. Table 3 provides an overview of the status of the technologies presented in

Table 3 Summary of concentrate minimization technologies: status, costs, and limitations

	Process							
	Thermal		Cation control	Anion control	Separation			
	Vapor compression	Freeze desalination	ICD	Biological sulfate control	EDR or ED	FO	MD	CDI
Industrial status	Commercial	Commercial	Demonstration-scale tested	Bench-scale tested	Commercial	Bench-scale tested	Bench-scale tested	Pilot-scale tested
Applied feed TDS	Brackish water, seawater, and brine with TDS > 300 g/L	Brackish water, seawater, and brine with TDS > 300 g/L	TDS 0.5–10 g/L	TDS 0.5 g/L	EDR: TDS < 8 g/L; ED: wide TDS range	Brackish water, and seawater	Brackish water, and seawater	TDS < 5 g/L
Salt rejection	~ > 99%	66–73% with product water TDS > 500 mg/L	> 94%[b]	> 94%[b]	~86.6% in treating RO concentrate	~95%	~99.5%	Rejection increases with CDI stages
Achieved recovery	40–50% for seawater desalination, can achieve zero liquid discharge (ZLD)	60–80%	~95% for brackish water desalination	~95% for brackish water desalination	~85% to 96% for brackish water desalination, can further reduce 50–86% of RO concentrate	Can further reduce ~70% of RO concentrate	Can further reduce 60–65% of RO concentrate	33% in treating TDS 5 g/L brackish water

Energy[a]	~100 to 250 kWh/kgal for ZLD	42–80 kWh/kgal	NA	NA	~3 kWh/kgal	0.21	NA	2.3–4.2 kWh/kgal
Estimated total cost[a]	~\$12 to 13 $kgal^{-1}$ for concentrate recovered (ZLD)	NA		NA	~\$0.7 to 2.1 $kgal^{-1}$	NA	~\$3 $kgal^{-1}$	~\$1.1 to 7.3 $kgal^{-1}$
Advantages	Achieving ZLD	Commercial technology	Mature technology	Reduced scaling potential	Low fouling/scaling potential	Low energy demand; low fouling and easy chemical cleaning	Low fouling and easy chemical cleaning	Low fouling and requires minimum pretreatment
Challenges	High energy demand and costs	Incomplete separation of salts, fouling, handling of ice residuals	Chemical and sludge handling	Developmental stage: technical challenges; chemical and sludge handling	Poor removal of organic matter	Developmental stage: lack of appropriate FO membranes, and draw solutions	Developmental stage: process optimization	Developmental stage: low recovery, high operating cost, module optimization, and so on
References	[61,87,88]	[17]	[61]		[75]	[83,51]	[11,51,79,83]	[5,59]

[a] Costs and energy are typically very site-specific and depending upon capacity, feedwater chemistry and salinity, targeted product water quality, and many other factors.
[b] When used in tandem with a secondary RO process.

this chapter, as well as relative energy consumption and costs. It should be noted that both energy consumption and treatment costs are highly site-specific; nonetheless, these broad ranges are presented for general guidance and comparative purposes. In selecting potential concentrate minimization technologies, the end user must select based on water quality characteristics, concentrate water recovery goals, disposal options available, permitting requirements, and site-specific characteristics such as available infrastructure, space, and skilled workforce.

REFERENCES

[1] T. Pankratz, The 19th IDA Worldwide Desalting Plant Inventory, Global Water Intelligence, Houston, TX, 2006.
[2] M. Mickley, Concentrate management, in: M. Wilf (Ed.), The Guidebook to Membrane Desalination Technology: Reverese Osmosis, Nanofiltration and Hybrid Systems Process, Design, Applications and Economics, Balaban Desalination Publications, L'Aquila, Italy, 2007, p. 524.
[3] B. van der Bruggen, L. Lejon, C. Vandecasteele, Reuse, treatment, and discharge of the concentrate of pressure-driven membrane processes, Environ. Sci. Technol. 37 (2003) 3733–3738.
[4] J.S. Taylor, Reverse osmosis and nanofiltration, in: J. Mallevialle, P.E. Odendaal, M.R. Wiesner (Eds.), Water Treatment: Membrane Processes, American Water Works Association Research Foundation, Lyonnaise des Eaux, Water Research Commission of South Africa, McGraw-Hill, New York, 1996, p. 9.1.
[5] A. Rahardianto, B. McCool, Y. Cohen, Reverse osmosis desalting of inland brackish water of high gypsum scaling propensity: kinetics and mitigation of membrane mineral scaling, Environ. Sci. Technol. 42 (2008) 4292–4297.
[6] A. Rahardianto, W.-Y. Shih, R.-W. Lee, Y. Cohen, Diagnositc characterization of gypsum scale formation and control in RO membrane desalination of brackish water, J. Membr. Sci. 279 (2006) 655–668.
[7] A. Burbano, S. Adham, W. Pearce, The state of full-scale RO/NF desalination-results from a worldwide survey, J. Am. Water Works Assoc. 99(4) (2007) 116–127.
[8] C. Gabelich, M. Williams, A. Rahardianto, J. Franklin, Y. Cohen, High-recovery reverse osmosis using intermediate chemical demineralization, J. Membr. Sci. 301 (2007) 131–141.
[9] A. Rahardianto, J. Gao, C. Gabelich, M. Williams, Y. Cohen, Accelerated precipitation softening for high-recovery desalination of brackish surface water, J. Membr. Sci. 289 (2007) 123–137.
[10] M. Pirbazari, V. Ravindran, J. Jung, M. Samee, A. Vahdati, Biological Sulfate Reduction for Recovering Reverse Osmosis Concentrate, United States Environmental Protection Agencey Region IX, San Francisco, CA, 2006.
[11] M. Turek, Electrodialytic desalination and concentration of coal-mine brine, Desalination 162 (2004) 355–359.
[12] M. Andelman, US Patent No. 5,779,891, 1998a.
[13] M. Andelman, US Patent No. 6,32590, 1998b.
[14] B. Ericsson, B. Hallmans, Treatment and disposal of saline wastewater from coal mines in Poland, Desalination 98 (1994) 239–248.
[15] T.A. Cath, V.D. Adams, A.E. Childress, Experimental study of desalination using direct contact membrane distillation: a new approach to flux enhancement, J. Membr. Sci. 228 (2004) 5–16.

[16] A. Mofidi, C. Gabelich, B. Coffey, J. Green, Freeze Desalting of Reverse-Osmosis Brine, Electric Power Research Insitute, Water and Wastewater Technology Demonstration Projects, Palo Alto, CA, 2000.
[17] T.Y. Cath, V.D. Adams, A.E. Childress, Membrane contactor processes for wastewater reclamation in space: II. Combined direct osmosis, osmotic distillation, and memebrane distillation for treatment of metabolic wastewater, J. Membr. Sci. 257 (2005) 111–119.
[18] T.Y. Cath, S. Gormly, E.G. Beaudry, M.T. Flynn, V.D. Adams, A.E. Childress, Membrane contactor processes for wastewater reclamation in space: part I. Direct osmotic concentration as pretreatment for reverse osmosis, J. Membr. Sci. 257 (2005) 85–98.
[19] B.M. Hamiah, J. Beckman, M. Ybarra, Brackish and seawater desalination using a 20 ft dewvaporation tower, Desalination 140 (2001) 217–226.
[20] J. Madole, J. Peterson, Concentrate management using lime softening and vibratory microfiltration, in: Proceedings of the 20th Anuual WaterReuse Symposium, 2005.
[21] C.J. Gabelich, Concentrate minimization for reverse osmosis treatment: demonstration-scale testing and economic feasibility analysis, Final Report for California Department of Water Resources, in: Interim Water Supply Grant Commitment – Safe Drinking Water, Clean Water, Watershed Protection and Flood Protection Act (Proposition 13, Chapter 9, Article 4), January 2009.
[22] Y. Le Gouellec, M. Elimelech, Calcium sulfate (gypsum) scaling in nanofiltration of agricultural drainage water, J. Membr. Sci. 205 (2002) 387–397.
[23] W.-Y. Shih, A. Rahardianto, R.-W. Lee, Y. Cohen, Morphometric characterization of calcium sulfate dihydrate (gypsum) scale on reverse osmosis membranes, J. Membr. Sci. 252 (2005) 253–263.
[24] A. Almulla, M. Eid, P. Cote, J. Coburn, Developments in high recovery brackish water desalination plants as part of the solution to water quantity problems, Desalination 153 (2003) 237–243.
[25] R. Bond, S. Veerapaneni, Zero Liquid Discharge for Inland Desalination, AWWA, & IWA Publishing, Denver, CO, 2007.
[26] I.M. Bremere, M. Kennedy, P. Michel, R. van Emmerik, G.-J. Witkamp, J.C. Schippers, Controlling scaling in membrane filtration systems using a desupersaturation unit to prevent scaling, Desalination 124 (1999) 51–62.
[27] I.M. Bremere, M. Kennedy, A. Johnson, R. van Emmerik, G.-J. Witkamp, J.C. Schippers, Increasing conversion in membrane filtration systems using a desupresaturation unit to prevent scaling, Desalination 119 (1998) 199–204.
[28] J. Gilron, D. Chaikin, N. Daltrophe, Demonstration of CAPS pretreatment of surface water for RO, Desalination 127 (2000) 271–282.
[29] A.J. Graveland, Developments in water softening by means of pellet reactors, J. Am. Water Works Assoc. 75 (1983) 619.
[30] R.C. Harries, A field trial of seeded reverse-osmosis for the desalination of a scaling-type mine water, Desalination 56 (1985) 237–243.
[31] G.J.G. Juby, C.F. Schutte, Membrane life in a seeded-slurry reverse osmosis system, Water SA 26 (2000) 239.
[32] O. Kedem, G. Zalmon, Compact accelerated precipitation softening (CAPS) as a pretreatment for membrane desalination. 1. Softening by NaOH, Desalination 113 (1997) 65–71.
[33] Y. Oren, V. Katz, N.C. Daltrophe, Improved compact accelerated precipitation suftening (CAPS), Desalination 139 (2001) 155–159.
[34] S. Seewoo, R. Van Hille, A. Lewis, Aspects of $CaSO_4$ precipitation in scaling waters, Hydrometallurgy 75 (2004) 135–146.
[35] S. Sethi, J. Drewes, P. Xu, Desalination Product Water Recovery and Concentrate Volume Minimization, AWWA, & IWA Publishing, Denver, CO, 2008.

[36] J. Rybicki, W. Alda, A. Rybicka, S. Feliziani, T.G. Kawecki, S.S. Chappie, D.R. Mahony, J.T.M. Sluys, D. Verdoes, J.H. Hanemaaijer, Water treatment in a membrane-assisted crystallizer (MAC), Desalination 104 (1996) 135–139.
[37] T. Yun, C. Gabelich, F. Gerringer, A. Mofidi, M.D. Norris, Evaluation of large diameter reverse osmosis elements and novel concentrate reduction technologies for colorado river water desalting, in: Proceedings of AMTA Biennial Conference, Anaheim, 2006.
[38] S.Y. Matsui, Kinetics of glucose decomposition with sulfate reduction in the anaerobic fluidized bed reactor, Water Sci. Technol. 28(2) (1993) 135–144.
[39] S.C. Nagpal, Ethanol utilization by sulfate-reducing bacteria: an experimental and modeling study, Biotechnol. Bioeng. 70 (2000) 533–543.
[40] S.C. Nagpal, Microbial sulfate reduction in a liquid-solid fluidized bed reactor, Biotechnol. Bioeng. 70 (2000) 370–380.
[41] S. Okabe, T. Itoh, H. Satoh, Y. Watanabe, Analyses of spatial distributions of sulfate-reducing bacteria and their activity in aerobic wastewater biofilms, Appl. Environ. Microbiol. 65(11) (1999) 5107–5116.
[42] J. Perry, J. Staley, S. Lory, Microbial Life, Sinauer Associates, New York, NY, 2002.
[43] T.T. Van Houten, Biological sulphate reduction using gas-lift reactor fed with hydrogen and carbon dioxide as energy and carbon source, Biotechnol. Bioeng. 44 (1994) 586–594.
[44] J. Postgate, The Sulfate-Reducing Bacteria, Cambridge University Press, New York, 1993.
[45] A.P. Annachhatre, Biological sulfate reduction using molasses as a carbon source, Water Environ. Res. 73 (2001) 118–126.
[46] D.C. Freedman, Biotransformaiton of explosive-grade nitrocellulose under denitrifying and sulfidogenic conditions, Waste Manage. 22 (2002) 283–292.
[47] B.W. Lee, Biological and chemical interaction with U(VI) during anaerobic enrichment in the presence of iron oxide coated quartz, Water Res. 39 (2005) 4363–4374.
[48] M.M. Madigan, Brock Biology of Microorganisms, Prentice Hall, New Jersey, NJ, 2002, p. 1140.
[49] J.S. Dries, High rate biological treatment of slufate-rich wastewater in an acetate-fed EGSB reactor, Biodegradation 9 (1998) 103–111.
[50] G.S. Parkin, Anaerobic filter treatment of sulfate-containing wastewaters, Water Sci. Technol. 23 (1991) 1283.
[51] D.H. Zitomer, High-sulfate, high-chemical oxygen demand wastewater treatment using aerated methanogenic fluidized beds, Water Environ. Res. 72 (2000) 90–97.
[52] R. Speece, Anaerobic Biotechnology for Industrial Wastewaters, Archae Press, Nashville, 1996.
[53] M. McFarland, W.J. Jewell, In situ control of sulfide emissions during the thermophilic anaerobic digestion of process water, Water Res. 23 (1989) 118–126.
[54] E. Reahl, Half a Century of Desalination with Electodialysis, GE Water & Process Technologies, Technical Paper, 2006.
[55] E. Reahl, Reclaiming reverse osmosis wastewater, in: Industrial Water Treatment, January–February 2002, pp. 36–39.
[56] P. Murray, Electrodialysis and Electrodialysis Reversal, AWWA, Denver, CO, 1995.
[57] S.E. Letant, C.M. Schaldach, M.R. Johnson, A. Sawvel, W.L. Bourcier, W.D. Wilson, Evidence of gating in hundred nanometer diameter pores: an experimental and theoretical study, Small 2 (2006) 1504–1510.
[58] C.M. Schaldach, Dielectrophoretic forces on the nanoscale, Langmuir 20 (2004) 10744–10750.
[59] C.M. Schaldach, Electrostatic fields in the vicinity of engineered nanostructures, J. Colloid. Interface Sci. 275 (2004) 601–611.

[60] W.L. Bourcier, Molecular Engineering of Electrodialysis Membranes, Lawrence Livermore National Laboratory, Livermore, CA, 2005.

[61] A.M. Johnson, J. Newman, Desalting by means of porous carbon electrodes, J. Electrochem. Soc. 118 (1971) 510–517.

[62] M. Matlosz, J. Newman, Experimental investigation of a porous carbon electrode for the removal of mercury from contaminated brine, J. Electrochem. Soc. 133 (1986) 1850–1859.

[63] T.S. Welgemoed, C.F. Schutte, Capacitive deionization technology: an alternative desalination solution, Desalination 183 (2005) 327–340.

[64] E.C. Ayranci, Removal of phenol, phenoxide and chlorophenols from wastewater by adsorption and electrosorption at high area carbon felt electreodes, Electroanal. Chem. 513 (2001) 100–110.

[65] K. Dai, L. Shi, J. Fang, D. Zhang, B. Yu, NaCl Adsorption in multi-walled carbon nanotubes, Mater. Lett. 59 (2005) 1989–1992.

[66] J.C. Farmer, D.V. Fix, G.V. Mack, R.W. Pekala, J. Poco, Capacitive deionization of NaCl and $NaNO_3$ solutions with carbon aerogel electrodes, J. Electrochem. Soc. 143 (1996) 159–169.

[67] S.-W. Hwang, S.-H. Hyun, Capacitance control of carbon aerogel electrodes, J. Non-Crystalline Solids 347 (2004) 238–245.

[68] R.W. Pekala, S.T. Mayer, J.F. Poco, J.L. Kaschmitter, Structure and performance of carbon aerogel electrodes, in: Spring Meeting of the Materials Research Society, San Francisco, CA, April 4–8, 1994.

[69] C. Gabelich, T.D. Tran, I.H. Suffet, Electrosorption of inorganic salts from aqueous solution using carbon aerogels, Environ. Sci. Technol. 36(13) (2002) 3010–3019.

[70] P. Xu, J. Drewes, D. Heil, G. Wang, Treatment of brackish produced water using carbon aerogel-based capacitive deionization technology, Water Res. 42(10 and 11) (2008) 2605–2617.

[71] C. Gabelich, T. Yun, C.R. Bartels, J. Green, Nonthermal Technologies for Salinity Removal, AWWA, Denver, CO, 2001.

[72] J. Farmer, J. Richardson, D.V. Fix, S.L. Thomson, S.C. May, Desalination with carbon aerogel electrodes, UCRL-ID-125298, December 4, 1996b.

[73] Lawrence Livermore National Laboratory (2005), Desalination Using Electrostatic Ion Pumping. Retrieved from http://www.owue.water.ca.gov/recycle/DesalPSP/FinalFundingAwards2005.pdf

[74] G. Kronenberg, F. Lokiec, Low-temperature distillation processes in single-and-dual-purpose plants, Desalination 136 (2001) 189–197.

[75] H. Aybar, Analysis of mechanical vapor compression desalination system, Desalination 142 (2002) 181–186.

[76] A.A. Mabrouk, A.S. Nafey, H.E.S. Fath, Thermoeconomic analysis of some existing desalination processes, Desalination 205 (2007) 354–373.

[77] L.A. Liu, Dewatering reverse osmosis concentrate using forward osmosis, in: Proceedings of Membrane Technology Conference, AWWA, Tampa, FL, 2007.

[78] T.S. Bryant, J. Stuart, I. Fergus, R. Lesan, The use of reverse osmosis as a 35,600 m^3/day concentrator in the waste water management scheme at 4640 MW Bayswater/LIddell power sation complex – Australia, Desalination 67 (1987) 327–353.

[79] B. Tleimat, M. Tleimat, Developments in saline water distillation technology, Desalination 93 (1993) 15–42.

[80] C.C. Martinetti, Novel Application of membrane distillation and forward osmosis for brackish water desalination, in: 2007 Membrane Technology Conference, American Water Works Association, Tampa, FL, 2007.

[81] K. a. Skirar, Novel membrane and device for direct contact membrane distillation-based desalination process – Phase II, Report No. 96, U.S. Department of Interior, Bureau of Reclamation, 2003.

[82] R. Halde, Concentration of impurities by progressive freezing, Water Res. 14 (1980) 575–580.
[83] L.a. Zhu, The new technology of desalination by vacuum freezing-vapor condensation, IDA World Congress on Desalination and Water Science, Abu Dhabi, 1995.
[84] W.E. Johnson, The story of freeze desalting, Int. Desalination Water Reuse Quar.3/4 (1994) 20–27.
[85] T.C. Cath, A.E. Childress, M. Elimelech, Forward osmosis: principles, applications, and recent developments, J. Membr. Sci. 281 (2006) 70–87.
[86] J.R. McCutcheon, R.L. McGinnis, M. Elimelech, A novel ammonia-carbon dioxide forward (direct) osmosis desalination process, Desalination 174 (2005) 1–11.
[87] S.-W. Hwang, S.-H. Hyun, Capacitive deionization technology: an alternative desalination solution, Desalination 183 (2005) 238–340.
[88] WHO, WHO Offers Guidance on Desalination for Safe Water Supply www.who.int, 2007.

CHAPTER 11

Inland Desalination: Current Practices, Environmental Implications, and Case Studies in Las Vegas, NV

Benjamin D. Stanford[1], **Joseph F. Leising**[1], **Rick G. Bond**[2] *and* **Shane A. Snyder**[1,*]
[1]Applied Research and Development Center, Southern Nevada Water Authority, Las Vegas, NV, USA
[2]Black and Veatch

Contents

1. INTRODUCTION

As worldwide human populations continue to rise and clean water sources become more stressed from increased demand, alternative water sources must be made available to meet those demands. Freshwater drinking sources are a precious commodity and, in many arid regions of the world, can be scarce or completely unavailable. In areas where freshwater sources are available, the biological and/or chemical pollutant load may make the waters unsuitable for conventional drinking water treatment, adding to the scramble for potable water sources. Industrial, agricultural, and home uses all compete for the available water supplies. According to the United States

*Corresponding author.
E-mail address: shane.snyder@snwa.com

Sustainability Science and Engineering, Volume 2 ISSN 1871-2711, DOI 10.1016/S1871-2711(09)00211-6

Department of Agriculture, 80% of water consumption in the United States is related to agricultural use (i.e., irrigation) but that number may be as high as 90% in several western states where droughts are common and water supplies scarce.[1] Such demands further stretch the resources available to a given population. As continued population growth and expansion in these arid and semiarid areas place heavy burden on the available water supplies, utilities are examining innovative ways to stretch water supplies to ensure their ability to meet the demand for safe, clean, and sustainable drinking water.

Not all end uses of potable water (e.g., industrial, agricultural) require the level of treatment necessary for human consumption. As such, many utilities in the United States and around the world have also been turning to the reuse of municipal wastewater, either directly or indirectly, to help meet this specific demand. In fact, recycled water is currently an indispensable fraction of numerous western communities' water resource portfolio. Water reuse projects range from turf irrigation to recharge of groundwater supplies for drinking water (i.e., indirect potable reuse [IPR]). In the United States, IPR and direct potable use of reclaimed wastewaters has not been widely accepted. While water reuse is a viable tool for utilities, it has yet to reach its full potential, in part due to public perceptions about the use of "sewage" for drinking water and a range of other misconceptions. Much of this opposition stems from the discovery of various pharmaceutically active compounds (PhACs), endocrine-disrupting compounds (EDCs), and other trace contaminants in treated wastewater [1–6], and the potential impact that such compounds have on wildlife species [7–10]. Furthermore, certain compounds such as caffeine, DEET, iopromide, and sulfamethoxazole have been shown to be highly pervasive in water systems and resistant to multiple treatment barriers [11]. Conventional drinking water treatment processes typically involve treatment options such as coagulation, filtration (e.g., sand, activated carbon, and anthracite), and disinfection, and are only moderately able to remove trace contaminants. More advanced treatment options, including ozone, UV disinfection, advanced oxidation processes (e.g., UV/peroxide, ozone/peroxide), and membranes have been shown to be better suited for removal of PhACs and EDCs. Membrane systems, including nanofiltration (NF) and reverse osmosis (RO) offer the highest potential for removal of trace contaminants via a physical barrier capable of removing large and small

[1] http://www.ers.usda.gov/Briefing/WaterUse/.

molecules, depending on the properties of the membranes. Membrane systems have been shown to be highly effective at removing trace contaminants, though even RO membranes are vulnerable to breakthrough by a few PhACs, EDCs, and other compounds such as *N*-nitrosodimethylamine (NDMA), bromide, and boron [11–14]. However, state-regulated multibarrier approaches such as RO followed by UV/peroxide has allowed IPR projects to expand in areas such as California where West Basin and Orange County utilities have groundwater recharge/IPR systems in place.

Beyond the application of membranes in IPR systems, desalination also is experiencing tremendous growth. This is especially evident in the Middle East, where access to seawater is plentiful while freshwater sources are scarce. In the United Arab Emirates, desalination may account for 98% of the total domestic water use, as reported in a recent review [15]. Several coastal communities in the United States have been turning to desalination as well, including Tampa, Florida, and Long Beach, California. As of 2005, over 230 municipal membrane-desalination plants for potable water production had been built in the United States [16]. Several options for desalination of brackish water and seawater exist including NF and RO membrane treatment, multistage flash distillation (MSF), multiple-effect distillation (MED), vapor compression (VC), and electrodialysis reversal (EDR). Each process has distinct advantages and disadvantages associated with it, though they all share in common a relatively high energy cost. Older desalination plants that are still in operation may require up to 20 kWh/m^3 for operation [17], though energy recovery devices coupled with modern RO systems have aided in reducing that value to 2–4 kWh/m^3, depending upon desired water recovery rates [18,19]. Despite these important gains in energy recovery and improvements in membrane efficiency, energy use remains a primary issue in the consideration of desalination as a source of potable water. As energy costs continue to rise globally and carbon footprint concerns escalate, the future of desalination will depend on continued improvements in reducing energy requirements, this fact is exponentially more critical for inland applications where brine disposal and zero-liquid discharge (ZLD) may be required.

Even with the considerable energy expense associated with desalination versus freshwater treatment,[2] inland desalination is another potential

[2]The current cost of water treatment in southern Nevada is approximately \$0.0132 m^{-3} (including personnel, treatment energy, and chemicals, but not including pumping costs), whereas at a modest electricity rate of \$0.03 kWh^{-1}, even the best RO desalination process running at 2 kWh/m^3 would cost \$0.06 m^{-3} for the RO treatment alone. *Source*: Personal contact, SNWA.

resource for water utilities not located near an ocean source, though brine disposal is typically the major limiting factor in implementation of such projects. Depending on the salinity of the concentrate, inland desalination plants have considerably higher costs associated with brine disposal than those proximal to oceans. Brine disposal is generally limited to one or two options for any given area and is directly related to land cost, energy costs, regulations, and the type and quantity of salts in the brine stream [16,20,17]. Inland brine disposal options include the reuse of concentrates, further treatment, surface water discharge, sewer disposal, deep well injection, land applications, and evaporation ponds (followed by landfilling of solids) [16,20]. The brine stream is generally considered a waste product that is incompatible with most available inland bodies of water and may have enormous environmental consequences if disposed of improperly. This makes discharge to surface waters and sewers infeasible for most arid areas without access to adequate dilution water. As such, significant logistical and fiscal planning must be coupled with environmental cost–benefit analysis well before any potential inland desalination site can move forward with permitting and construction.

2. STRATEGIES FOR INLAND BRINE DISPOSAL: ZLD AND FLUIDIZED BED CRYSTALLIZERS

Innovative strategies have been suggested to address the issue of brine disposal with inland desalination. One inland plant constructed a 30 km pipeline to send concentrate to the sea [21] while another proposed project will send, by gravity, concentrate from Red Sea desalination plants to the Dead Sea, producing energy, disposing of the brine, and returning much-needed volume of liquid to the Dead Sea [22]. Such innovative solutions are more a rarity than the norm, however. As such, considerable effort must be spent toward engineering a cost-effective solution with the minimum environmental impact possible and potential sites must evaluate other options such as ZLD and/or beneficial uses and applications of the brine by-products.

Many ZLD applications in operation today treat power plant cooling water with thermal desalination and/or evaporation ponds to gain maximum energy use of the heated water. Thermal desalination (distillation) has been practiced for over 30 years, particularly in the Middle East, and it is a mature technology unlikely to result in any major technological improvements. Although there have been a few design innovations over the years to optimize energy efficiency, thermal desalination remains an expensive, energy-intensive process. Therefore, membrane-based ZLD operations are gaining favor as the technology expands for improved membranes and energy recovery, though evaporative systems may be viable in regions with low humidity and large tracts of undeveloped land.

Enhanced evaporation systems use mechanical energy to increase the surface area of water in contact with air and thereby achieve evaporation rates several times that of conventional evaporation ponds. Such evaporation systems can be used alone or in combination with multistage membrane ZLD processes. Several enhanced evaporation systems have been tested, which may have the potential to significantly reduce the costs and area required for evaporation [23]. Most rely on water lines with spray nozzles to disperse brine to the atmosphere as a mist, into a high-speed air stream, countercurrent air streams, or onto vertical surfaces to augment evaporation. A portion of the water is evaporated and residual salts fall to the ground. These systems may have the potential to achieve efficient evaporation in arid climates. A disadvantage to these systems is that windborne salt drift could potentially contaminate a sizeable surrounding area, which is especially of concern if the system is located near an urban area.

One possible membrane-based ZLD treatment process is illustrated in Fig. 1 [24]. This process schematic shows a primary RO system for desalination of an inland source of water. Concentrate from the primary RO is treated to reduce its precipitation potential and then desalinated in a secondary RO system. The concentrate from the secondary RO, typically

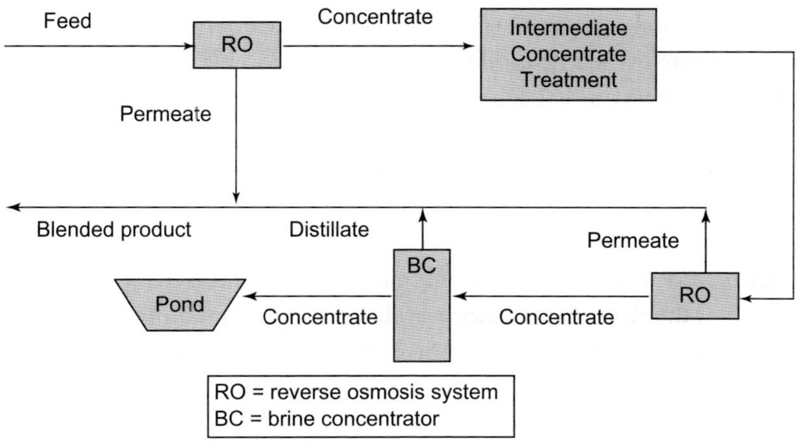

Figure 1 Process schematic for desalination with zero-liquid discharge.

2–6% of the feed to the primary RO, is treated using a thermal desalination process, the brine concentrator. Brine concentrators typically recover about 95% of RO concentrate from brackish water desalination as a distillate with very low total dissolved solids (TDS). The residual concentrate, 0.1–0.3% of the feed to the primary RO, is discharged to an evaporation pond. Consequently, no liquid waste is discharged from the site.

Precipitation of solutes within membrane systems limits the amount of treated water that can be produced without some form of enhanced recovery. As recovery is increased, water on the feed side of the membrane becomes increasingly supersaturated with sparingly soluble salts such as calcium carbonate ($CaCO_3$), calcium sulfate ($CaSO_4$), and barium sulfate ($BaSO_4$). Consequently, recovery in RO systems is limited by the precipitation potential of sparingly soluble salts. Crystallizers can be used to enhance recovery of water from the concentrated brine stream, which often is supersaturated [24].

An understanding of the factors that affect precipitation from a supersaturated solution is essential to development of a well-designed, functional ZLD process. The presence of foreign particles enhances precipitation kinetics by reducing the amount of free energy required for solids formation. Consequently, precipitation is made more energetically favorable. Induction time, the time required for precipitation to begin, decreases as the fit between the foreign particle and the crystal to be formed increases. The best fit between the two solid phases occurs when the particle is a seed crystal of the same salt (secondary nucleation). One group studied the effect of brushite ($CaHPO_4 \cdot 2H_2O$) on gypsum ($CaSO_4 \cdot 2H_2O$) precipitation [25]. The researchers noted that the ability of one crystalline phase to grow on the surface of another is strongly dependent on the compatibility of their surface characteristics, and they observed a close fit in lattice structure between brushite and gypsum. Their results showed that brushite crystals served as effective nuclei for the growth of gypsum crystals. Another group studied the kinetics of seeded growth of gypsum in both the presence and the absence of antiscalant additives and concluded that seed concentration greatly influenced induction time [26]. At the same antiscalant concentration and solution temperature, the induction time was 73 min for a seed concentration of 110 mg/mL and 31 min for a seed concentration of 193 mg/mL. Without the antiscalant, precipitation was immediate at each seed concentration. Without seed addition, the supersaturated solution was stable.

The schematic of a fluidized bed crystallization system optimized with these considerations in mind is illustrated in Fig. 2. The process is initiated,

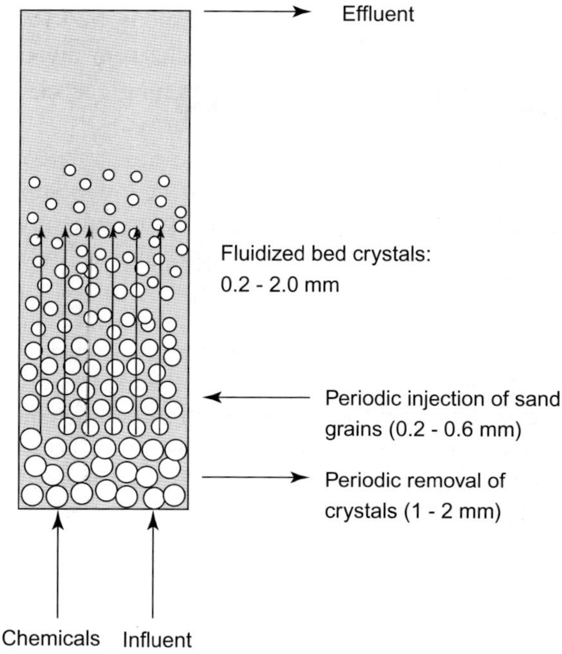

Figure 2 Process Schematic for fluidized bed crystallization.

by introducing flow at the bottom of the reactor at a rate sufficient to fluidize the media without causing it to be washed out in the effluent. Sodium hydroxide or lime is fed at the bottom of the bed to achieve supersaturation with respect to $CaCO_3$, and precipitation occurs as calcium and carbonate ions leave the solution and are adsorbed to the sand to form calcium carbonate pellets. Calcium removal continues with crystal growth of the pellets, and as the pellet diameter increases, the crystal surface area per unit volume of reactor decreases. The process is controlled by periodically removing larger crystals from the bottom of the column and adding new sand.

Crystallization was first applied to water softening in 1938 [27] with the invention of the Spiractor®, a conical shaped upflow reactor. In 1971, a fluidized bed crystallizer, the Crystalactor®, was developed in the Netherlands with a cylindrical shape and water and chemical feed nozzles designed to enhance vertical plug flow and improve initial chemical mixing. This reactor was selected to provide water softening at the main water treatment plants in the Netherlands [28] and is currently in use at 25 treatment plants in Europe, Asia, and Australia. Fluidized bed crystallization has not been widely used for concentrate treatment, however [24]. Several features of an optimized fluidized bed crystallization are expected to be advantageous compared to conventional softening for treatment of RO concentrate:

- Fluidized bed crystallization provides a large surface area of seed crystals for precipitation. Such precipitation occurs at lower supersaturation in the presence of crystals of the precipitate. Consequently, precipitation can be achieved with smaller chemical doses and at lower pH.
- The fluidized bed crystallizer produces near anhydrous pellets that are approximately 90% solid by weight, and these pellets drain rapidly under gravity to a solids content of 99%. Sludge produced in conventional softening has a solids content of 3–15%. Consequently, the fluidized bed crystallizer generates approximately 10% of the solids volume generated by conventional softening. Furthermore, calcium carbonate crystals have beneficial uses in agriculture and industry. All of the solids generated by fluidized bed crystallization softening in the Netherlands are reused. These calcite crystals have been used for treatment of aggressive groundwater, neutralization of acid wastewater, for road construction, cement manufacture, and in the metal industry [27].
- Crystallizers are designed with upflow velocities as high as 120 m/h (49 gallons per minute [gpm]/ft^2) [28]. Consequently, the fluidized bed crystallizer footprint is much smaller than that required for conventional softening.

For optimal performance in a fluidized bed reactor, it is important that calcium be removed by crystal growth rather than by spontaneous nucleation of calcium carbonate. According to Graveland et al. [27], good crystallizer design should include the following features:

- Proper water and chemical distribution to produce plug flow and avoid short-circuiting.
- Intensive mixing of the chemical to avoid locally high supersaturation and spontaneous nucleation of $CaCO_3$ rather than crystal growth.
- Chemical mixing in the presence of a high seed surface area to promote crystal growth.
- Sufficient turbulence in the reactor to prevent scaling of nozzles and the reactor wall.

3. BENEFICIAL USES OF BRINE BY-PRODUCTS

Beneficial use of waste by-products requires innovative consideration of brine as a product as opposed to waste. Such beneficial use may hold potential to generate income from products that could offset some portion of the treatment cost. Such a treatment process might involve alternating applications of RO with previously discussed precipitation processes designed to remove the least soluble salts under conditions of controlled mixing, salt seeding, chemical addition, residence time, temperature, and pH. After removal of the salts that limited recovery in the primary RO, the brine could be treated again in a secondary RO for further product water recovery and brine volume reduction. Factors that must be considered in evaluating beneficial use of desalination by-products include the following:

- technical feasibility of isolating salts of the requisite morphology and purity,
- additional cost required to produce salt products,
- size and location of the market for specific salt products, and
- transportation costs to deliver salt products to market.

The salts that could potentially be produced from desalinated brine depend on the relative ion concentrations and the treatment processes used. For example, potential products from a shallow groundwater might include calcium carbonate, gypsum, magnesium hydroxide, calcium chloride, and sodium chloride. These salt products could potentially be generated for use in the manufacture of paper, ink, plastics, and paint products, as fertilizers and soil conditioners, as fillers for lightweight and fireproof building products, as chemical agents in water and wastewater treatment, and as dust suppressants. In order for a salt product recapture system to work, a demonstrated market for the sale of the salts must be available and proximal to the desalination site. Presently, the technical and economic feasibility of generating salt products from drinking water desalination still needs to be demonstrated, so enhanced evaporation systems remain a potentially viable disposal option.

4. LAS VEGAS VALLEY SHALLOW GROUNDWATER STUDY

Given the water scarcities present in the southwestern United States, many areas are looking to find new water resources to meet growing demand. The city of Las Vegas, Nevada, has experienced a population boom from approximately 1200 residents in 1905 to nearly 2 million residents in 2007. The main water source for the city is Lake Mead, which has seen a 100 ft decline in water levels during the past 8 years, reducing the volume from >27 million acre-feet to approximately 12.5 million acre feet, leaving the lake at well below half capacity.[3] Furthermore, Lake Mead (fed by the Colorado River) has a relatively high TDS content of around 700 mg/L. As such, the Southern Nevada Water Authority (SNWA) has been researching

[3]U.S. Bureau of Reclamation: http://www.usbr.gov/lc/region/g4000/hourly/levels.html.

(1) methods to soften the water through systems with high recovery and no addition of chemical softeners (e.g., ZLD RO systems) and (2) other sources of water, including brackish surface and groundwaters. In the remainder of this chapter we present two studies: one on the brackish groundwater found in and around Las Vegas, Nevada, and the other examining a ZLD sidestream softening plant for the Lake Mead surface water and several other water sources in the desert southwest.

The Las Vegas hydrographic basin, situated in southern Nevada, includes the Las Vegas Valley and portions of its flanking mountain blocks. The basin covers approximately 1600 square miles and contains the cities of Las Vegas, North Las Vegas, and Henderson in addition to contiguous unincorporated urbanized areas and Nellis Air Force Base. Groundwater in the Las Vegas Valley derives from the areas of greatest natural recharge in the adjoining Spring and Sheep Mountains and enters the Las Vegas Springs Aquifer, formerly termed the "principal" aquifer system, which contains the primary water producing alluvial horizons [29]. The Las Vegas Springs Aquifer is capped by a confining layer, the Las Vegas Wash aquitard, often mislabeled the "shallow aquifer." The shallow groundwater system lies within the upper 100 ft of the Las Vegas Wash aquitard [23]. In predevelopment times, the shallow groundwater in the central and eastern portions of the Las Vegas Valley originated though upward migration of water from the underlying aquifers through playa sediments that contain soluble sodium, calcium, and sulfate salts. In recent decades, the shallow system has been augmented by retention of the leaching fraction from irrigation applications.

Groundwater within the shallow system is brackish, with a salinity that varies by location from 1000 to about 10,000 mg/L TDS [23,30]. The State of Nevada has designated the shallow groundwater as a geologic hazard because of its high salinity, its impact on structures, the need to dewater during construction, and its potential water quality impact on the Las Vegas Wash and Lake Mead, home to several endangered species. Nevada water law permits beneficial use of water from shallow groundwater systems when the groundwater poses a geologic hazard (NRS 534.050 [2c] [31]). As part of its integrated resources planning beginning in 1995, SNWA sought to develop means of beneficial use that would add as much as 20,000 acre-feet per year of shallow groundwater to a diversified resource portfolio.

In 1999, Black & Veatch, Inc. conducted a prepilot study (Research and Development of the Saline Shallow Aquifer in the Southeastern Las Vegas Valley) that compared treatment and brine disposal alternatives for the shallow groundwater in an area on the southeast side of the Las Vegas

Valley. The following membrane treatment processes were compared by desktop (i.e., simulated) evaluation in the prepilot study:

- RO with disposal of the RO brine to evaporation ponds,
- lime/soda ash softening followed by RO with disposal of the RO brine to evaporation ponds,
- EDR with disposal of the EDR brine to evaporation ponds,
- RO with thermal desalination of the RO brine,
- lime/soda ash softening followed by RO and thermal desalination of the RO brine, and
- EDR with thermal desalination of the EDR brine.

The prepilot study concluded that RO membrane treatment followed by thermal desalination of the RO brine was the most cost-effective treatment process for obtaining potable water in that area. The study also recommended that a pilot study based on the RO process be conducted to confirm the technical feasibility of RO treatment, optimize RO operating parameters, and verify water quality predictions.

In late 2002, a contract was awarded to Black & Veatch to carry out a groundwater desalination pilot investigation, located at the City of Las Vegas Advanced Wastewater Treatment Facility. This study, completed in 2004, tested the performance of RO and NF as the primary treatment technologies on groundwater extracted at the site [24]. Pilot test results were used to develop full-scale design criteria and treatment costs for wells in the vicinity of the pilot site. The cost estimates did not include infrastructure, such as wells and pipelines. A desktop study was performed to evaluate the potential of four pretreatment options to increase RO membrane recovery and reduce treatment cost. Another desktop study applied the pilot RO and NF investigation results to computer models to evaluate treatment and costs at two additional Las Vegas locations.

Both pilot and modeling results indicated that irrespective of pretreatment techniques, a maximum of about 65% of the water could be recovered using membrane processes alone, leaving a substantial brine stream. Because brine disposal options such as deep well injection or discharge to a wastewater treatment plant are not permissible in the Las Vegas Valley, disposal options for the waste stream necessitated a ZLD treatment. To achieve ZLD, the brine generated during desalination must be treated to produce near dry salts and additional potable water. This has the benefit of raising the proportion of total recovered water to near 100%, but at a high cost. Options evaluated to treat the membrane waste brine were

- evaporation ponds,
- brine concentrator followed by a crystallizer, and
- brine concentrator followed by an evaporation pond.

Evaluation of these options was based on computer simulations and bench-scale testing results for thermal desalination. Disposal options for the brine concentrator waste stream included further thermal treatment using a crystallizer or discharge to an evaporation pond. The primary comparison criteria were cost and product water recovery. The quantity of solids produced and the eventual disposal of the solids are essentially the same for the three options.

The bench-scale testing was conducted by Resource Conservation Corporation (RCC) in conjunction with a pilot study to evaluate thermal treatment of the brine using a brine concentrator and a crystallizer. As tested, the membrane waste stream was determined to be fully compatible with evaporative concentration in a seeded slurry brine concentrator, and would feasibly recover water in excess of 95%, which at 1 mgd (million gallons per day) production would produce about 28 gpm feed to a crystallizer or evaporation pond. A crystallizer utilizing this product water would produce solids with 20% water content by weight. Overall water recovery with the concentrator and crystallizer was estimated to be 99.6%. These tests not only indicated the technical feasibility of thermal treatment, but also confirmed the high energy requirements and high costs that had been estimated previously by computer modeling. These results were used to evaluate full-scale costs for brine treatment.

Treating all of the RO brine in evaporation ponds was the most expensive option, required the most land, and had the least product recovery. An evaporation pond area of approximately 344 acres was required for a 1.02 mgd RO brine flow. Although operation and maintenance costs for ponds were low, the pond capital cost, including land and pond construction, was excessive. Furthermore, approximately 37% of the well production would be lost. The estimated annual cost of evaporation ponds, including amortized land acquisition but not infrastructure, was about \$11.29 per 1000 gal in excess of membrane treatment expenses, which totaled approximately \$1.61 $kgal^{-1}$.[4]

The second option, thermal treatment, produced the highest potable water recovery and had smallest footprint of the three options. Virtually all of the well water was recovered as product, with the only losses being a small purge stream from the crystallizer (approximately 1 gpm) and the water in the crystallized salts. The relative expense of the crystallizer was high: flow to the crystallizer was 6% of the flow to the brine concentrator, yet its capital cost was 58% and its operating cost was 91% of those for the

[4]All dollar amounts have been adjusted to 2007 values based on the United States Consumer Price Index. http://www.bls.gov/CPI/.

brine concentrator. Under this option, brine treatment costs were estimated to be about \$4.15 $kgal^{-1}$.

Brine concentration followed by pond evaporation proved the least expensive of the three options at about \$3.11 $kgal^{-1}$. At 98%, water recovery was only slightly less than that for the second option, and so this option was determined to be the most realistic brine treatment alternative of the three. As such, the estimated cost of such a system including RO treatment would be \$4.73 $kgal^{-1}$.

5. ZERO-LIQUID DISCHARGE WITH FLUIDIZED BED CRYSTALLIZER STUDY

A second research project involving the ZLD softening of Lake Mead surface water and several other sites was conducted for the American Water Works Association Research Foundation (AwwaRF) to evaluate an updated ZLD approach using fluidized bed crystallizers [24]. Testing was conducted on five brackish water sources that included three groundwaters, one surface water, and one reclaimed water.

The water sources and the ZLD applications evaluated in this research are listed in Table 1. TDS of the source waters ranged from 690 to 3500 mg/L. Applications included development of a new brackish source to reducing the salinity of a finished water supply.

Bench-scale testing was conducted at each site to evaluate chemical doses required to achieve target removals of calcium, barium, and silica. A pilot-scale study was conducted at one site, SNWA, to evaluate fluidized bed crystallizer performance under continuous-flow conditions. Modeling with computer software programs was performed to help predict sustainable RO recovery with various antiscalant products and to establish treatment goals for removal of calcium, barium, and silica from the primary RO concentrate.

The treatment goals established for each site are summarized in Table 2. For the treatment goals listed, the combined recovery of the two RO systems was projected to range from 94% to 98%. The analyses indicated that calcium removal would be required at each site, barium removal would be required at each site except San Antonio Water System (SAWS), and silica removal would be required at each site except SNWA.

6. TEST RESULTS

Calcium removal in fluidized bed crystallization bench-scale tests is exemplified by the Beverly Hills, California, results shown in Fig. 3.

Table 1 Source water characteristics and ZLD applications evaluated

	TDS (mg/L)	Hardness (mg/L as $CaCO_3$)	Silica (mg/L)	Water type	Application
SNWA, NV	690	290	9	Surface	Desalination of treated water to reduce TDS and hardness
Beverly Hills, CA	720	300	41	Groundwater	Recovery of concentrate from existing desalination facility
Phoenix, AZ	1300	68	27	Groundwater	Development of new water source
Scottsdale, AZ	1100	340	34	Reclaimed	Recovery of concentrate from existing water reclamation facility
SAWS, TX	3500	1720	57	Groundwater	Development of new water source

Calcium removal was greater than 60% at a NaOH dose of 100 mg/L, and greater than 80% at a NaOH dose of 250 mg/L. The NaOH dose required for meeting the treatment goals varied among the water sources tested, but effective calcium removal was achieved at each site with fluidized bed crystallization.

Barium was also effectively removed in fluidized bed crystallization tests and there was a correlation between barium removal and calcium removal as illustrated in Fig. 4. Barium fraction remaining is plotted against calcium fraction remaining in this figure for the SAWS, Scottsdale, and SNWA sites.

Table 2 Summary of concentrate treatment goals and projected recovery by reverse osmosis

	Primary RO recovery (%)	Ca removal (%)	Ba removal (%)	SiO_2 removal (%)	Secondary RO recovery (%)	Total RO recovery (%)
Beverly Hills, CA	73	26	33	84	85	96
SNWA, NV	85	20	63	0	76	96
Scottsdale, AZ	73	69	67	64	76	94
Phoenix, AZ	85	72	50	77	85	98
SAWS, TX	66	77	0	87	88	96

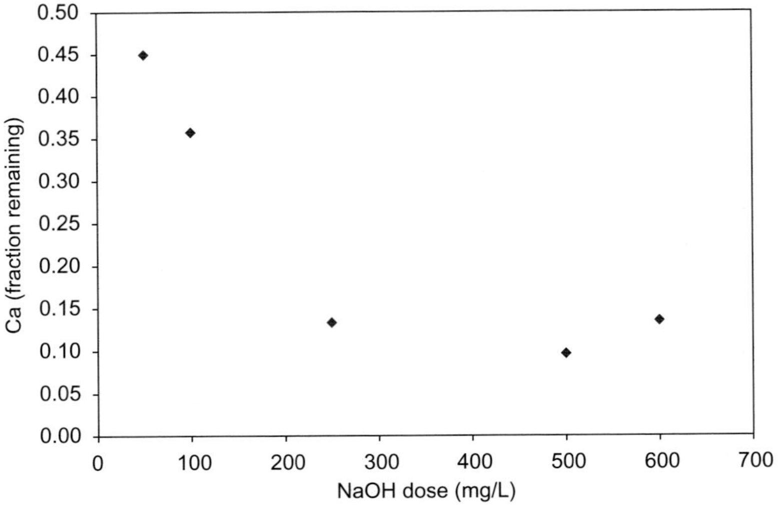

Figure 3 Calcium in fluidized bed crystallization tests at Beverly Hills, California.

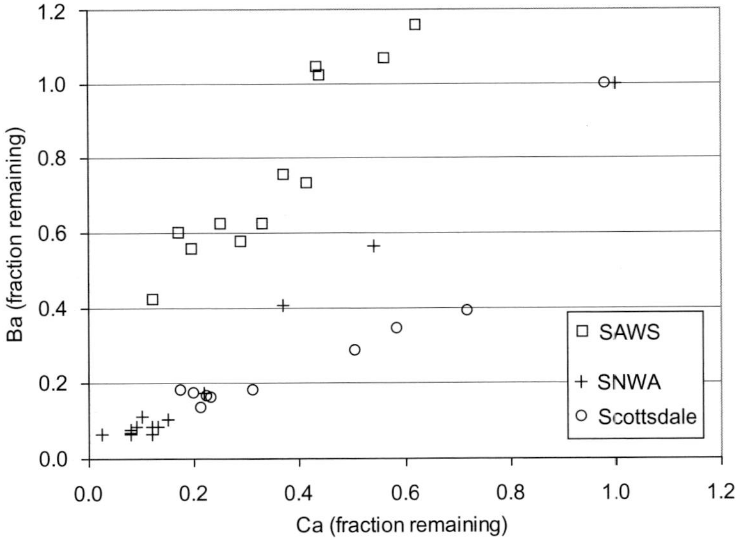

Figure 4 Correlation between barium and calcium in fluidized bed crystallization test.

It was theorized that the removal mechanism was substitution of barium for calcium in the calcium carbonate crystal lattice.

Silica, however, was not effectively removed in the fluidized bed crystallization test at pH below 10, as illustrated in Fig. 5, and it was

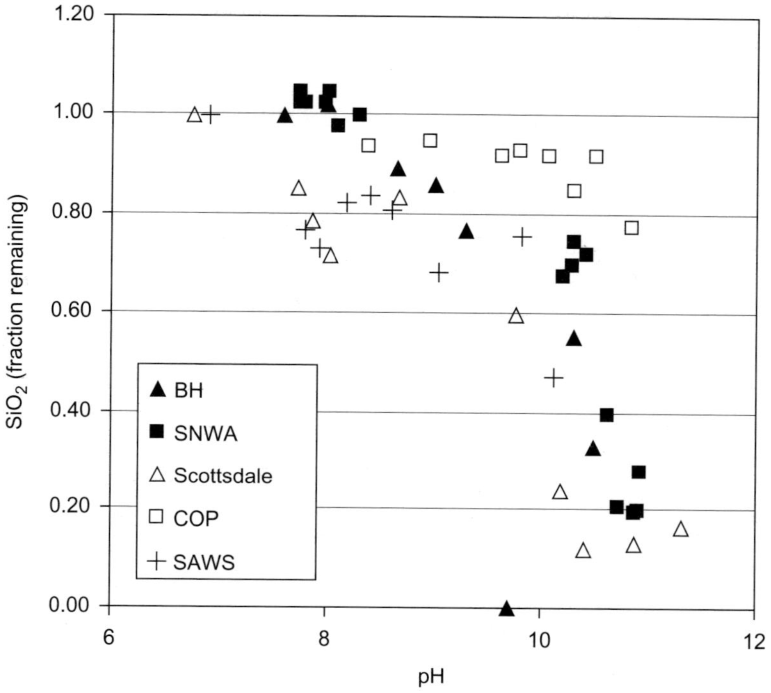

Figure 5 Silica in fluidized bed crystallization test.

concluded that effective silica removal in each test occurred only after the pH was high enough to cause $Mg(OH)_2$ precipitation. The need for $Mg(OH)_2$ precipitation for effective silica removal in chemical softening is a long-recognized phenomenon, and as illustrated in Fig. 6, correlations between silica and magnesium in the fluidized bed crystallization tests confirmed that silica removal occurred due to magnesium hydroxide precipitation.

Rather than increasing pH for silica removal, tests were conducted with alum and sodium aluminate to evaluate silica removal by adsorption to aluminum hydroxide. The tests showed that effective silica removal could be achieved in the pH range of 8–9 by adding either alum or aluminum hydroxide to the fluidized bed crystallization process in addition to NaOH or lime.

7. TREATMENT COSTS AND ENERGY REQUIREMENTS

The test results were used to establish treatment parameters and costs for each site. The goal of this research was to reduce the costs and energy

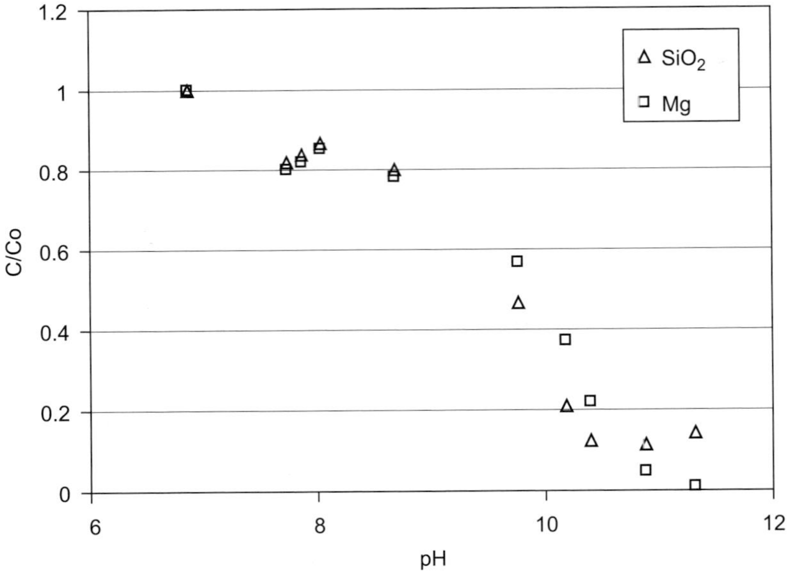

Figure 6 SiO_2 and Mg fractions remaining versus pH in Scottsdale fluidized bed crystallization test.

consumption for desalination with ZLD. To evaluate this, it was necessary to determine benchmark costs and energy consumption to serve as basis for comparison. The established technologies for treatment of RO concentrate to achieve ZLD are thermal desalination and evaporation ponds.

Consequently a brine concentrator followed by an evaporation pond was used as the benchmark ZLD process, and treatment costs were estimated for two process trains at each site: (1) the ZLD treatment process investigated in the AwwaRF study and (2) the benchmark process. The primary RO is identical for the two process trains. In the benchmark train, however, primary RO concentrate is discharged directly to a brine concentrator (Fig. 1), and in the evaluated process the primary RO concentrate is first treated by a fluidized bed crystallizer (Fig. 2), microfiltration (MF), and secondary RO before discharge to a brine concentrator. Blended product water quality and the total recovery for the benchmark and evaluated ZLD process are equivalent.

Fluidized bed crystallizer chemical doses required to meet concentrate treatment goals at each site for the evaluated ZLD process are listed in Table 3. Alum and sodium aluminate ($Na_2Al_2O_4$) were used for silica removal. The use of one or both was governed by the goal to maintain fluidized bed crystallizer effluent pH below 8.5. Alum suppresses pH while

Table 3 Summary of chemical doses calculated for fluidized bed crystallization

	Alum (mg/L)	$Na_2Al_2O_4$ (mg/L)	NaOH (mg/L)	Na_2CO_3 (mg/L)
Beverly Hills, CA	220	400	0	0
SNWA, NV	0	375	0	0
Scottsdale, AZ	0	431	170	350
Phoenix, AZ	10	280	0	0
SAWS, TX	0	844	600	550

sodium aluminate adds alkalinity. Sodium hydroxide was used to raise pH and increase calcium carbonate supersaturation. Soda ash (Na_2CO_3) was used where additional carbonate was needed for calcium carbonate precipitation.

Unit prices used in estimating treatment costs are listed in Table 4. Capital costs included equipment, installation, mechanical, electrical and instrumentation, engineering, and a 20% contingency. Because the objective was to evaluate treatment costs based on process selection, water quality, and the type of application, one set of unit prices was used for all sites. The goal was to ensure that cost differences were a reflection of water quality characteristics and not local economic factors.

Treatment costs were calculated only for the concentrate treatment processes. For the evaluated ZLD process, these were the fluidized bed crystallizer, secondary RO, brine concentrator, and evaporation pond. The processes for the benchmark ZLD process were the brine concentrator and evaporation pond.

An annual treatment cost was calculated for the evaluated ZLD process and the benchmark process at each site. The annual treatment costs included capital costs amortized over 20 years at a 6% annual rate of return, and all operation and maintenance costs. The annual treatment cost was used to calculate a unit treatment cost per kilogallon of concentrate recovered.

The treatment cost comparison is summarized in Table 5. Costs for the benchmark ZLD process ranged from \$12.05 to \$12.96 $kgal^{-1}$ concentrate recovered compared to \$3.92–6.63 $kgal^{-1}$ concentrate recovered for the evaluated ZLD process.

Energy requirements for the benchmark and evaluated ZLD processes were compared by calculating the energy requirements for the RO

Table 4 Unit cost and design factors used in cost projections

Item	Unit cost
Capital recovery factor (based on 0.06 rate of return and 20 year return period)	0.09
Energy	\$0.12 kWh^{-1}
NaOH	\$400 t^{-1}
$Na_2Al_2O_4$	\$200 t^{-1}
Na_2CO_3	\$400 t^{-1}
Alum	\$300 t^{-1}
Antiscalant	\$1 lb^{-1}
RO capital cost, pressure < 600 psi	\$1.35 gpd^{-1}
RO capital cost, pressure > 600 psi	\$4.00 gpd^{-1}
RO element cost	\$500
RO element life	5 years
MF capital cost	\$0.57 gpd^{-1}
MF element cost	\$500
MF element life	5 years
Fluidized bed crystallizer capital cost	$a \times Q^{0.7}$, where Q is capacity (gpd) and a is a coefficient that increased with chemical dose $a = 70$ for Beverly Hills, SNWA, and Phoenix $a = 75$ for Scottsdale $a = 80$ for SAWS
Fluidized bed crystallizer power consumption (kWh/day)	$2.84 \times Q$, where Q is capacity (mgd)
Pond capital cost	\$200,000 $acre^{-1}$ (includes land cost)
Brine concentrator capital cost	8130Q+3,000,000, where Q is capacity (gpm)
Brine concentrator energy (kW)	4.85Q−192, where Q is capacity (gpm)
Building area	2000 ft^2/mgd capacity RO plus 1000 ft^2/mgd capacity of MF, and fluidized bed crystallizer
Building capital cost	\$150 ft^{-2}

Table 5 Summary of comparison of concentrate treatment costs ($/kgal concentrate recovered)

	Benchmark ZLD process ($)	Evaluated ZLD process ($)	Percent difference
Beverly Hills, CA	12.86	5.34	58
SNWA, NV	12.96	4.51	65
Phoenix, AZ	12.05	3.92	67
Scottsdale, AZ	12.57	5.29	58
SAWS, TX	12.83	6.63	48

high-pressure pumps and the brine concentrator for each. All other energy uses are insignificant, relative to the energy required for RO and the brine concentrator.

The energy comparison is summarized in Table 6. The greatest energy sink in the treatment process is the brine concentrator. The evaluated ZLD process requires significantly less energy because it uses less brine concentrator capacity. ZLD desalination is a new endeavor in the drinking water industry. ZLD has been practiced in other industries, most notably at power plants, but the established technologies are expensive and energy intensive. The ZLD research summarized here demonstrates that the costs and energy requirements for ZLD can be reduced significantly through development of new ideas and practices.

8. OUTCOMES AND FUTURE CONSIDERATIONS

The pilot, bench-scale, and desktop studies undertaken for the Las Vegas Valley indicate that desalination, while technically feasible, is costly. The initial estimates were, at best, $4.73 kgal^{-1} with substantial water loss involved. The follow-up study indicated that an optimized ZLD system could be set in operation at a cost of $4.51 kgal^{-1} with improved water recovery and reduced energy consumption. However, these costs still compare unfavorably to production costs of less than $1.23 kgal^{-1} for water from other sources. As such, one must carefully weigh the costs and benefits of implementing a desalination system, especially in inland settings where high recovery and adequate means for brine disposal are paramount to success.

Table 6 Summary of energy requirements (kWh/kgal concentrate recovered)

	Benchmark ZLD process	Evaluated ZLD process
Beverly Hills, CA	12.1	64.9
SNWA, NV	20.6	68.6
Phoenix, AZ	12.0	74.2
Scottsdale, AZ	21.2	75.9
SAWS, TX	16.9	75.3

In a critical discussion of the interrelationship between increasing supply and rising demand, and the distinction between true needs and perceived needs, von Medeazza cautions against unchecked expansion into desalination and production of water simply because the resources exist to continue society's water-intensive lifestyles [17]. Such considerations must be included in future discussion of the value of tapping into unexploited water resources. Energy requirements, environmental impacts, monetary costs, and societal benefits must all be weighed in the final decision about using desalination for drinking water production at inland sites. If desalination of Las Vegas Valley shallow groundwater is to be a realistic option, a system must be designed that accounts for true population needs, local conditions, and takes advantage of as many factors as possible to optimize energy consumption and maximize profit potential. As a general rule, such an approach will be necessary anywhere – each facility will need to be tailored to match site-specific conditions.

REFERENCES

[1] H.-R. Aerni, B. Kobler, B. Rutishauser, F. Wettstein, R. Fischer, W. Giger, A. Hungerbühler, M.D. Marazuela, A. Peter, R. Schönenberger, A.C. Vögeli, M.F. Suter, R.L. Eggen, Combined biological and chemical assessment of estrogenic activities in wastewater treatment plant effluents, Anal. Bioanal. Chem. 378(3) (2004) 688–696.

[2] M. Ahel, E. Molnar, S. Ibric, W. Giger, Estrogenic metabolites of alkylphenol polyethoxylates in secondary sewage effluents and rivers, Water Sci. Technol. 42(7–8, Hazard Assessment and Control of Environmental Contaminants (Ecohazard 99)) (2000) 15–22.

[3] J.H. Al-Rifai, C.L. Gabelish, A.I. Schafer, Occurrence of pharmaceutically active and non-steroidal estrogenic compounds in three different wastewater recycling schemes in Australia, Chemosphere 69(5) (2007) 803–815.

[4] S.A. Snyder, T.L. Keith, D.A. Verbrugge, E.M. Snyder, T.S. Gross, K. Kannan, J.P. Giesy, Analytical methods for detection of selected estrogenic compounds in aqueous mixtures, Environ. Sci. Technol. 33(16) (1999) 2814–2820.
[5] S.A. Snyder, D.L. Villeneuve, E.M. Snyder, J.P. Giesy, Identification and quantification of estrogen receptor agonists in wastewater effluents, Environ. Sci. Technol. 35(18) (2001) 3620–3625.
[6] D. Voutsa, P. Hartmann, C. Schaffner, W. Giger, Benzotriazoles, alkylphenols and bisphenol a in municipal wastewaters and in the Glatt River, Switzerland, Environ. Sci. Pollut. Res. Int. 13(5) (2006) 333–341.
[7] J.V. Brian, C.A. Harris, M. Scholze, A. Kortenkamp, P. Booy, M. Lamoree, G. Pojana, N. Jonkers, A. Marcomini, J.P. Sumpter, Evidence of estrogenic mixture effects on the reproductive performance of fish, Environ. Sci. Technol. 41(1) (2007) 337–344.
[8] F. Brion, C.R. Tyler, X. Palazzi, B. Laillet, J.M. Porcher, J. Garric, P. Flammarion, Impacts of 17[Beta]-estradiol, including environmentally relevant concentrations, on reproduction after exposure during embryo-larval-, juvenile- and adult-life stages in zebrafish (*Danio rerio*), Aquat. Toxicol. 68(3) (2004) 193–217.
[9] R. Gibson, M.D. Smith, C.J. Spary, C.R. Tyler, E.M. Hill, Mixtures of estrogenic contaminants in bile of fish exposed to wastewater treatment works effluents, Environ. Sci. Technol. 39(8) (2005) 2461–2471.
[10] P. Labadie, H. Budzinski, Alteration of steroid hormone profile in juvenile turbot (*Psetta maxima*) as a consequence of short-term exposure to 17[alpha]-ethynylestradiol, Chemosphere 64(8) (2006) 1274–1286.
[11] S.A. Snyder, E.C. Wert, H.D. Lei, P. Westerhoff, Y. Yoon, Removal of EDCs and Pharmaceuticals in Drinking and Reuse Treatment Processes, Awwa Research Foundation, Denver, CO, 2007.
[12] P. Andrzejewski, B. Kasprzyk-Hordern, J. Nawrocki, *N*-Nitrosodimethylamine (NDMA) formation during ozonation of dimethylamine-containing waters, Water Res. 42(4–5) (2008) 863–870.
[13] W. Lee, P. Westerhoff, J.P. Croue, Dissolved organic nitrogen as a precursor for chloroform, dichloroacetonitrile, *N*-nitrosodimethylamine, and trichloronitromethane, Environ. Sci. Technol. 41(15) (2007) 5485–5490.
[14] E. Steinle-Darling, M. Zedda, M.H. Plumlee, H.F. Ridgway, M. Reinhard, Evaluating the impacts of membrane type, coating, fouling, chemical properties and water chemistry on reverse osmosis rejection of seven nitrosoalklyamines, including NDMA, Water Res. 41(17) (2007) 3959–3967.
[15] A.M.O. Mohamed, M. Maraqa, J. Al Handhaly, Impact of land disposal of reject brine from desalination plants on soil and groundwater, Desalination 182(1–3) (2005) 411–433.
[16] M. Mickley, Concentrate management, in: M. Wilf (Ed.), The Guidebook to Membrane Desalination Technology: Reverse Osmosis, Nanofiltration and Hybrid Systems Process, Design, Applications and Economics, Balaban Desalination Publications, L'Aquila, Italy, 2007, p. 524.
[17] G.L.M. von Medeazza, "Direct" and socially-induced environmental impacts of desalination, Desalination 185(1–3) (2005) 57–70.
[18] J.P. MacHarg, The evolution of SWRO-energy recovery systems, Desalination and Water Reuse 11(3) (2001) 49–53.
[19] S.V. Veerapaneni, B. Long, S. Freeman, R. Bond, Reducing energy consumption for seawater desalination, J. Am. Water Works Assoc. 99(6) (2007) 95–106.
[20] B. van der Bruggen, L. Lejon, C. Vandecasteele, Reuse, treatment, and discharge of the concentrate of pressure-driven membrane processes, Environ. Sci. Technol. 37(17) (2003) 3733–3738.
[21] R. Semiat, D. Hasson, G. Zelmanov, I. Hemo, Threshold scaling limits of RO concentrates flowing in a long waste disposal pipeline, Water Sci. Technol. 49(2) (2004) 211–219.

[22] H. Abu Qdais, Environmental impacts of the mega desalination project: the Red-Dead Sea conveyor, Desalination 220(1–3) (2008) 16–23.
[23] J.F. Leising, Chemical Conditions in the Primary Producing Aquifers and Portions of the Shallow Groundwater System of the Las Vegas Valley in 2000, Southern Nevada Water Authority in cooperation with the Las Vegas Valley Water District, Las Vegas, NV, 2004.
[24] R.G. Bond, S. Veerapaneni, Zero Liquid Discharge and Volume Minimization for Inland Desalination, AwwaRF and AWWA, Denver, CO, 2007.
[25] A. Hina, G.H. Nancollas, M. Grynpas, Surface induced constant composition crystal growth kinetics studies, J. Cryst. Growth 223 (2001) 213–214.
[26] S.T. Liu, G.H. Nancollas, A kinetic and morphological study of the seeded growth of calcium sulfate dihydrate in the presence of additives, J. Colloid. Interface Sci. 52 (1975) 593–601.
[27] J.C. Graveland, P.J. de Moel, J.H. Oomen, Developments in water softening by means of pellet reactors, J. Am. Water Works Assoc. 75(12) (1983) 619.
[28] C. van der Veen, A. Graveland, Central softening by crystallization in a fluidized bed process, J. Am. Water Works Assoc. 80(6) (1988) 51.
[29] D.J. Donovan, Hydrostratigraphy and Allostratigraphy of the Cenozoic Alluvium in the Northwestern Part of Las Vegas Valley, Thesis, Department of Geosciences, University of Nevada, Las Vegas, 1996, 197 pp.
[30] K.S. Zikmund, Extent and Potenital Use of the Shallow Aquifer and Wash Flow in Las Vegas Nevada, Southern Nevada Water Authority, Las Vegas Valley Water District, Las Vegas, NV, 1996.
[31] R.M. Turnipseed, Regulations for Water Well and Related Drilling, Nevada Department of Conservation and Natural Resources, Division of Water Resources, Las Vegas, NV, 1998.

PART Five

Comparison of Desalination versus Water Recycling

CHAPTER 12

Renewable Energy Powered Water Treatment Systems

Bryce S. Richards[1] *and* **Andrea I. Schäfer**[2]
[1]School of Engineering and Physical Sciences, Heriot-Watt University, Edinburgh EH14 4AS, UK
[2]School of Engineering, University of Edinburgh, Edinburgh EH9 3JL, UK

Contents

1. INTRODUCTION

Renewable energy is defined as energy that is generated from natural resources that are replenished both naturally and constantly – thus, from sunlight, wind, rain, tides, waves, and geothermal heat. The International Energy Agency (IEA) defines renewable energy as "electricity and heat generated from solar, wind, ocean, hydropower, biomass, geothermal resources, and biofuels and hydrogen derived from renewable resources" [1].

In the present age, there are a wide range of motivations for powering water treatment systems from renewable energy resources, such as the following:

- There is an increasing demand for water around the world.
- The cost of traditional fuels for power and/or thermal energy generation such gas, oil and coal are increasing.

Sustainability Science and Engineering, Volume 2 ISSN 1871-2711, DOI 10.1016/S1871-2711(09)00212-8

- Security of supply – some of the traditional fuels, such as gas, are expected to be mostly used up within the next 50 years, while supplies of other fuels can depend on political stability.
- The cost of both desalination systems and RE technologies is falling.
- The need for small systems to operate in remote areas that often do not have an electricity grid.
- And finally, climate change is an increasing motivation – as per the Intergovernmental Panel on Climate Change:

  Warming of the climate system is unequivocal, as is now evident from observations of increases in global average air and ocean temperatures, widespread melting of snow and ice and rising global average sea level [2]

The majority of renewable energy technologies are powered by the sun, as outlined below, except for geothermal energy and tidal energy.

1.1 Solar energy

Solar energy refers to both electricity and thermal energy that is harnessed from the sun. Solar thermal energy can be harnessed either passively – where no moving parts are required such as for passive solar building design – or actively for a wide range of applications, including domestic hot water and space heating via solar thermal collectors. Direct current (DC) electricity can be generated directly from sunlight using photovoltaic (PV) or solar cells that contain no moving parts. In addition, photoelectrochemical cells can also be used to generate solar hydrogen, and while this offers interesting potential for energy storage, it is neither a mature nor widespread technology and will therefore not be mentioned further in this chapter.

1.2 Wind energy

The uneven heating of the earth by the sun results in the poles receiving less solar energy than that received on the equator. In addition, land is able to both heat up and cool down more quickly than the oceans do. This differential heating results in the jet stream in the upper atmosphere and the characteristic winds on earth: midlatitude westerlies, polar easterlies, and the trade winds. Wind energy is most commonly harvested via a windmill for generating mechanical energy or a wind turbine for generating electrical energy.

1.3 Wave energy

Wave energy is in effect a stored and concentrated form of solar energy, since the waves are generated by wind passing over them, and as long as the waves propagate slower than the speed of the wind speed (just above the waves), then a transfer of energy occurs from the wind to the most energetic waves.

1.4 Tidal Energy

Tidal energy devices exploit the natural rise and fall of coastal tidal waters resulting from the interaction of the gravitational fields of the sun and the moon. In some estuaries, the difference between high and low tide is accentuated and can create tidal ranges of up to 11 m. While several demonstration projects exist worldwide – notably a 240 MW barrage in La Rance, France, that has been operational since 1967 – there currently no major expansions of this technology anticipated. In addition, given that this resource is more geographically restricted than those above, it will not be discussed in more detail in this chapter.

1.5 Small-scale hydroelectric energy

Hydro schemes convert the energy available in flowing water (rivers, canals, or streams) into electricity. The technology is commercially and technically mature, with small-scale hydro projects being defined as having an installed capacity of less than 10 MW. Importantly, it also has a greatly reduced environmental impact compared to the flooding of valleys required for large-scale hydro projects. The majority of the world's small-scale hydro projects are found in China.

1.6 Geothermal energy

Geothermal energy originates from heat generated deep within the earth. While naturally occurring water from aquifers with a temperature of 50–150 °C can be used for district heating, temperatures of over 150 °C are required for electricity generation. The primary disadvantage of geothermal energy is that the geological conditions that determine the quality of the resource – such as formation fluid temperature and flow rate – are difficult to predict in advance without significant capital investment in drilling and tests. Consequently, geothermal energy is regarded as a high-risk investment relative to other forms of energy production.

2. RENEWABLE ENERGY IN THE WORLD TODAY

In 2006, about 12.3% of world total primary energy consumption (TPES, includes all forms of energy) came from renewables, with the largest large fraction (10.1%) coming from traditional biomass sources, such as wood-burning. Hydroelectricity was the next largest renewable source, providing 2.2% of TPES in 2008, as well accounting for 16% of global electricity

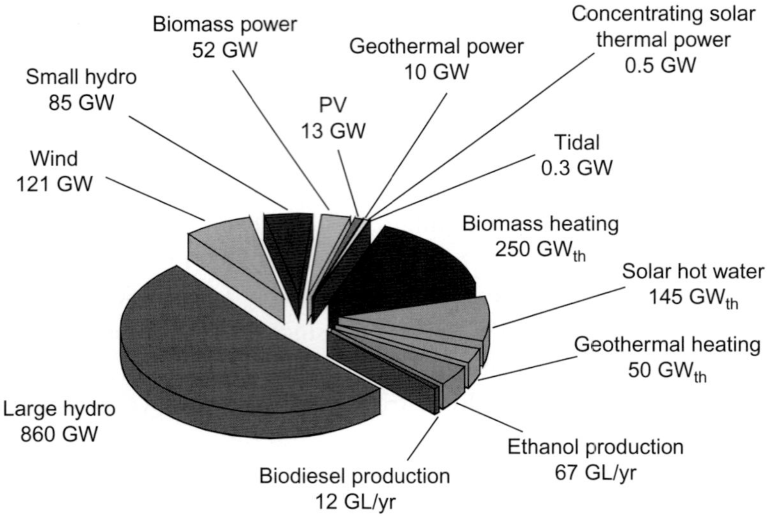

Figure 1 World renewable energy supply capacity by the end of 2008, broken down in electricity generation (units: GW), thermal generation (GW_{th}) and bio-fuel production (units: GL/year) [4].

generation [3]. Fig. 1 below shows the world renewable energy supply capacity by the end of 2008 [4].

From Fig. 1, it can be seen that wind power has an installed capacity of 121 MW worldwide. The wind industry this is growing at annual rate of 30%, with widespread use in Europe and the USA [4]. The annual manufacturing output of the photovoltaics (PV) industry reached a record 6.9 GW in 2008, bringing the installed capacity to 13 MW with the largest "solar farms" operating Germany, Spain, and Portugal. Several large solar thermal power plants operate in USA and Spain with the largest of these being the 354 MW SEGS power plant in the Mojave Desert. While being more geographically restricted than other renewable energy sources, the world's largest geothermal power installation is located in California, with a rated capacity of 750 MW [4]. Hydroelectricity is also, by its nature, more geographically limited and growth in this area has been less, about 8% for small-scale hydro projects and 3% in large-scale hydro projects. Brazil is leading the world in ethanol production from sugar cane, with ethanol now accounting for 18% of that country's automotive fuel consumption [4].

While many of the above renewable energy projects are designed for large-scale power generation for the electricity grid, several of these technologies are also very well-suited to small off-grid applications, especially in remote areas. For example, "solar homes systems" – consisting of a PV panel (20–100 W), battery, charge controller, and DC lights – are very popular in countries such as China, Sri Lanka, India, Bangladesh, and Kenya.

Therefore, in order to understand the potential for powering a water treatment system – whether water recycling or desalination – from renewable energy, it is important to have a good understanding of local resource availability. Fig. 2 shows the world average availability of solar irradiance, measured as the daily number of peak sunshine hours incident on a horizontal surface at an intensity of 1 kW/m^2 [5].

It can be seen that the solar radiation resource is very good throughout the North America, South America, and much of Asia, while an excellent solar resource is found in Africa, the Middle East, and Australia. While this serves as a rough guide as to where solar-powered systems would be a good choice, a system designer would also need to consider seasonal variation in the solar resource and how well demand (in this case clean water) matches the supply of energy. For a critical application where, for example, the system would be the sole source of clean water for a community, solar energy systems are typically sized for the month with the least solar irradiance. This is often winter at greater latitudes, but in the tropics, this usually coincides with wet seasons.

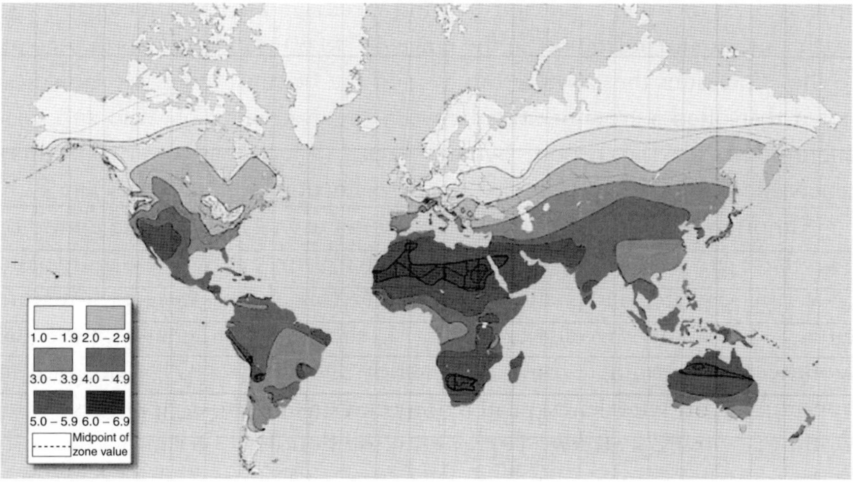

Figure 2 World solar irradiance, plotted as the number the daily number of peak sunshine hours incident on a horizontal surface at an intensity of 1 kW/m^2 (adapted from Ref. [5]) (see plate 6 in color plate section at the end of this book).

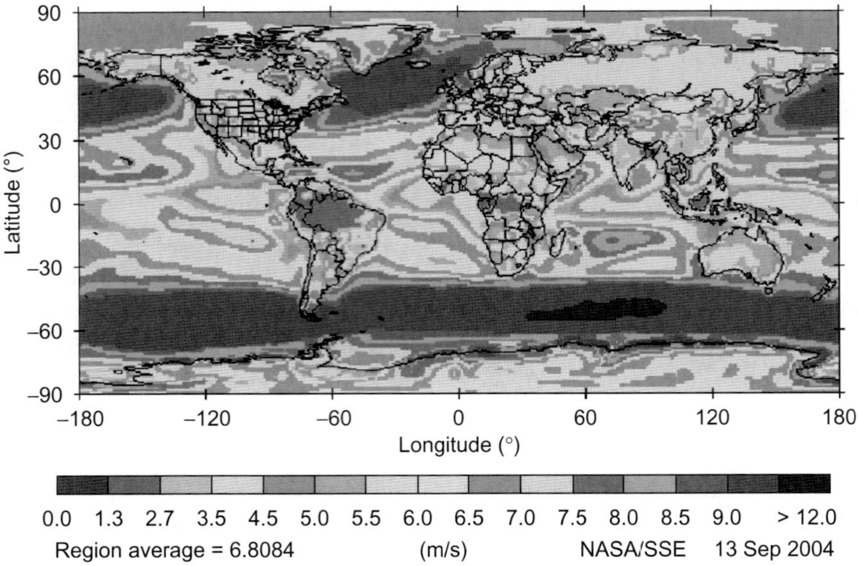

Figure 3 World map of average annual wind speed at a height of 50 m based on 10 years of data (July 1983 to June 1993). Adapted from Ref. [6] (see plate 7 in color plate section at the end of this book).

Fig. 3 shows a world map of average annual wind resource, depicted as wind speed at a height of 50 m based on 10 years of data (July 1983 to June 1993) [6]. As mentioned previously, wind speeds are less near the equator and reach a maximum at latitudes in the range 40° to 60° south and north, leading to the expressions the "roaring forties" and the "furious fifties." It should be noted that often a synergy exists between the availability of wind and solar energy and, for this reason, hybrid systems, which rely on two sources renewable electricity to maximize water production over all four seasons.

Although marine energy is a much less mature technology than wind or PV, there are obvious synergies between marine energy availability and powering seawater desalination plants. Therefore, as an example of marine energy availability worldwide, Fig. 4 shows wave energy potential worldwide, determined from 15 years of satellite data [7]. The greatest opportunity for wave energy harvesting exists along those coastlines in the world that possess a western exposure to the Southern Ocean (Chile, parts of Australia and New Zealand) as well as parts of Europe (Ireland, Scotland, Iceland) and as well as western Canada and South Africa. Additional factors when selecting a site include how steady the resource is – both in strength and direction – and therefore the most promising areas are probably the

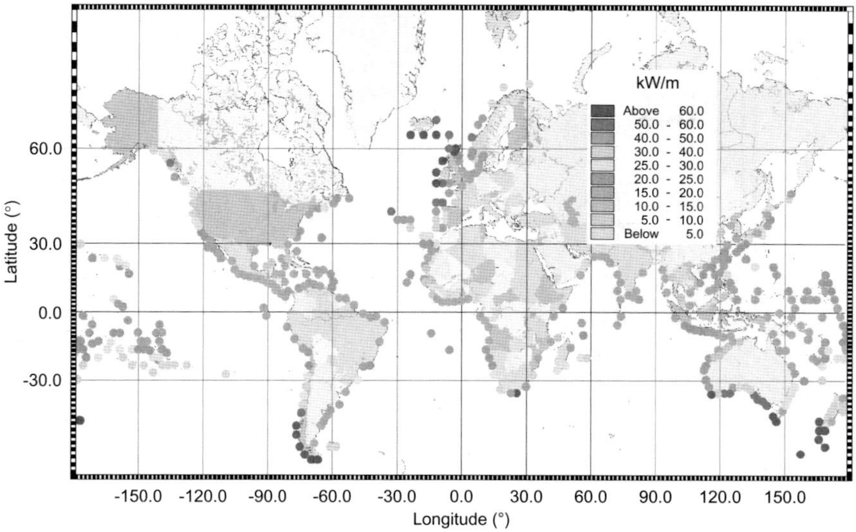

Figure 4 Wave energy estimates (in units of kW/m) along global coastlines determined via satellite data. Adapted from Ref. [7] (see plate 8 in color plate section at the end of this book).

islands in the trade wind belt of the Pacific [7]. A further impetus for such development is that high cost of imported diesel that is used for power generation on these islands.

Fig. 5 compares the cost of electricity (CoE) generated from renewable energy sources (PV, wave, wind, tidal, biomass) compared to coal, gas, and nuclear. For some technologies, there is quite a spread in the CoE, which is influenced by the scale of the generation system and the quality of the renewable resource.

With the majority of renewable energy sources being variable in nature, the traditional design approach is that renewable energy powered systems will require some form of energy storage to accommodate variability in the resource availability. The only renewable energy resource that can sometimes be regarded as continuous is small-scale hydro projects, for streams and rivers that do not experience a dry season.

Energy storage can take many forms, including mechanical energy devices such as pressure accumulators and flywheels, and electrical energy storage devices such as batteries, supercapacitors, and fuel cells. Alternatively, if the water treatment system is connected to the electricity grid, electricity can be exported during times of excess generation and imported when the renewable resource provides insufficient power.

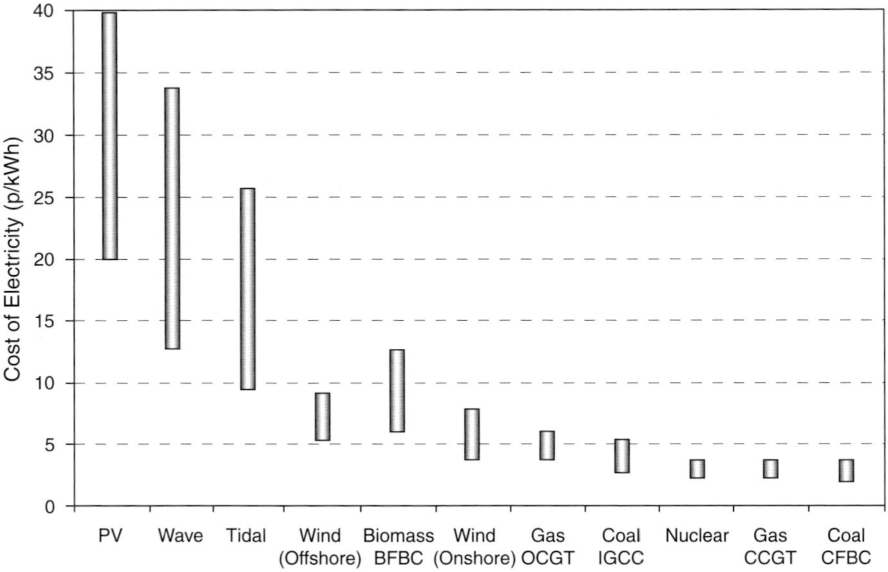

Figure 5 Cost of electricity (in British pence per kWh) generated from both renewable and traditional energy sources for the UK. BFBC = bubbling fluidized bed combustion, OCGT = open cycle gas turbine, IGCC = integrated gasification combined cycle, CFBC = circulating fluidized bed combustion. Adapted from Ref. [8].

3. RENEWABLE ENERGY POWERED WATER TREATMENT TECHNOLOGIES

The most common renewable energy technologies for powering water treatment systems in the past have been PV, solar thermal energy, and wind energy. This is shown in Fig. 6 for the following desalination technologies: reverse osmosis (RO) including nanofiltration, multieffect distillation (MED), electrodialysis (ED), multistage flash (MSF), and mechanical vapor compression (MVC). Although, no examples of renewable energy powered water recycling schemes exist at the present time, there are plans to develop such schemes. An Australian scheme is described in a section below.

Before considering the energy consumption of various technologies, it is important to understand the implications of the chosen desalination technology. Desalination can be accomplished via phase change (including MED, MSF, MVC) or membrane separation (including RO, ED, NF) processes. The former all involve a phase change of the feedwater (either to vapour or solid), whereas technologies like RO or ED rely on the filtration properties of polymeric membranes. The figure of merit for clean water production is the specific energy consumption (SEC), defined as how many

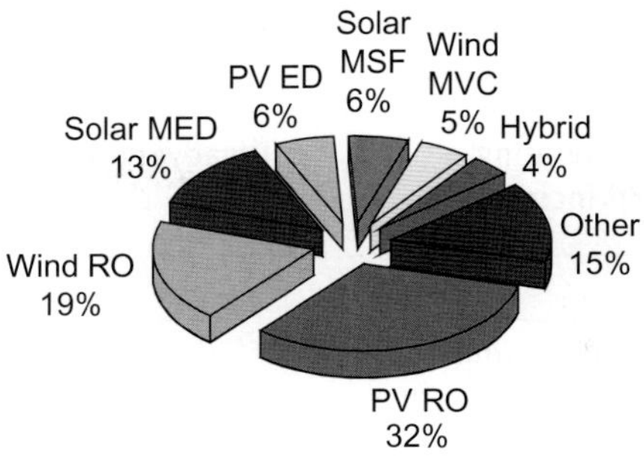

Figure 6 Breakdown of renewable energy powered desalination system technologies implemented worldwide Ref. [10].

units of clean drinking water can be produced for each unit of energy consumed (units: kWh/m^3). The SEC of a phase change process is proportional to the amount of water produced, whereas the energy requirements for a membrane separation process are proportional to the salinity of the feedwater.

Table 1 compares all existing renewable energy powered desalination technologies, highlighting the energy consumption and disadvantages of each technology. For RO and NF systems, the major energy requirement is for pressurising the feedwater, with brackish water systems typically operating at pressures of about 5–15 bar, while seawater desalination ranges from about 40 to 60 bar. It should be noted that many technologies are still going through a learning curve: overall, the SEC of Spanish seawater desalination plants has decreased from $22\,kWh/m^3$ in 1970 to $8\,kWh/m^3$ by 1990, and is presently at $4\,kWh/m^3$ [9].

Given that the capital cost of installing a renewable energy system is high, it is naturally desirable to couple this with the desalination technology with the lowest SEC. However, this is not the only consideration. For example, if significant amounts of low-grade heat are available, then perhaps one of the phase change processes could offer a lower lifecycle water cost if operation and maintenance costs are less than for RO systems.

4. SYNERGY BETWEEN RENEWABLE ENERGY RESOURCE AND WATER SUPPLY

It is critical to recognize that there can be synergies between the availability of a renewable energy and water resources.

One interesting example that illustrates this is the provision of clean drinking water to remote communities in outback Australia. While arid

Table 1 Comparison of all existing renewable energy powered desalination technologies

Technology	Operating principle	Disadvantages	References
Phase change processes			
Solar still	Solar thermal energy evaporates the water, which condenses onto the sloping glass surface and it then drains into a collection trough	• High SEC of 639 kWh/m^3 • Low daily production • High maintenance costs • Glass sheets vulnerable to storms and vandalism	[11]
Multistage flash (MSF)	Saline water held under pressure at ~120 °C and "flashed" into vapor in a series of ~50 chambers, which then condenses and is collected.	• Both thermal and electric energy required • High SEC of 20–64 kWh/m^3 (electrical component ~4 kWh/m^3)	[12–14]
Multiple effect distillation (MED)	Thin film evaporation process where vapour formed one chamber condenses in the next, providing a heat source for further evaporation.	• Both thermal and electric energy required • High SEC of 15 kWh/m^3 (electrical component ~2 kWh/m^3)	[12]
Mechanical vapor compression (MVC)	Evaporative system where vapor boiled off in the evaporator is mechanically compressed and reused as the heating medium	• Both thermal and electric energy required • High SEC of 11–16 kWh/m^3	[12]

Table 1 *(Continued)*

Technology	Operating principle	Disadvantages	References
Freeze separation (FS)	Ice crystals formed in feed-water are then separated and subsequently melted to form the product water	• High SEC of 97 kWh/m^3 • Separating ice crystals from the brine; operation in vacuum required due to lower freezing point of saline water	[13–15]
Membrane separation processes			
Reverse osmosis (RO)	Pressure-driven separation of two solutions with differing salt concentrations across a semipermeable membrane	• Low SEC of 4 kWh/m^3 for seawater • Specialized chemicals not available in remote locations • Chemicals required to control fouling: increases system complexity and cost, and reduces system reliability • Membrane life 3–5 years	[13,16]
Nanofiltration (NF)	As above	• As above, but reduced SEC for brackish water (2 kWh/m^3) • Not suitable for seawater	[13,17]
Electrodialysis (ED)	Electromigration of ions through cation and anion exchange membranes	• Low SEC of 2 kWh/m^3 for brackish water • Chemical cleaning required • No pretreatment for removing particulates • Not suitable for seawater	[13,18, 19–21]

Table 1 *(Continued)*

Technology	Operating principle	Disadvantages	References
Electrodialysis reversal (EDR)	As above, however, electrode polarity is periodically reversed to facilitate cleaning of ED membrane	• As above, however reduced chemical cleaning required	[22]

countries experience minimal rainfall and hence limited freshwater availability, there is often an abundance of solar radiation received at such locations. In addition, there are often significant groundwater reserves available, although these are often of marginal (total dissolved solids [TDS] 0.5–1.5 g/L) or brackish (TDS 1.5–5 g/L) quality. This is the situation in Australia, where the majority of the rainfall occurs along the coastline, compared to 200–300 mm annual rainfall in central and Western Australia, as shown in Fig. 7a [23]. The arid region in central Australia receives a daily average of at least 6.7 h of full sunshine (kW/m^2) [24] as shown in Fig. 7b, which is 20–50% more than is received along the wetter coastline. While this climate and freshwater availability reflects very strongly in the country's population distribution, as shown Fig. 7c, a large fraction of Central Australia is farmland or home to indigenous communities that often rely on poor water resources, with many communities being too small to have controlled and monitored water supplies. Drinking water for these communities is generally supplied from groundwater bores, which are of varying quality ranging from drinkable water to inconsumable brackish water. Fig. 7d indicates that brackish groundwater can be found in significant volumes throughout the majority of Australia, mostly with good extraction rates [26]. Given that the consumption of brackish water has been linked to poor health and that many of the central regions of Australia are not serviced by the national electricity grid, communities are often drinking water of substandard quality, as they do not possess the electrical power or appropriate technology to purify the water. Therefore, application and feasibility of PV-powered desalination systems, both on a small [17] a large scale [27] have been investigated as a sustainable technology for the provision of clean water in remote areas of outback

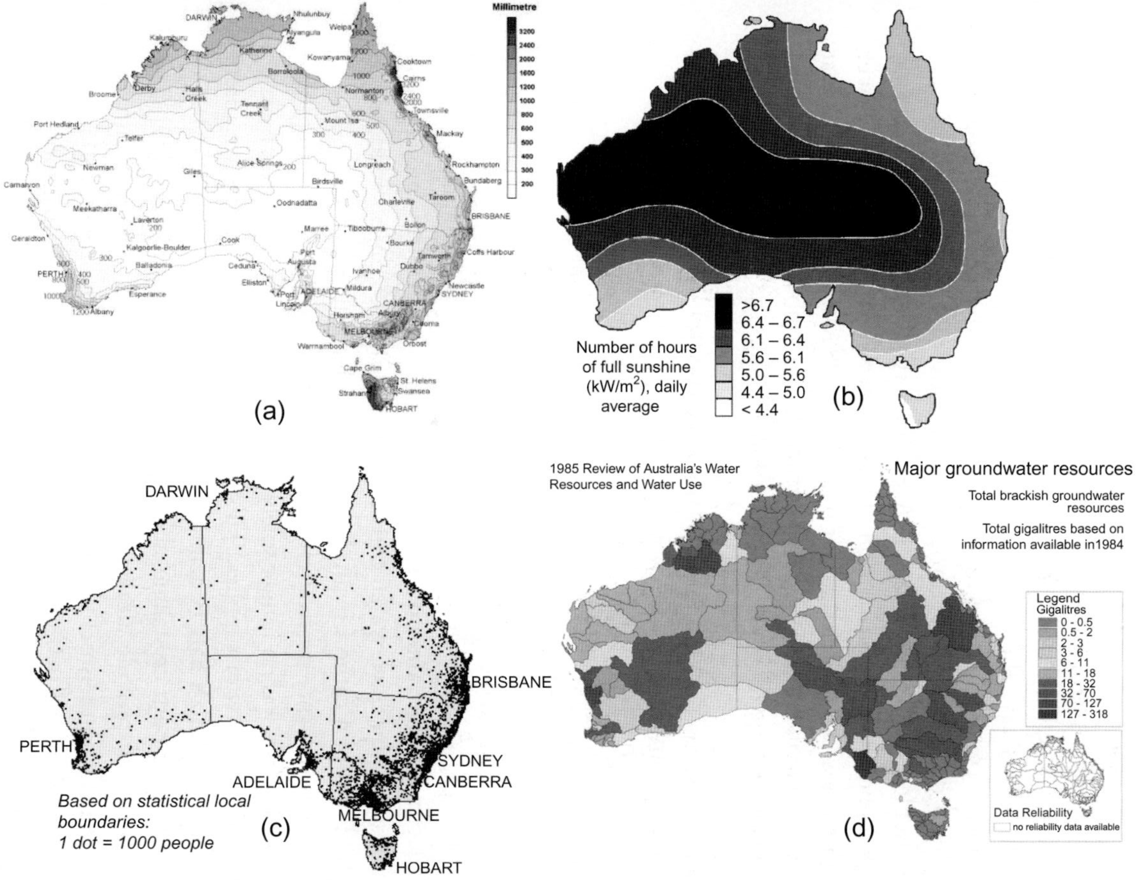

Figure 7 Australian (a) annual rainfall [23]; (b) solar radiation resource [24]; (c) population distribution [25]; (d) major brackish groundwater resources [26].

Australia. A further advantage of solar technologies is that peak energy production in the summer months coincides with peak water demand.

A second example, in Townsville, Australia, involves the addition of a water recycling aspect to the existing Cleveland Bay Purification Plant, enabling up to 20 ML of water per day to be recycled from the main treatment plant. Currently, the treatment plant discharges the treated waters into Cleveland Bay, however, future limits on water and nutrient disposal necessitate the utility to develop a water recycling program in conjunction with private sector partners [28]. A further driver is the security of clean drinking water throughout periods of drought by reducing Townsville's raw water demand. A preliminary study indicated that both the cost and energy consumption of water recycling were about 10% lower than the only other alternative of pumping in 28 ML of water over a great distance and allowing 8 ML losses incurred via evaporation. The potential renewable energy sources capable of powering this large-scale project include:

- wind power (in the form of two 2 MW REpower MM70 wind turbines); and
- methane, sourced from
  - the wastewater treatment plant itself,
  - a nearby landfill, and
  - a meatworks settling pond.

This average wind speed at the site is about 6.9 m/s, which, while low for Australia, is nonetheless a valuable resource potential yielding 4.3 GWh of electricity per annum. If all proceeds to plan, this project will demonstrate that raw water consumption of 28 ML can be reduced via the addition of a carbon neutral water recycling plant.

Sometimes, the motivation is purely financial, demonstrating that renewable energy is no longer solely applicable for niche applications, such as remote area power supplies. This is demonstrated by a water treatment plant in California's San Fernando Valley that is powered by 1.6 MW of PV, including both crystalline silicon and thin-film technologies [29]. The solar farm will provide almost all of the power needs for the South San Joaquin Irrigation District water treatment plant, which provides 40 million gallons/day for 155,000 residents and businesses, as well as irrigation water for 55,000 farm acres. The main goal of the project is to stabilize electrical costs, which can spike in summer months because of time-of-use metering implemented in California, which result in the cost of grid electricity reaching US$0.32 kWh^{-1}, however the peak times for water demand also coincide well when solar output is at a maximum.

5. SMALL-SCALE RENEWABLE ENERGY POWERED MEMBRANE FILTRATION PLANTS

Membrane-driven processes account for over half the existing renewable energy powered desalination in existence. Some of the reasons for this include that they are a modular technology, easy to install, compact in size, and simple to operate. Many of these advantages are also mirrored by renewable energy microgenerators, such as PV modules. These are also modular, contain no moving parts, have a long life (>20-year warranty) and involve low maintenance. The modularity of both of these technologies has also assisted in cost reduction being achieved via economies-of-scale. Wind turbines are also available in a wide variety of sizes (from 100 W up to MW scale) and multiple turbines can be included in a system design. Therefore, it is possible to scale a renewable energy powered membrane system to almost any size. These factors, combined with RO and NF exhibiting a very low SEC for seawater and brackish water, respectively, makes membranes an obvious choice when powering such systems with renewable energy.

Further advantages for small-scale systems can be realized coupling the DC output of PV modules and small wind turbines to power the necessary DC pump(s) and electronics, as well as possibly storing some energy in batteries. A DC only system increases system efficiency by 5–10% due to the avoidance of losses in power conversion (DC–AC) and rectification (AC–DC). In addition, the majority of renewable energy powered membrane filtration systems tend to use batteries to avoid energy fluctuations to enable continuous operation and avoid variations in pressure and flows. While energy storage enables a membrane system to produce a known amount of water at the desired quality, the use of batteries results in several problems:

1. The charge-in/charge-out efficiency of a typical deep-cycle lead acid battery is 75–80% [30], which results in a loss in system efficiency on the order of 20–25%. In order to overcome this loss, a 20–25% larger PV array is needed, substantially increasing the system cost.
2. Batteries both perform worse and degrade faster at higher temperatures, which is likely to coincide with arid regions where PV technology will be implemented. Specifically, with increasing operating temperature, the battery capacity decreases, followed by the charge efficiency decreasing and the self-discharge rate increasing [30]. This has resulted in battery banks requiring replacement in as little as 2 years after installation, thus adding considerably to maintenance costs [31].

3. Even for a "long" battery-life of 5 years – representing over 1500 charge–discharge cycles – the battery bank will require replacement on average four times during the life of the system, since PV systems are designed to have at least a 20-year life, thus further adding to the lifecycle cost of the system.

4. A follow-on problem is that lack adequate disposal/recycling facilities rarely exist in remote regions, and improper disposal can create further environmental hazards [32].

For these reasons, renewable energy powered membrane systems are being investigated where the energy is stored in the form of the product water. This means that the system may have to be slightly oversized to account for variations in the energy resource availability, for example, to store enough water to account for a very cloudy day with minimal clean water production. However, as long as the water stored in the permeate tank remains free from biological contamination, this approach can lead to a lower lifecycle cost – and hence cost of water – as well as a much more robust system design that facilitates autonomous operation.

Therefore, it is interesting to investigate the performance of batteryless RE membrane systems. While PV-powered water pumping systems, which are directly DC-coupled between the PV panel and pump motor, operate very successfully without any form of energy storage [33], relatively little is known about the consequences of variable operation (flow, pressure) on NF and RO membrane systems [17,34]. This research is being pursued for both PV- and wind-powered membrane filtration systems [17,35].

Field trials performed in outback Australia have demonstrated that while relatively large variations in solar irradiance occur, due to large clouds passing overhead, the system still produces good quality water. This is demonstrated in the graphs in Fig. 8, which detail the performance of a 300 W PV-powered RO filtration system when treating brackish feedwater with an electrical conductivity (EC) of 8.2 mS/cm during October 2005 (spring) [17]. The two gray curves in Fig. 8a show the incident solar irradiance measured on the horizontal (dashed line) as well as that falling on the PV panels attached to a single-axis (east-west) solar tracker (solid line). This clearly indicates the advantage of having the PV modules track the path of the sun throughout the day, producing 36% more electricity throughout the day [9.5 kWh/(m^2 day) instead of 7.0 kWh/(m^2 day)]. Fig. 8a also plots the power output from the PV panels, which closely matches the solar resource availability. The maximum occurs at slightly less than the 300 W rating of the PV module due to temperature effects.

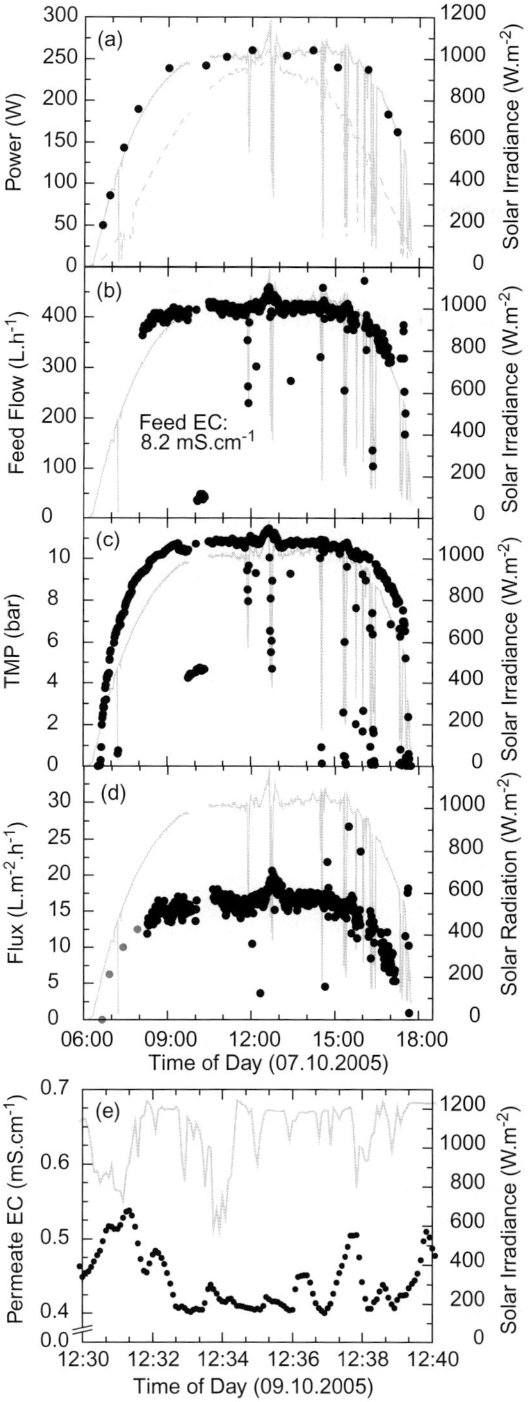

Figure 8 (a) Pump power (black) as fluctuation of tracked solar irradiance (solid gray) throughout the solar day on 07.10.2005. The solar irradiance measured on a horizontal surface is also plotted (dotted gray). This results in (b) a varying feed flow, (c) transmembrane pressure (TMP), and (d) flux when using the BW30 membrane. Higher resolution permeate EC data (e) is plotted over short period of high solar irradiance fluctuation for NF90 membrane between 12:30 and 12:40 on 09.10.2005.

The DC power produced by the modules closely is electronically optimized to power the positive displacement pump. Feedwater is sucked through an ultrafiltration (UF) prefilter at a pressure of about −0.6 bar. The resulting feedflow reaches a maximum of about 400 L/h between 10:00 and 16:00 h (Fig. 8b) while the transmembrane pressure (TMP) is typically around 10.5 bar during this period, as shown in Fig. 8c. Under full sunlight, the flux is around 16 L/(m^2 h) (Fig. 8d), which corresponds to a daily permeate production of 1.1 m^3 with the Dow Filmtec BW30 membrane at an average permeate EC of 0.28 mS/cm. The retention was over 96% on average, while recovery was 28%. The average SEC for this experiment over the whole solar day was 2.3 kWh/m.

Similar experiments performed with other RO membranes including Dow Filmtec NF90, Hydranautics ESPA4, and Koch TFC-S yielded interesting results. Under similar solar conditions, the system produced 1.4 m^3 with the NF90 membrane, albeit at a slightly higher permeate EC (0.52 mS/cm). The performance with the ESPA4 membrane looked very promising, as even on a rainy and overcast day, the system still produced 0.85 m^3 of permeate that exhibited a permeate EC of 0.81 mS/cm, which is only fractionally over the Australian Drinking Water Guideline value of 0.78 mS/cm, which is equivalent to 500 mg/L TDS (ref). When using the TFC-S membrane, the system was not able to produce was of good quality (permeate EC = 2.1 mS/cm).

It was noted, however, that during periods of low power availability, the permeate EC value occasionally exceeds the guidelines. This is a result of stagnated water being flushing out of the system during periods of cloud cover. Fig. 8e examines this effect in more detail, showing 10 min period that was recorded with higher resolution conductivity data. This shows that despite a sudden drop in solar irradiance by 50%, the fluctuation in permeate quality is minimal. These encouraging results indicate show the ability of the system to perform well under partial cloud coverage. Further research is currently underway to characterize the system further under a wide range of fluctuating power conditions.

6. CONCLUSIONS

There are many motivations for choosing renewable energy technologies to provide the necessary energy to power water treatment systems for reuse and desalination. These range from the lack of an existing electricity grid, particularly in remote areas, to securing future energy and water supplies, to purely financial incentives. While many renewable energy technologies exist, the two dominant ones used for powering desalination systems are PV modules and wind turbines. While wave power devices are a less mature technology, there are definitely synergies for desalination if these systems can be demonstrated to last 20 years in the harsh marine environment. Wind energy exhibits the lowest cost of electricity produced, while solar electricity is the highest. However, PV modules have a definite advantage as they contain no moving parts, thus enabling them to operate well in harsh conditions for over 20 years.

Solar technologies are particularly promising for powering water treatment schemes, given that the amount of power produced in summer will also coincide with increased water demand. The hypothesis was presented that energy storage devices may not be required, and that variations in the supply of energy could be absorbed by storing enough product water. Finally, the performance of a PV-powered membrane system filtering very brackish in outback Australia over a solar day was described. The system was able to tolerate large fluctuations in solar irradiance availability, however, more extensive testing is required before a more conclusive answer can be provided as to whether the use of batteries in such systems can be truly avoided.

REFERENCES

[1] IEA, International Energy Agency: Renewable Energy Policy... into the Mainstream, 2002, Novem, SITTARD, The Netherlands, October 2002.

[2] IPCC, Intergovernmental Panel on Climate Change, Climate Change 2007: The Physical Science Basis, Cambridge University Press, Cambridge, UK, 2007, http://www.ipcc.ch/ipccreports/ar4-wg1.htm, accessed on 07.01.2009.

[3] IEA., International Energy Agency: Key World Energy Statistics 2008, IEA, Paris, France, 2008.

[4] RGSP, Renewables Global Status Report 2009 Update, Ren21, Paris, France, 2009, http://www.ren21.net/pdf/RE_GSR_2009_Update.pdf, accessed on 18.08.2009.

[5] OK Solar, World solar radiation map, 2009, http://www.oksolar.com/abctech/images/world_solar_radiation_large.gif, accessed on 07.01.2009.

[6] NASA, North American Space Association, NASA Surface Meteorology and Solar Energy: Methodology, 2004, http://www.ceoe.udel.edu/WindPower/ResourceMap/SSE_Methodology.pdf, accessed on 07.01.2009.

[7] H.E. Krogstad, S.F. Barstow, Satellite wave measurements for coastal engineering applications. WAVES '97 conference, Virginia Beach, USA, November 1997.

[8] RAEng, The Royal Academy of Engineering, The Cost of Generating Electricity: A Commentary, The Royal Academy of Engineering, London, UK, http://www.raeng.org.uk/news/publications/list/reports/Cost_Generation_Commentary.pdf, 2004, accessed on 18.08.2009.

[9] J.J. Sadhwani, J.M. Veza, Desalination and energy consumption in Canary Islands, Desalination 221 (2008) 143–150.

[10] E. Tzen, Overview of the desalination technologies powered by renewable energies (RE), in: Proceedings of International Seminar Desalination Systems Powered by

Renewable Energy, Amman, Jordan, 2006, http://www.adu-res.org/pdf/ADU-RES_D1_4_Proceedings_Jordan%20Seminar.pdf, accessed on 18.08.2009.
[11] W.D. Childs, A.E. Dabiri, H.A. Al-Hinai, H.A. Abdullah, VARI-RO solar-powered desalting technology, Desalination 125 (1999) 155–166.
[12] M. Abdel-Jawad, Energy sources for coupling with desalination plants in the GCC countries. consultancy report prepared for ESCWA, http://faculty.ksu.edu.sa/Almutaz/Documents/Energy%20Consumption%20and%20Performance%20for%20Various%20Desalination%20Processes.pdf, accessed on 18.08.2009.
[13] D.L. Block, I.I. Melody, Assessment of solar desalination processes, in: Proc. Solar '89, 1989, pp. 62–71.
[14] A. Joyce, D. Loureiro, C. Rodrigues, S. Castro, Small reverse osmosis units using PV systems for water purification in rural places, Desalination 137 (2001) 39–44.
[15] B.W. Tleimat, Principles of Desalination, Academic Press, New York, 1980.
[16] M. Thomson, D. Infield, Laboratory demonstration of a photovoltaic-powered seawater reverse-osmosis system without batteries, Desalination 183 (2005) 105–111.
[17] B.S. Richards, D.P.S. Capão, A.I. Schäfer, Renewable energy powered membrane technology. 2. The effect of solar fluctuations on performance of a photovoltaic hybrid membrane system, Environ. Sci. Technol. 42 (2008) 4563–4569.
[18] M.R. Adiga, S.K. Adhikary, P.K. Narayanan, W.P. Harkare, S.D. Gomkale, K.P. Govindan, Performance analysis of photovoltaic electrodialysis desalination plant at Tanote in Thar desert, Desalination 67 (1987) 59–66.
[19] N. Ishimaru, Solar photovoltaic desalination of brackish water in remote areas by electrodialysis, Desalination 98 (1994) 485–493.
[20] T. Tablawi, R. Botros, H. Fotouh, Mahmoud Taha, Mohamed Faheed, J. F. Hoelsche, R. Quinn, Design of a PV-powered desalination plant in Egypt, in: Proc. 19th IEEE PVSC, 1987, pp. 1045–1049.
[21] J.M. Ortiz, E. Expósito, F. Gallud, V. García-García, V. Montiel, A. Aldaz, Electrodialysis of brackish water photovoltaic energy without batteries: direct connection behaviour, Desalination 208 (2007) 89–100.
[22] J.E. Lundstrom, Water desalting by solar powered electrodialysis, Desalination 31 (1979) 439–488.
[23] BOM, Commonwealth Bureau of Meteorology, Australian Annual Average Rainfall 1961–1990, http://www.bom.gov.au/cgi-bin/climate/cgi_bin_scripts/annual_rnfall.cgi, accessed on 18.08.2009.
[24] ABM Australia Bureau of Meteorology. Laboratory and Radiation Sub-Section Solar Exposure Data 1947–2001 (NCCSOL), v2.209 (2002).
[25] ABS, Australian Bureau of Statistics, Year Book Australia 2008, http://www.abs.gov.au/ausstats/abs@.nsf/0/5A717784C2562A99CA2573D20010FF17?opendocument, accessed on 18.08.2009.
[26] AWRC. Australian Water Resource Council. 1985 Review of Australia's Water Resources and Water Use, Canberra, 1987.
[27] A. De Munari, D.P.S. Capão, B.S. Richards, A.I. Schäfer, Application of solar-powered desalination in a remote town in South Australia, Desalination 251 (2010) 72–82.
[28] Townsville Sate of the Environment, Citiwater Carbon Neutral Water Recycling Project, http://www.soe-townsville.org/sustainable/water/, accessed on 06.30.2009.
[29] V. Boyd, Solar Power Leader, EnerG Magazine http://www.altenerg.com/back_issues/index.php?content_id = 113, accessed on 06.30.2009.
[30] D. Linden, T.B. Reddy, Handbook of Batteries, 3rd ed., New York, 2002.
[31] M. Thomson, M.S. Miranda, D. Infield, A small-scale seawater reverse-osmosis system with excellent energy efficiency over a wide operating range, Desalination 153 (2002) 229–236.

[32] E.A. Alsema, Environmental Life Cycle Assessment of Solar Home Systems, NWS-E-2000-15, Utrecht University, 2000.
[33] M.G. Thomas, Water Pumping: The Solar Alternative, SAND 87-0804, Sandia National Laboratories, Albuquerque, NM,1987.
[34] A.I. Schäfer, A. Broeckmann, B.S. Richards, Renewable energy powered membrane technology. 1. Development and characterization of a photovoltaic hybrid membrane system, Environ. Sci. Technol. 41 (2007) 998–1003.
[35] G.L. Park, A.I. Schäfer, B.S. Richards, Potential of wind-powered renewable energy membrane systems for Ghana, Desalination 251 (2010) 169–176.

CHAPTER 13

Desalinated Versus Recycled Water: What Does the Public Think?

Sara Dolnicar[1,*] *and* **Anna Hurlimann**[2]

[1]Marketing Research Innovation Centre, School of Management and Marketing, University of Wollongong, Northfields Ave., 2522 Wollongong, NSW, Australia

[2]Faculty of Architecture Building and Planning, The University of Melbourne, Melbourne 3010, Australia

Contents

1. INTRODUCTION

Australia, like many other regions worldwide, is facing a serious water crisis. The main source of water supply in Australia is rainwater stored in dams. Over the past decade, rainfall has decreased in many parts of the country. Thus the reliability of the supply of water harvested in dams is decreasing. Information regarding future rainfall projections for some parts of the country (such as Victoria) indicates that this shortfall will continue in future [1]. This pressure on supply is in addition to pressure from population growth. Thus if per capita water demand remains constant, pressure on supplies is expected to increase significantly.

*Corresponding author.
E-mail address: sarad@uow.edu.au

Sustainability Science and Engineering, Volume 2 ISSN 1871-2711, DOI 10.1016/S1871-2711(09)00213-X

As a consequence, many states in Australia affected by the drought have commenced planning and at times construction and implementation of large-scale water augmentation projects such as water recycling and desalination [2–6]. The technical solutions for such water augmentation projects are available and have been used in other countries for many years. The challenge in Australia, however, has been public resistance to alternative water sources. However, this challenge is not new. Dishman et al. [7] concluded 20 years ago that – while technical aspects of potable water reuse can be resolved – "the issue of public acceptance could kill the proposal" (p. 158). Many researchers since Dishman have called for more research on community acceptance of water from alternative sources [8–10].

Increased understanding of the public's attitudes can help in the process of successfully implementing sustainable water augmentation projects. Additionally, it may have a positive influence on policy, to be developed in line with community aspirations. Past work has focused mainly on recycled water, but other sources of water should be included in social research on water alternatives in the future, most importantly desalination, but also other options such as gray water, stormwater etc.

In this chapter, we report on a recent study comparing the Australian public's attitude toward recycled and desalinated water. In so doing we contribute to the understanding of public acceptance of water from alternative sources and provide possible explanations for the substantial differences in people's willingness to adopt recycled and desalinated water.

2. PRIOR WORK

As indicated above, the vast majority of studies investigating public acceptance of water from alternative sources focused on recycled water. Pioneering work in the area [11,12] concluded more than 30 years ago that people differentiate between the kinds of uses and show the highest level of opposition when asked about close to body uses, such as drinking and bathing. This finding has been replicated in all subsequent studies on public acceptance of recycled water in Australia [10,13–17] and beyond.

Very few studies have conducted comparisons of public acceptance of water from different alternative sources. One study (Nancarrow et al. [18]) found the highest public acceptance was for treated stormwater for use in parks and gardens (96%) and gray water for gardening (87%). However, this comparison included a limited number of alternative water sources. Also, and with implications for the proposed study, reasons behind the identified difference in attitudes between sources were not investigated. A second study was conducted by Dolnicar and Schäfer [17] and compared recycled water with desalinated water, finding that public acceptance – while generally higher for desalinated water – varied by the intended water use purpose.

There have been many factors that have been investigated with regard to influence on willingness to use recycled water. Past research has found that some demographic characteristics such as gender, age, and education, influence attitudes toward recycled water use. However Marks [19] in a recent review article found that there is little evidence that demographic factors, apart from gender, can predict acceptance of recycled water use. Factors, which have been found to influence willingness to use recycled water, include but are not limited to: trust (including [19–21]), information/knowledge [21,22], and concerns about quality and aesthetic attributes [23–26]. However, there has been limited research regarding attitudinal factors influencing willingness to use desalinated water.

3. METHODOLOGY

3.1 Fieldwork administration

A survey study was conducted in January 2009 with 1495 Australian respondents aged 18 or above who were quota sampled to be representative of the Australian population. Data was collected using a permission-based research-only Internet panel. Respondents were compensated for their time with a payment following standard rates used by the Internet panel for a survey of this length.

3.2 Questionnaire

The 30-min questionnaire was developed on the basis of (1) a previous survey instrument measuring public acceptance of recycled and desalinated water [17] as well as (2) an extensive qualitative research phase in which one focus group and up to 10 in-depth interviews were conducted at eight locations in Australia, which distinctly different water supply situations [27].

The key question respondents were asked was, what their likelihood of using recycled and desalinated water is, for a range of uses. Because this question is by its very nature hypothetical given that recycled water is not widely available to Australian residents, and because the validity of such hypothetical questions has been criticized in the past [28], we chose to provide respondents with the following scenario that made the question setting more realistic:

"For the following question, imagine that you live in a town where:

- *Dams supplying household water currently hold 20 per cent of capacity*
- *Level 5 Mandatory Water Restrictions are in place for the use of tap water (no outside watering of gardens, no watering systems, no refilling swimming pools, no washing vehicles except for windows and headlights).*
- *Recycled water/desalinated water is readily available without restrictions.*

Under these circumstances, please indicate how likely you would be to use RECYCLED WATER/DESALINATED WATER for the following purposes by placing the slider in the respective position along the line. The extreme points of the line indicate "very likely" and "very unlikely". Some of these behaviors may not apply to you, e.g., because you do not have a swimming pool. In this case please tick the "not applicable" option."

Respondents were then presented with 14 typical water uses, for 7 of which a "not applicable" option was provided.

Note that the choice of the horizontal line (also referred to as visual analogue scale) is uncommon. We have chosen this answer format because it has a number of advantages over traditionally used 5- or 7-point multicategory scales: (1) the distance between answer options is clearly and unambiguously defined, thus enabling metric data analysis to be performed, (2) there is no need for verbal labeling of options, which may be interpreted in different ways by different respondents, thus introducing bias.

Additional questions asked related to people's attitudes toward recycled and desalinated water, and water conservation; their level of proenvironmental attitude as well as proenvironmental behaviour; and basic sociodemographic criteria.

3.3 Data analysis

Frequency counts, χ^2 tests, analyses of variance, and *t*-tests for proportions were computed to assess differences in the stated likelihood of use between recycled and desalinated water as perceived by the Australian population.

4. RESULTS

The results of the question about people's likelihood of using recycled and desalinated water, respectively, for the 14 uses presented in the questionnaire are provided in Fig. 1. Higher figures indicate higher levels of stated likelihood of use, with 100 indicating "very likely" and 0 indicating "very unlikely."

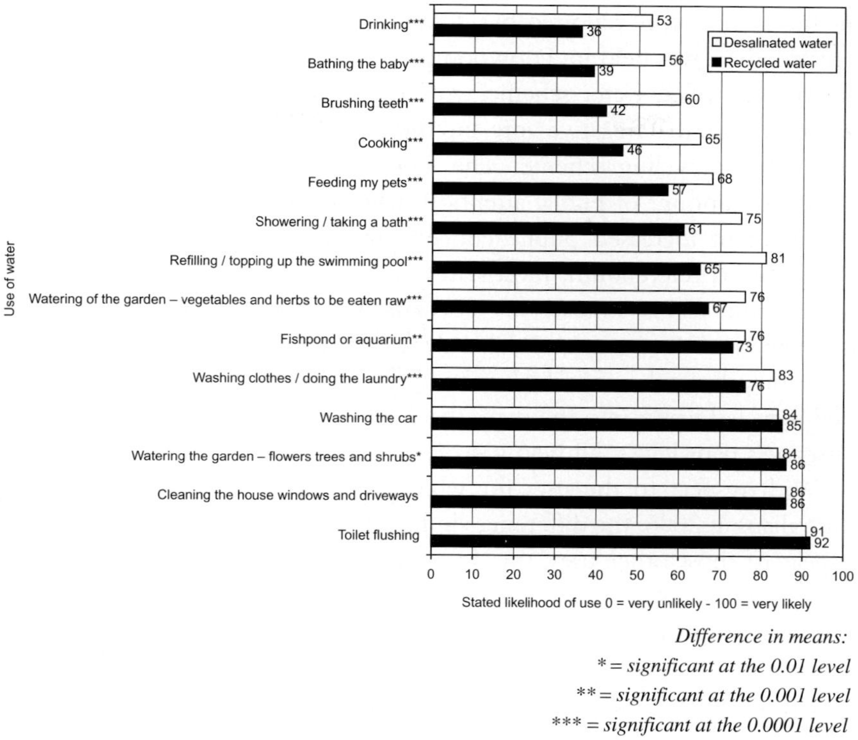

Figure 1 Stated likelihood of using recycled and desalinated water for a range of 14 purposes. Difference in means: (*) significant at the 0.01 level; (**) significant at the 0.001 level; (***) significant at the 0.0001 level. *Source:* Hurlimann and Dolnicar [27].

As can be seen, desalinated water outperforms recycled water in all but three uses, and only one at a significant level. This is a change from the 2006 survey results reported by Dolnicar and Schäfer [17] where a number of uses has been identified for which recycled water was preferred. These preferred uses included watering flowers and shrubs, toilet flushing, and washing the house, windows, driveways, and car. In the 2009 survey (the subject of this chapter), only one use can be identified with respect to which Australians state a significantly higher likelihood of adoption for recycled water (watering the garden - flowers, trees and shrubs). This comparison with Dolnicar and Schäfer's findings indicates that attitudes are in a constant state of flux.

There may be many possible explanations for this difference over time. One possible influence is the increased media coverage and community discussion and awareness regarding recycled water since the data collected for the 2006 survey reported by Dolnicar and Schäfer. On July 29, 2006, a referendum was held in Toowoomba, a city in Queensland, on whether or not a recycled water scheme would be introduced. Significant public opposition led to a negative vote and the postponement of the introduction of a water recycling scheme. The Toowoomba referendum attracted a significant amount of public attention nationwide and may well have had the effect of increased public concern about recycled water. Since the Toowoomba referendum, most states in Australia have chosen to develop desalination plants instead of recycling schemes for large-scale water augmentation. Interestingly, the only exception currently is a large-scale water recycling scheme, which is being developed in Queensland and will feed recycled water into the dams that supply Toowoomba with water – if Brisbane's dam levels fall below 40%.

With respect to the uses for which recycled water was preferred by Australians in 2006, the differences in the 2009 study are insignificant. With respect to most other uses, the differences in stated likelihood of use are highly significant. For example, the average stated likelihood of the Australian population to drink desalinated water is 53 on a 100-point scale, whereas it is only 36 for recycled water.

Fig. 2 shows some of the statements that were made by respondents in the qualitative phase of the study (phase I). Those most frequently stated were subsequently included in phase II of the research – the 2009 survey. For each of those statements, respondents were asked to state whether they agreed or disagreed. Fig. 2 shows the percentage of respondents who agreed with these statements for recycled and desalinated water, respectively.

χ^2 tests were undertaken to assess whether there was a significant difference in the agreement with the attitudinal statements between water sources. The results indicate that there was a significant difference (at the significance = 0.0001 level) for every single statement. As can be seen, when compared to desalinated water, recycled water is generally perceived by a larger percentage of respondents as having a health risk; tasting/smelling bad; and as disgusting. More people state that they are cautious of what is actually in recycled water, and express skepticism about how safe and clean it is. For both sources of water, a very high percentage of respondents want reassurance about its quality: 71% (recycled) and 77% (desalinated) would feel comfortable about its quality if it would be approved for human consumption by scientists. With respect to cost, Australians are more concerned about the implementation of desalination solutions: 58% believe that desalinated water is too expensive to implement/operate, whereas only 35% of respondents feel the same about

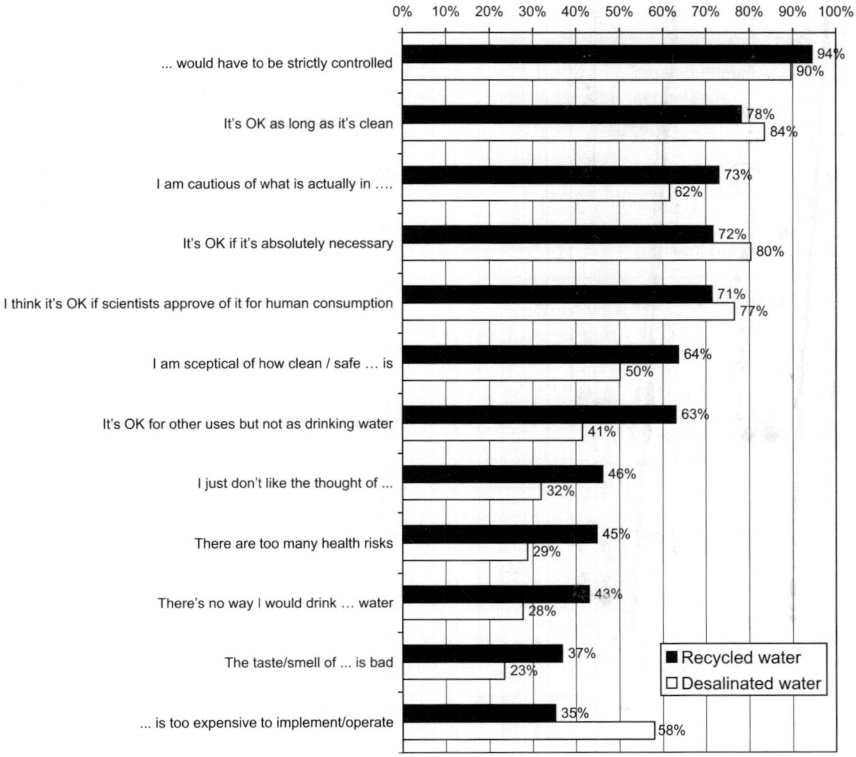

Figure 2 Percent of respondents agreeing with statements about recycled and desalinated water.

recycled water. Given the stated likelihoods of use in Fig. 1, health concerns outweigh cost concerns. It should also be noted, however, that the vast majority of Australians state that recycled (72%) and desalinated (80%) water, is OK if it is absolutely necessary. This mirrors the findings from the qualitative study where respondents were able to produce a number of arguments for and against various alternatives for securing Australia's future water supply. However, when confronted with a worst case scenario, all understood the need for water from alternative sources and were mostly willing to accept these solutions.

Statistical analysis (ANOVA) was undertaken to establish if agreement with attitudinal statements was significantly associated with the stated likelihood of using recycled water and desalinated water. The results of this analysis are displayed in Table 1 for recycled water and Table 2 for desalinated water. As can be seen from these tables, there were a number of significant results.

Table 1 Mean differences in stated likelihood of using recycled water between various attitudinal groups

Attitudinal statement/Use of recycled water– mean likelihood to use (scale: 0 = very unlikely - 100 = very likely)		Watering the garden – flowers, trees, shrubs	Washing clothes/ laundry	Cooking	Showering/taking a bath	Drinking	Brushing teeth	Bathing the baby	Filling up the fishpond/aquarium	Toilet flushing	Cleaning the house, windows, driveway	Watering the garden – vegetables , herbs eaten raw	Washing the car	Refilling the swimming pool	Feeding my pets
Recycled water would have to be strictly controlled	Disagree	77	73	54	65	47★	50	47	69	89	80	70	76	65	61
	Agree	86★	76	46	60	36	41	38	71	92	86	67	85★	64	56
It's OK if it is absolutely necessary	Disagree	85	65	34	49	27	30	26	67	90	82	56	81	50	45
	Agree	86	80★★★	51★★★	66★★★	41★★★	47★★★	44★★★	73★★	92	87★★★	72★★★	86★	70★★★	61★★★
The taste / smell of recycled water is bad	Disagree	86	82★★★	57★★★	70★★★	47★★★	52★★★	50★★★	77★★★	93★	88★	75★★★	87★	73★★★	67★★★
	Agree	86	65	27	45	18	24	22	64	90	83	54	82	51	40
It is OK as long as it is clean	Disagree	84	61	20	40	12	16	14	61	88	81	44	80	42	31
	Agree	86	80★★★	54★★★	67★★★	43★★★	49★★★	46★★★	75★★★	92★★★	87★★★	74★★★	86★★★	71★★★	64★★★
I am skeptical of how clean and safe recycled water is	Disagree	86	87★★★	69★★★	79★★★	60★★★	65★★★	61★★★	80★★★	93	88	81★★★	87	78★★★	76★★★
	Agree	86	70	33	50	23	28	27	67	91	85	59	84	57	46
There are too many health risks	Disagree	86	83★★★	58★★★	71★★★	47★★★	52★★★	49★★★	76★★★	92★	88★★★	76★★★	86★	73★★★	66★★★
	Agree	85	64	25	41	17	22	21	64	90	82	51	82	48	40
I am cautious of what is actually in recycled water	Disagree	87	87★★★	70★★★	79★★★	62★★★	66★★★	62★★★	80★★★	93	89★	83★★★	86	77★★★	77★★★
	Agree	85	72	38	54	27	33	31	69	91	85	61	85	60	49
It is OK for other uses but not as drinking water	Disagree	84	85★★★	70★★★	78★★★	62★★★	67★★★	63★★★	80★★★	92	87	84★★★	85	77★★★	76★★★
	Agree	87	71	32	50	21	27	26	67	91	85	60	85	58	45
I just don't like the thought of recycled water	Disagree	86	83★★★	61★★★	73★★★	52★★★	57★★★	53★★★	77★★★	92	87	79★★★	87★	75★★★	69★★★
	Agree	86	68	29	47	19	24	24	66	91	84	54	83	54	43
Recycled water is too expensive to implement	Disagree	86	81★★★	53★★★	67★★★	43★★★	48★★★	46★★★	75★★★	93★★	87★	73★★★	87★★★	70★★★	63★★★
	Agree	84	67	33	49	24	30	27	67	90	84	57	81	55	45
I think it is OK if scientists approve it for human consumption	Disagree	86	63	24	42	13	18	18	62	90	83	51	82	46	35
	Agree	86	81★★★	56★★★	69★★★	46★★★	51★★★	48★★★	76★★★	92	87★	74★★★	86★	72★★★	65★★★
There's no way I would drink recycled water	Disagree	86	85★★★	62★★★	74★★★	53★★★	57★★★	55★★★	77★★★	92	88★★	77★★★	86	75★★★	70★★★
	Agree	86	65	25	43	15	21	19	64	91	83	54	84	50	39

Note: Significances always apply to the whole "agree-disagree" pair, but are market only for either "agree" or "disagree" in this table to highlight associations with higher stated likelihood of use. Difference in means: (★★★) significant at the 0.0001 level, (★★) significant at the 0.001 level, (★) significant at the 0.01 level.

Table 2 Mean differences in stated likelihood of using desalinated water between various attitudinal groups

Attitudinal statement/Use of desalinated water – mean likelihood to use (scale: 0 = very unlikely - 100 = very likely)		Watering the garden – flowers, trees, shrubs	Washing clothes/ laundry	Cooking	Showering/taking a bath	Drinking	Brushing teeth	Bathing the baby	Filling up the fishpond/aquarium	Toilet flushing	Cleaning the house, windows, driveway	Watering the garden – vegetables, herbs eaten raw	Washing the car	Refilling the swimming pool	Feeding my pets
Desalinated water would have to be strictly controlled	Disagree	79	82	72	79	63**	71***	65**	76	88	82	79	78	78	69
	Agree	85*	83	65	75	53	59	55	75	91	87	76	85**	81	68
It's OK if it is absolutely necessary	Disagree	77	72	48	61	37	42	39	63	85	79	61	75	67	51
	Agree	86***	86***	70***	79***	58***	65***	60***	78***	92***	88***	80***	86***	84***	72***
The taste / smell of desalinated water is bad	Disagree	86***	87***	73***	81***	62***	68***	64***	80***	93***	88***	81***	86***	84***	75***
	Agree	78	70	42	55	27	35	31	62	86	80	58	77	68	46
It is OK as long as it is clean	Disagree	74	65	37	51	26	31	28	59	82	76	54	72	64	40
	Agree	86***	86***	71***	80***	59***	66***	62***	79***	93***	88***	80***	86***	84***	73***
I am skeptical of how clean and safe desalinated water is	Disagree	87***	91***	80***	87***	71***	76***	74***	82***	93***	89***	86***	87***	88***	81***
	Agree	81	75	51	63	36	44	40	69	89	83	65	81	73	55
There are too many health risks	Disagree	87***	89***	76***	84***	65***	71***	67***	81***	93***	89***	83***	87***	87***	78***
	Agree	78	68	39	53	26	33	30	63	86	80	57	77	66	45
I am cautious of what is actually in desalinated water	Disagree	87*	90***	80***	86***	73***	77***	73***	81***	92	88	85***	85	86***	82***
	Agree	83	79	56	68	42	49	46	72	91	85	70	83	77	60
It is OK for other uses but not as drinking water	Disagree	86	88***	77***	83***	70***	74***	70***	79***	92	87	83***	85	84***	79***
	Agree	83	76	49	63	31	41	38	70	90	85	66	83	75	53
I just don't like the thought of desalinated water	Disagree	87***	88***	76***	83***	65***	71***	66***	79***	93***	89***	83***	86***	85***	76***
	Agree	79	72	44	58	30	37	35	68	87	81	61	79	71	51
Desalinated water is too expensive to implement	Disagree	86	86***	71***	80***	60***	66***	61***	78	92	88	80***	86*	83	72***
	Agree	83	80	61	72	49	56	52	74	90	85	73	82	79	65
I think it is OK if scientists approve it for human consumption	Disagree	79	70	42	56	26	32	31	62	85	80	58	76	66	46
	Agree	87***	87***	73***	81***	62***	69***	64***	80***	93***	88***	81***	86***	85***	75***
There's no way I would drink desalinated water	Disagree	87***	88***	76***	83***	66***	72***	68***	80***	92***	88***	83***	86***	86***	77***
	Agree	78	69	37	54	21	30	28	63	87	81	58	78	67	44

Note: Significances always apply to the whole "agree-disagree" pair, but are market only for either "agree" or "disagree" in this table to highlight associations with HIGHER stated likelihood of use. Difference in means: (***) significant at the 0.0001 level, (**) significant at the 0.001 level, (*) significant at the 0.01 level.

For recycled water, there was only one statement that had a significant difference between agreement groups for likelihood of using recycled water to water the garden (flowers, trees, and shrubs). This was agreement with the statement "recycled water should be strictly controlled." The following attitudes were found to be significantly associated with a higher stated likelihood of using recycled water for all of the other 13 uses investigated:

- those who disagree that the taste/smell of recycled water is bad;
- those who agree that recycled water is OK as long as it is clean;
- those who disagree that there are too many health risks associated with recycled water use;
- those who disagree that recycled water is too expensive to implement.

Other attitudes were found to be significantly associated with the stated likelihood of using recycled water, but for fewer that 13 uses – the details of this can be found in Table 1.

For desalinated water, the following attitudes were significantly associated with stated likelihood of use for all 14 uses investigated:

- those who agreed that "desalination is OK if absolutely necessary";
- those who disagreed that the taste/smell of desalinated water is bad;
- those who agreed that desalinated water is OK as long as it is clean;
- those who disagreed that are skeptical of how clean and safe desalinated water is;
- those who disagree that there are too many health risks associated with recycled water;
- those who disagree that they "just don't like the thought of desalinated water";
- those who think it is OK as long as scientists approve it for human consumption;
- those who disagree that there is no way they would drink recycled water.

Details for other associations between attitudes and stated likelihood of use can be found in Table 2.

The results above indicate to water policy officers, attitudes which may facilitate higher likelihood of using recycled water and desalinated water. This information could thus be the focus of any public communication plan regarding recycled water or desalinated water to help increase likelihood of use.

5. CONCLUSIONS

In this chapter, results from an Australian survey study conducted in January 2009 have been reported. The public's stated likelihood of using recycled and desalinated water for a range of 14 purposes was compared and attitudinal factors associated with different levels of stated likelihood of use were explored.

Results indicate that, as opposed to previous research [16,17], Australians express a higher likelihood of using desalinated water then recycled water for all household uses included in the survey. Differences were insignificant for low contact uses such as toilet flushing, but were highly significant for high contact uses, such as drinking, bathing the baby, brushing teeth, and cooking.

Significant differences between recycled and desalinated water were also found with respect to attitude statements. A number of attitudinal statements were also found to be significantly associated with higher levels of stated likelihood of use. For both water sources, the following three attitudes had significantly higher mean ratings for likelihood of use: those who disagree that the taste/smell of the water source is bad; those who agree that the alternative water source is OK as long as it is clean; and those who disagree that there are too many health risks associated with the source of water's use.

When asked about their attitudes toward recycled and desalinated water, it becomes evident from the responses that the primary concern of people relating to recycled water remains public health, whereas the main weakness perceived in relation to desalinated water appears to be its cost (58%) and caution about what is actually in the water (62%). Also, 48% of respondents raised environmental concerns (this question was not asked for recycled water). The cost and environmental concern appear to be outweighed by health concerns, because the stated likelihood of use levels are consistently higher for desalinated water than for recycled water. Finally, the vast majority of Australians are willing to accept water from alternative sources if it is absolutely necessary.

The findings from this study have major implications for water policy: first, it appears that the fertile ground for public resistance is the perception of choice. When the public feels that introducing water from alternative sources into their tap water is a choice they make (or a public policy decision they want to boycott) resistance is more likely to occur than in the situation where people are aware that there are no other viable options and using water from alternative sources is not actually an option but a necessity. To date, public policy makers in Australia have not used this line of argument toward the public much, leaving the impression that indeed it is a choice.

Second, given that the Australian public does appear to view water augmentation as a matter of choice, it may be necessary to provide the public with more factual information about water alternatives. The attitudinal factors found in our study to be significantly associated with higher stated likelihood of using recycled and desalinated water, could inform public communication plans. For example, sources of potential health hazards should be outlined clearly, clarifying also that health risks are inherent wherever any kind of water is transported over significant distances. It appears that factual information may be the best counter-measure against people developing unreasonable health concerns. To date, little factual information has been provided to the Australian public. Some water authorities add flyers with messages about water conservation and water augmentation projects to their bill mailouts. However there is currently no wide-scale independent source of information that would enable Australians to inform themselves about facts relating to current tap, recycled, desalinated water as well as other water options, which are widely used in Australia. These other sources may also be perceived as unhealthy but are typically not perceived as such by the population because they are "close to home" (e.g., rainwater tanks).

Finally, the uses of alternative water sources for which people have a higher level of acceptance could be used to increase people's experience with these kinds of water and increase acceptance for other uses. This recommendation was made a long time ago by Baumann and Kasperson [29] who suggested to "put the reclaimed water in an attractive setting and invite the public to look at it, sniff it, picnic around it, fish in it, and swim in it" (p. 670).

In sum, results indicate that Australians have a differentiated view of different kinds of water from alternative sources. As such, a range of public policy measures could be taken to provide the public with factual information and experience to increase their acceptance and improve their attitudes.

ACKNOWLEDGMENTS

This study was funded through Australian Research Council (ARC) Discovery Grant (DP0878338). We thank our Research Assistant, Sarah Oberklaid, for her help with formatting the tables.

REFERENCES

[1] Government of Victoria, Sustainable Water Strategy, Central Region Action to 2055. Melbourne, Department of Sustainability and Environment, available online at: www.dse.vic.gov.au, 2006, Date accessed: 19 August 2009.

[2] Government of South Australia, Securing Tomorrow's Water Today Our four-way strategy. Adelaide, available online at http://www.water.sa.gov.au/, 2008, Date accessed: 22 February 2009.

[3] Government of Victoria, Our Water Our Future. Department of Sustainability and Environment, Melbourne, available online at: www.ourwater.vic.gov.au/, 2007, Date accessed 11 April 2008.
[4] Government of Western Australia, State Water Plan 2007. Department of the Premier and Cabinet, Perth, available online at: http://portal.water.wa.gov.au/portal/page/portal/PlanningWaterFuture/StateWaterPlan/Content/State%20Water%20Plan%202007%20text%20only.pdf, 2007, Date accessed: 22 February 2009.
[5] Government of New South Wales, 2006 Metropolitan Water Plan, Sydney, available online at: www.waterforlife.nsw.gov.au/data/assets/pdf_file/0015/1446/wfl2006.pdf, 2006, Date accessed: 31 May 2005.
[6] Queensland Water Commission, Water for today, water for tomorrow: South East Queensland water strategy draft-March 2008. Brisbane, available online at: http://www.qwc.qld.gov.au/myfiles/uploads/SEQWS/Foreword_Exec_Summary.pdf, 2008, Date accessed 19 August 2009.
[7] C. Dishman, J. Sherrard, M. Rebhun, Gaining public support for direct potable water reuse, J. Professional Issues Eng. 115(2) (1989) 154.
[8] P. Dillon, Water reuse in Australia: current, future and research, J. Australian Water Assoc. 28(3) (2001) 18–21.
[9] P. Jeffrey, Community and institutional engagement in agricultural water reuse projects, in: V. Lazarova (Ed.), Water Reuse for Irrigation Agriculture Landscapes and Turf Grass, CRC Press, Boca Raton, 2005, p. 285
[10] M. Po, J. Kaercher, B. Nancarrow, Literature Review of Factors Influencing Public Perceptions of Water Reuse, CSIRO, Perth, 2003.
[11] W. Bruvold, P. Ward, Public attitudes toward uses of reclaimed wastewater, Water Sew. Works 120 (1970) 120–122.
[12] W. Bruvold, Public Attitudes Towards Reuse of Reclaimed Water, University of California, California, USA, 1972.
[13] A. Hurlimann, Melbourne office worker attitudes to recycled water use, J. Australian Water Assoc. 33(7) (2006) 58–65.
[14] J. McKay, A. Hurlimann, Attitudes to reclaimed water for domestic use: part 1. Age, J. Australian Water Assoc. 30(5) (2003) 45–49.
[15] J.S. Marks, B. Martin, M. Zadoroznyj, Acceptance of water recycling in Australia: national baseline data, J. Aust. Water Assoc. 33(2) (2006) 151–157.
[16] S. Dolnicar, A.I. Schäfer, Public perception of desalinated versus recycled water in Australia, In: CD Proceedings of the AWWA Desalination Symposium 2006, 2006.
[17] S. Dolnicar, A.I. Schäfer, Desalinated versus recycled water – public perceptions and profiles of the accepters, J. Environ. Manage. 90 (2009) 888–900.
[18] B.E. Nancarrow, J.D. Kaercher, M. Po, Community Attitudes to Water Restrictions Policies and Alternative Sources, A Longitudinal Analysis 1988–2002, CSIRO, Perth, 2002.
[19] J. Marks, Advancing community acceptance of reclaimed water, J. Australian Water Assoc. 31(5) (2004) 46–51.
[20] A. Hurlimann, J. McKay, Attitudes to reclaimed water for domestic use: part 2. Trust, J. Australian Water Assoc. 31(5) (2004) 40–45.
[21] M. Po, B.E. Nancarrow, Z. Leviston, N.B. Poter, G.J. Syme, J.D. Kaercher, Predicting Community Behaviour in Relation to Wastewater Reuse: What Drives Decisions to Accept of Reject? CSIRO, Perth, 2005.
[22] A.C. Hurlimann, E. Hemphill, J. McKay, G. Geursen, Establishing components of community satisfaction with recycled water use through a structural equation model, J. Environ. Manage. 88(4) (2008) 1221–1232.
[23] H.J. Albrechtsen, Microbiological investigations of rainwater and graywater collected for toilet flushing, Water Sci. Technol. 46(6–7) (2002) 311–316.

[24] A. Hurlimann, J. McKay, Urban Australians using recycled water for domestic non-potable use – an evaluation of the attributes price, saltiness, colour and odour using conjoint analysis, J. Environ. Manage. 83(1) (2007) 93–104.
[25] P. Jeffrey, B. Jefferson, Public receptivity regarding "in-house" water recycling: results from a UK survey, Water Sci. Technol.: Water Supply 3(3) (2003) 109–116.
[26] J. Marks, N. Cromar, H. Fallowfield, D. Oemcke, M. Zadoroznyi, Community experience and perceptions of water reuse, in: Proceedings of 3rd World Water Congress of the International Water Association, Melbourne, International Water Association, 2002.
[27] S. Dolnicar, A. Hurlimann, Drinking water from alternative water sources: differences in beliefs, social norms and factors of perceived behavioural control across eight Australian locations. Water Science and Technology, in press.
[28] D. Baumann, Social acceptance of water reuse, Appl. Geogr. 3(1) (1983) 79–84.
[29] D. Baumann, R. Kasperson, Public acceptance of renovated waste water: myth and reality, Water Resour. Res. 10(4) (1974) 667.

CHAPTER 14

Conclusion: A Summary of Challenges still Facing Desalination and Water Reuse

Isabel C. Escobar*
Chemical and Environmental Engineering Department, The University of Toledo, 2801 West Bancroft Street, MS 305, Toledo, OH 43606-3390, USA

Contents

In water-scarce regions, two sources of water are normally available: sewage and seawater. In the last decade, the reclamation of effluents has developed rapidly as an alternative to seawater desalination for irrigation and indirect potable water reuse. The key in water reuse is to first treat the sewage biologically followed by membrane filtration to remove organic matter and suspended solids [1]. For seawater desalination, pretreatment must be provided if the source is open seawater [2]. Desalination has now been practiced on a large scale for more than 50 years [3], and in recent decades, membrane desalination has enjoyed tremendous success. Through continual improvements, the major technologies are now efficient and reliable. However, they are still too expensive to address the needs for additional supplies of freshwater in of many parts of the world [4].

On the other hand, water reclamation, recycling, and reuse address depleted water supply challenges by creating new sources of high-quality water supplies [5]. One of the most significant benefits of water reuse is the value created by the inclusion of water reuse in integrated water resources planning and other aspects of water policy and the implementation of water

*Corresponding author.
E-mail address: Isabel.Escobar@utoledo.edu

Sustainability Science and Engineering, Volume 2 ISSN 1871-2711, DOI 10.1016/S1871-2711(09)00214-1

projects, resulting in the long-term sustainability of water supplies. Membrane bioreactors (MBR) have been truly a revolutionary membrane technology that has increased the efficiency of water reuse facilities [6–9].

This concluding chapter will discuss some of the challenges still facing desalination and water reuse. Recently, the Joint Water Reuse and Desalination Task Force held a Water Innovation Symposium (San Diego, CA, October 17–25, 2005) [10] to develop a roadmap. The challenges discussed here can be considered to be adaptations from this roadmap.

1. CHALLENGES FACING DESALINATION

Despite its success, desalination, especially when using membranes, still faces numerous challenges, and among the major ones are the following:

1. *Membrane fouling*: Membrane fouling is one of the most important factors that limit greater use of desalination membranes [11]. Fouling occurs due to particulate matter, organic matter, microorganisms forming biofilms, and inorganic scaling [12–14]. Membrane replacement due to fouling is the single largest operating cost when membranes are used in water separation applications, and, thus, the greatest hindrance to the widespread use of membranes [11].
2. *Pretreatment for electrodialysis*: Electrodialysis reversal (EDR) is generally most attractive for the desalination of brackish water and in cases where organic matter removal and microbial control are not important. EDR is also of interest in treating brackish waters where silica is an important limitation. One of the principle claims of EDR is that its pretreatment requirements are significantly less than those of reverse osmosis (RO). However, in the presence of organic matter, EDR installations have experienced problems with fouling of the anion exchange membranes [15–18].
3. *Brine disposal*: The ideal process would recover all the water leaving behind only the salt. All current technologies produce a concentrate stream that requires discharge into the environment, or must undergo additional treatment to reduce the volume and remove the dissolved solids. Brine disposal is an additional cost and complication associated with desalination membranes.
4. *Water quality*: The information required to judge a water's suitability for the horticulture is sorely lacking and it is this information that is important if desalinated water is to be utilized for public use. Communities or parts of communities making a switch to desalinated

seawater are forced to take risks because of the lack of knowledge about water quality and its impact on horticulture.

5. *Contaminant removal*: Although desalination membranes are capable of removing dissolved solids, there are contaminants that are difficult to remove, or are inconsistently removed, with membrane processes that affect the product water quality, such as boron. Boron at relatively low levels can also have toxic effects on humans and animals, and there is almost no reliable literature on boron tolerance for horticultural plants such as those of interest to the general population in a municipality that might receive desalinated water. The World Health Organization recommends that boron in drinking water be limited to 0.5 mg/L. Various techniques have been considered to improve performance. Principal among these techniques are seeking membranes with higher boron rejection [19], [20], adding a second stage of RO treatment with or without pH adjustment [21–25], and ion exchange. Because boron, like silica, is not ionized at normal pHs, EDR does not remove it.
6. *Power consumption*: Membrane desalination has rapidly become the process of choice for water desalination due to reduced operational costs as compared to thermal processes. Although membrane processes benefit from not having to vaporize the entire flow, the energy required for membrane desalination is still greater than what is theoretically possible. For the RO process, new low-pressure membranes operate at only 4–10 times the osmotic pressure.

2. CHALLENGES FACING WATER REUSE

Although water reclamation and reuse is practiced in many countries around the world, current levels of reuse constitute a small fraction of the total volume of municipal and industrial effluent generated. Among the reasons for this are the following:

1. *Monitoring*: The number and variety of contaminants found in reuse source waters present challenges not only to the technologist, but also for the water provider. More information must be developed regarding in-line sensors capable of real-time assurance, monitoring, and control. Risk assessments must be conducted to determine requisite removal levels for unregulated contaminants and to contribute to public education campaigns and perception.
2. *Reverse osmosis pretreatment*: Prior to 1995, biologically treated wastewater was viewed as one of the most difficult waters to treat via RO. Most agreed that the natural organic material in wastewater was the most

likely cause of problems. Subsequent work in San Diego [1] and in Orange County, California [2], [26] demonstrated that extremely long runs could be accomplished between cleanings when membrane filtration and a chloramine residual were employed to control particulate and microbiological contaminants [27–31].

3. *Contaminant removal*: Removing contaminants from reuse source waters includes (1) removal/destruction of organics in combination with improved/reduced fouling of membrane-based, oxidative, thermal, and biological treatment processes; (2) removal of dissolved inorganic contaminants and materials, including metals, radioactive materials, emerging contaminants, and salinity reduction. Emerging organics are endocrine-disrupting compounds, pharmaceuticals, and personal care products commonly found in treated wastewater effluents. For indirect potable reuse applications, the removal of these compounds is required to assure that they are not present in significant concentrations in the potable water. Few studies have yet been conducted on the removal of emerging organics during water reuse. Kimura et al. [32] looked at the removal of 11 neutral endocrine disruptor compounds (EDCs) and pharmaceutically active compounds (PhACs) via nanofiltration (NF) and RO and observed incomplete removals, ranging from 57% to 91%. [33] looked at a much broader collection of EDCs, PhACs, and related compounds, observing complete removal in most instances. Only nine of several dozen compounds were observed downstream of RO and these exhibited removals ranging from 84% to 99.9%. In a recent study, Xu et al. [34] studied the rejection of six EDCs and pharmaceuticals on five different RO and NF membranes. Removals of 93–99% were observed with one membrane reducing two of the six compounds from 300 ng/L to below detection limits. Another challenge contaminant is *N*-nitrosodimethylamine (NDMA), which is present in most wastewater effluents. The California Department of Health Services requires drinking water sources with concentrations >10 ng/L to provide public notification. The RO process does not effectively reject NDMA (30–40%) and this poses a problem for indirect potable reuse applications where desired water quality meets or exceeds all criteria for drinking water. This means that additional treatment processes are required (photolysis) to achieve the desired water quality for potable reuse applications, thus significantly increasing the treatment cost. There is also the issue of understanding the behavior of new materials (nanoparticles, WMD-type materials) and the characterization of their physical and chemical separations properties.

3. CROSS-CUTTING NEEDS

There are numerous technologies and processes available today to desalinate and/or reuse water. Regardless of the technology or process (or combination of technologies and processes) chosen, there are several cross-cutting needs that must be addressed through some combination of basic research and applied development work. These include the following.

3.1 Implementation of novel processes (forward osmosis)

Osmosis, or as it is currently referred to as forward osmosis (FO), has new applications in separation processes for wastewater treatment, food processing, and seawater and brackish water desalination [35–40]. Like reverse osmosis (RO), FO uses a semipermeable membrane to separate water from dissolved solutes. Instead of employing hydraulic pressure as the driving force for the separation as in RO, FO uses the osmotic pressure gradient across the membrane to induce a net flow of water through the membrane into the draw solution, thus, efficiently separating the freshwater from its solutes. Driven by an osmotic pressure gradient, FO does not require significant energy input, as only stirring or pumping of the solutions are involved. Development of a desalination or a water reuse process technology using the FO route has the potential to achieve a significant decrease in costs and energy consumption in comparison to conventional RO processes.

3.2 Energy

The capital costs for a plant producing water from seawater are about twice the costs of a plant reusing secondary sewage. Both the pretreatment costs and RO costs are higher. In the case of pretreatment, this is due to the difference in recovery (75%for secondary effluent; 50% for seawater), which results in a larger seawater pretreatment system. The capital cost for the seawater RO process is higher than for the secondary effluent RO as it is operating at a much higher pressure, lower permeate flux, lower recovery, and must be made of materials that resist corrosion in seawater. Similarly, the operation and maintenance costs for producing RO water from seawater are twice the costs of reusing secondary sewage. The higher pretreatment costs are due to continuous dosage of a coagulant. The higher RO costs are due primarily to energy (the operating pressure is four times higher and the feedflow is 1.5 times higher), but also to membrane replacement. The total life cycle costs for producing RO water from secondary effluent and seawater are US$ 0.28 and US$ 0.62 m^{-3}, respectively, a ratio of 2.21 [41].

Desalination and water reuse processes tend to be energy-intensive, so reduction and recovery and alternative energy sources must be sought. Energy is used (and wasted) throughout desalination and water reuse processes. There is a potential to develop or apply energy recovery devices, and to devote time and resources to reducing the energy consumption of desalination- and water reuse-specific devices. Further, the use of renewable or alternative energy sources when constructing desalination and water reuse facilities should be considered. Worthy of mention are solar electricity, geothermal, and wind energy [4].

3.3 Public perception

Desalinated and/or reused waters are going to have to fight for acceptance, at least for the foreseeable future [42]. People generally favor reuse that promotes water conservation, provides environmental protection benefits, protects human health, and cost-effectively treats and distributes a valuable and limited resource. However, as the water options become more tangible to people with specific proposed projects in their communities and the likelihood of human contact increasing, attitudes change – the public's support wanes [43]. A survey study in 2000, in the UK, corroborated the existence of a "source" factor. It found people more willing to use recycled water from their own wastewater than from second parties or a common public source [44]. A survey in Australia in 2006 determined that the Australian population discriminated between recycled and desalinated water [45]. Respondents understood that recycled water was more environmentally friendly, whereas desalinated water was perceived as less risky. All of these studies show that acceptance levels of recycled water are lower than those of desalinated water.

To more rapidly gain public acceptance, it is vital that water providers be able to instantly and constantly monitor their processes and measure the quality of their product. This verification and assurance of quality will require the development of new technologies to monitor and evaluate the performance of membranes in real-time and without inhibiting the performance of the plant. It will also require development of monitoring technologies and measurement protocols suitable for application in water reuse scenarios.

In conclusion, there is a need for holistic, system-wide design of desalination and water reuse plants and processes. In particular, plants and processes should be optimized and integrated to minimize waste, recover by-products, and minimize energy consumption. Additionally, sustainability of the resource and the process should be integrated into all process optimization, planning, and public outreach activities.

REFERENCES

[1] P. Gagliardo, S. Adham, R. Trussell, A. Olivieri, Water repurification via reverse osmosis, Desalination 117 (1998) 73.
[2] B. Durham, Membrane pretreatment of reverse osmosis-long term experience on difficult waters, Desalination 110 (1997) 49.
[3] J. Glater, The early history of reverse osmosis membrane development, Desalination 117 (1998) 297.
[4] G. Filteau, P. Moss, Ultra-low pressure RO membranes: an analysis of performance and cost, Desalination 113 (1997) 147.
[5] Miller, W. Integrated concepts in water reuse: managing global water needs, in: Proceedings of International Conference on Integrated Concepts in Water Recycling, Wollongong, NSW, Australia, February 14, 2005.
[6] R.S. Trussell, S. Adham, R.R. Trussell, Process limits of municipal wastewater treatment with the submerged membrane bioreactor, J. Environ. Eng.-ASCE 131 (2005) 410.
[7] J. Lozier, A. Fernandez, B. Hines, K. Carns, Using a membrane bioreactor/reverse osmosis system for indirect potable reuse, Water Reuse (2000) .
[8] R. Merlo, S. Adham, P. Gagliardo, R.S. Trussell, R. Trussell, Application of Membrane Bioreactor (MBR) Technology for Water Reclamation, WEFTEC, Los Angeles, CA, 2000.
[9] R.S. Trussell, R.P. Merlo, S. Hermanowicz, D. Jenkins, The Effect of Organic Loading on Membrane Fouling in a Submerged Membrane Bioreactor Treating Municipal Wastewater, WEFTEC, New Orleans, LA, 2004.
[10] Joint Water Reuse and Desalination, Task Force Water Innovation Symposium, San Diego, CA, October 17–25, 2005.
[11] I. Escobar, E. Hoek, C. Gabelich, F. DiGiano, Y. Le Gouellec, P. Berube, K. Howe, J. Allen, K. Atasi, M. Benjamin, P. Brandhuber, J. Brant, Y. Chang, M. Chapman, A. Childress, W. Conlon, T. Cooke, I. Crossley, G. Crozes, P. Huck, S. Kommineni, J. Jacangelo, A. Karimi, J. Kim, D. Lawler, Q. Li, L. Schideman, S. Sethi, J. Tobiason, T. Tseng, S. Veerapaneni, A. Zander, American water works association membrane technology research committee report: membrane fouling – recent advances and research needs, J. Am. Water Works Assoc. 97(8) (2005) 79–89.
[12] J.-Y. Gal, J.C. Bollinger, H. Tolosa, N. Gache, Calcium carbonate solubility: a reappraisal of scale formation and inhibition, Talanta 43 (1996) 497–1509.
[13] R. Sheikholeslami, Composite fouling-Inorganic and biological: a review, Environ. Prog. 18 (1999) 113.
[14] R. Sheikholeslami, I.S. Al-Mutaz, S. Tan, S.D. Tan, Some aspects of silica polymerization and fouling and its pretreatment by sodium aluminate, lime and soda ash, Desalination 150 (2002) 85.
[15] W.E. Katz, Electrodialysis – the first twenty-five years, in: Proceedings of Fifth Annual Meeting, National Water Supply Improvement Association, San Diego, CA, 1977.
[16] W.E. Katz, Electrodialysis reversal (EDR) process, Desalination 28 (1979) 31.
[17] J.S. Park, H.J. Lee, S.J. Choi, K.E. Geckeler, J. Cho, S.H. Moon, Fouling mitigation of anion exchange membrane by zeta potential control, J. Colloid Interface Sci. 259 (2003) 293.
[18] A.J.M. von Gottberg, Advances in electrodialysis reversal systems, in: Proceedings of Membrane Technology Conference, Long Beach, CA, 1999.
[19] J. Redondo, M. Busch, J.P. De Witte, Boron removal from seawater using FILMTEC (TM) high rejection SWRO membranes, Desalination 156 (2003) 229.
[20] M. Taniguchi, Y. Fusaoka, T. Nishikawa, M. Kurihara, Boron removal in RO seawater desalination, Desalination 167 (2004) 419.
[21] N. Kabay, I. Yilmaz, S. Yamac, S. Samatya, M. Yuksel, U. Yuksel, M. Arda, M. Saglam, T. Iwanaga, K. Hirowatari, Removal and recovery of boron from geothermal

wastewater by selective ion exchange resins. I. Laboratory tests, React. Funct. Polym. 60 (2004) 163.
[22] N. Kabay, I. Yilmaz, S. Yamac, M. Yuksel, U. Yuksel, N. Yildirim, O. Aydogdu, T. Iwanaga, K. Hirowatari, Removal and recovery of boron from geothermal wastewater by selective ion-exchange resins. II. Field tests, Desalination 167 (2004) 427.
[23] N. Nadav, Boron removal from seawater reverse osmosis permeate utilizing selective ion exchange resin, Desalination 124 (1999) 131.
[24] M.R. Pastor, A.F. Ruiz, M.F. Chillon, D.P. Rico, Influence of pH in the elimination of boron by means of reverse osmosis, Desalination 140 (2001) 145.
[25] M. Wilf, C. Bartels, Optimization of seawater RO systems design, Desalination 173 (2005) 1.
[26] B. Durham, M.M. Bourbigot, T. Pankratz, Membranes as pretreatment to desalination in wastewater reuse: operating experience in the municipal and industrial sectors, Desalination 138 (2001) 83.
[27] S. BouHamad, M. AbdelJawad, S. Ebrahim, M. AlMansour, A. AlHijji, Performance evaluation of three different pretreatment systems for seawater reverse osmosis technique, Desalination 110 (1997) 85.
[28] K.T. Chua, M.N.A. Hawlader, A. Malek, Pretreatment of seawater: results of pilot trials in Singapore, Desalination 159 (2003) 225.
[29] N. Hilal, H. Al-Zoubi, N.A. Darwish, A.W. Mohammad, M. Abu Arabi, A comprehensive review of nanofiltration membranes: Treatment, pretreatment, modeling, and atomic force microscopy, Desalination 170 (2004) 281.
[30] S. van Hoof, J.G. Minnery, B. Mack, Dead-end ultrafiltration as alternative pre-treatment to reverse osmosis in seawater desalination: a case study, Desalination 139 (2001) 161.
[31] D. Vial, G. Doussau, The use of microfiltration membranes for seawater pre-treatment prior to reverse osmosis membranes, Desalination 153 (2003) 141.
[32] K. Kimura, S. Toshima, G. Amy, Y. Watanabe, Rejection of neutral endocrine disrupting compounds (EDCs) and pharmaceutical active compounds (PhACs) by RO membranes, J. Membr. Sci. 245 (2004) 71.
[33] S. Snyder, B. Vanderford, S. Adham, J. Oppenheimer, T. Gillogly, R. Bond, V. Veerapaneni, Membranes for the removal of EDCs and pharmaceuticals, in: Proceedings of WQTC, Denver, CO, 2004.
[34] P. Xu, J. Drewes, C. Bellona, G. Amy, T. Kim, M. Adam, T. Heberer, Rejection of emerging organic micropollutants in nanofiltration-reverse osmosis membrane applications, Water Environ. Res. 77 (2005) 40.
[35] R.L. McGinnis, M. Elimelech, Energy requirements of ammonia–carbon dioxide forward osmosis desalination, Desalination 207 (2007) 370–382.
[36] T.Y. Cath, A.E. Childress, M. Elimelech, Forward osmosis: principles, applications, and recent developments, review, J. Membr. Sci. 281 (2006) 70.
[37] J.R. McCutcheon, R.L. McGinnis, M. Elimelech, A novel ammonia–carbon dioxide forward (direct) osmosis desalination process, Desalination 174 (2005) 1–11.
[38] J.O. Kessler, C.D. Moody, Drinking water from sea water by forward osmosis, Desalination 18 (1976) 297.
[39] T.Y. Cath, S. Gormly, E.G. Beaudry, M.T. Flynn, V.D. Adams, A.E. Childress, Membrane contactor processes for wastewater reclamation in space Part 1. Direct osmosis concentration as pretreatment for reverse osmosis, J. Membr. Sci. 257 (2005) 85.
[40] T.Y. Cath, V.D. Adams, A.E. Childress, Membrane contactor processes for wastewater reclamation in space II. combined direct osmosis, osmotic distillation, and membrane distillation for treatment of metabolic wastewater, J. Membr. Sci. 257 (2005) 111.
[41] P. Côté, S. Siverns, S. Monti, Comparison of membrane-based solutions for water reclamation and desalination, Desalination 182 (2005) 251.

[42] T.W. Hartley, Public perception and participation in water reuse, Desalination 187 (2006) 115.
[43] W.H. Bruvold, Public opinion on water reuse options, J. Water Pollut. Control. Fed. 60(1) (1998) 45–50.
[44] P. Jeffery, Understanding Public Receptivity Issues Regarding "In-House" Water Recycling, Results from a UK Survey, Unpublished Manuscript, Cranfield University, UK, 2001.
[45] S. Dolnicar, A.L. Schäfer, Public perception of desalinated versus recycled water in Australia, CD Proceedings of the AWWA Desalination Symposium (2006).

SUBJECT INDEX

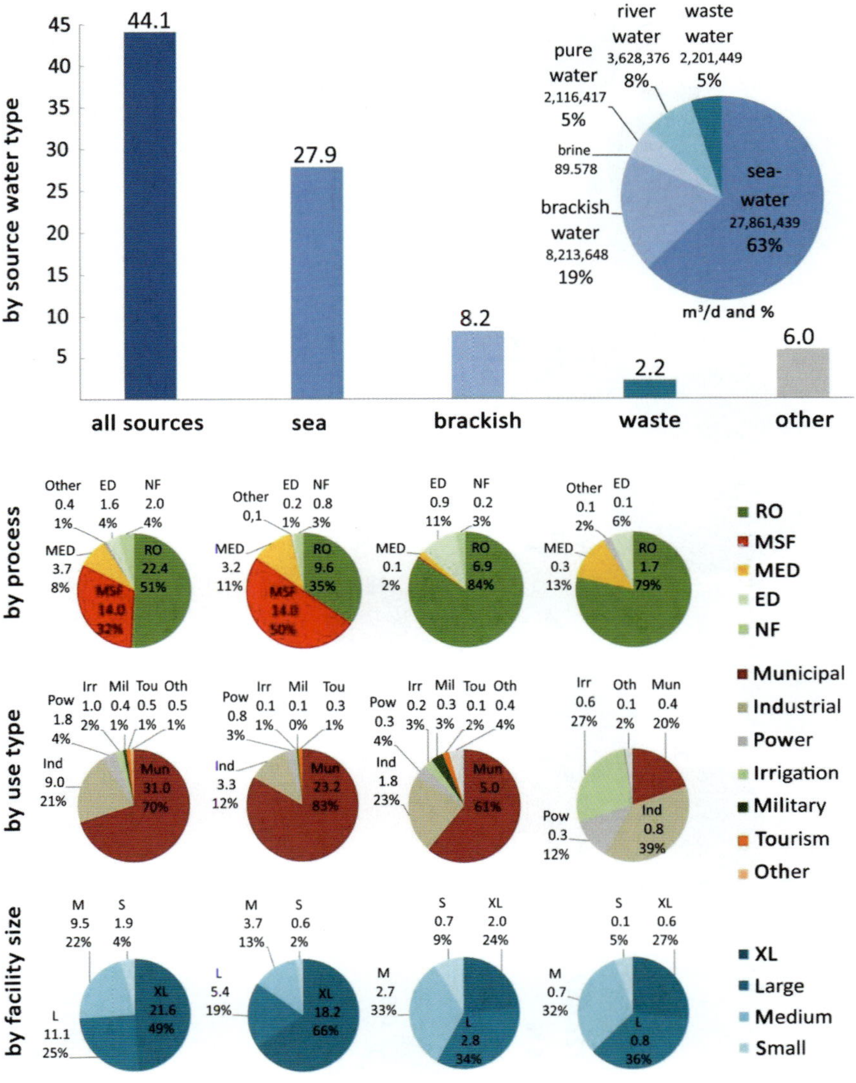

Plate 1 Global desalination capacities (in Mm^3/day and %) by source water type (top row), by process and source water type (second row), by use type and source water type (third row). Data analysis based on primary data from Ref. [5]. Abbreviations: RO, reverse osmosis; MSF, multistage flash distillation; MED, multieffect distillation; NF, nanofiltration; ED, electrodialysis; $XL \geq 50{,}000\ m^3/day > L \geq 10{,}000\ m^3/day > M \geq 1{,}000\ m^3/day > S$. (For b/w version, see page 13 in the volume.)

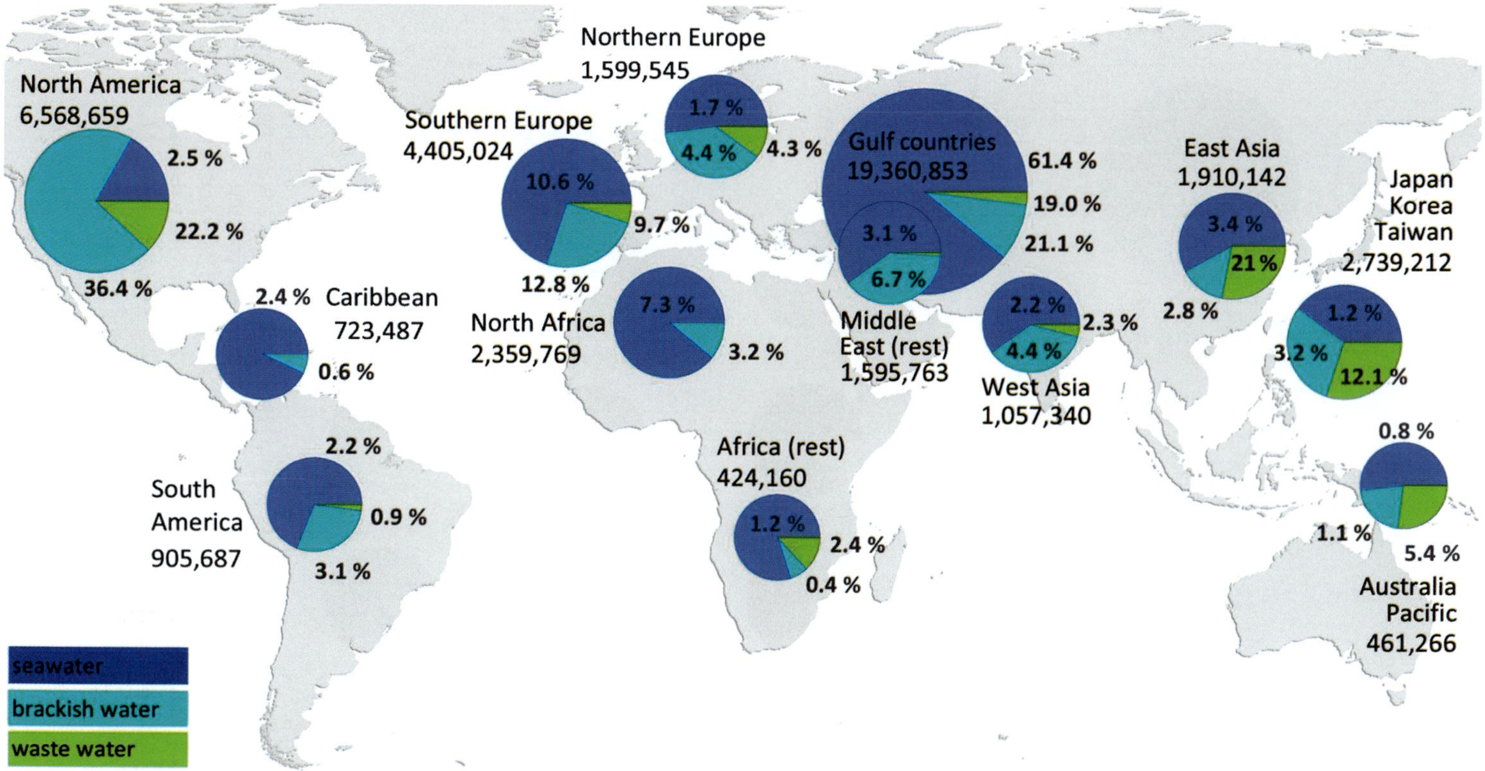

Plate 2 Global desalination capacities in cubic meters per day. For example, the installed capacity in Southern Europe is 4,405,024 m^3/day. This figure includes all source water types. Seawater desalination accounts for most of the production in Southern Europe, brackish water for about one-fourth of the production and wastewater desalination plays a relatively minor role (pie diagram). The figures next to the pie diagram give the contribution to the *global* production, that is, the seawater desalination capacity in Southern Europe represents 10.6% of the *global seawater* desalination capacity. The brackish water capacity – though it is less than half the seawater desalination capacity in Southern Europe – represents 12.8% of the *global*

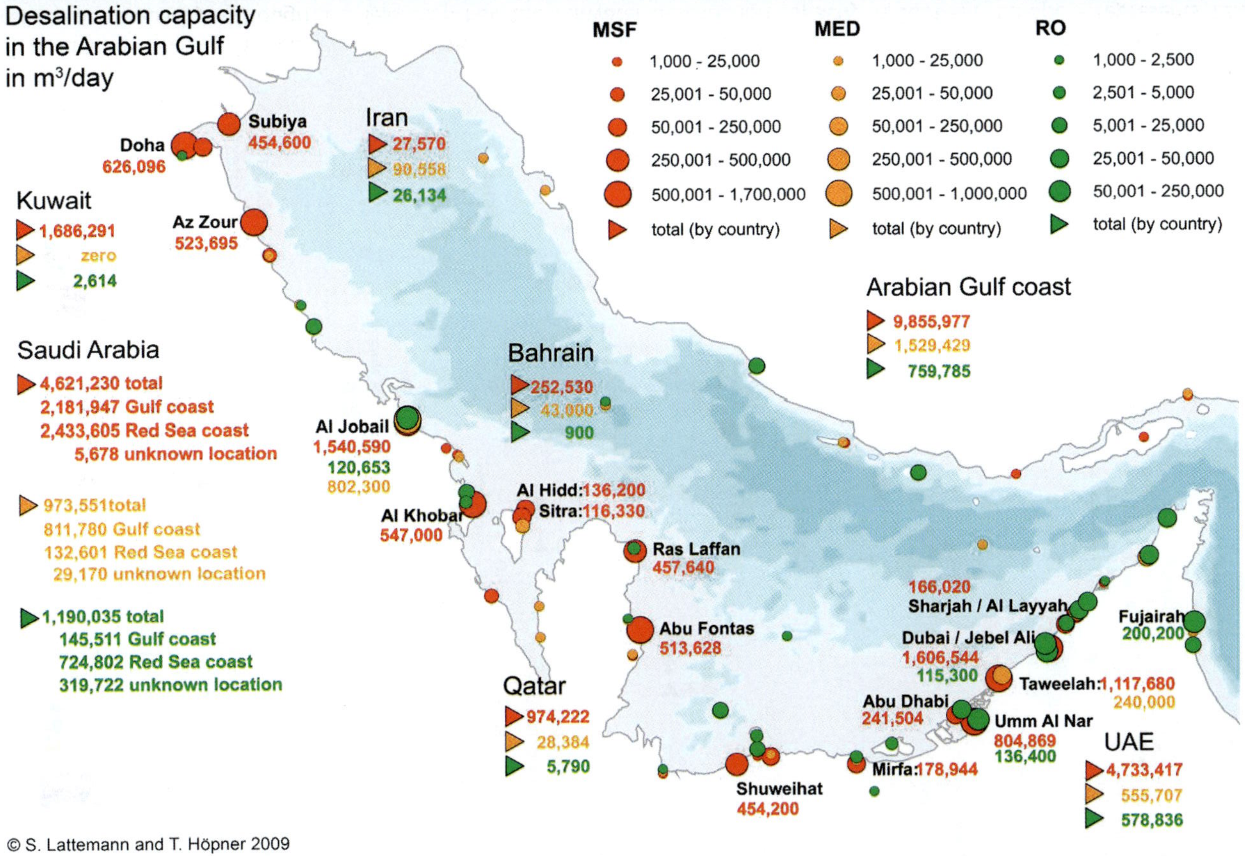

Plate 3 Cumulative MSF, MED, and RO capacities in the Arabian Gulf in cubic meters per day by site location (dots) and by country (triangles). The map shows all sites with an installed capacity $\geq 1000\ m^3$/day and displays sites with a capacity $\geq 100{,}000\ m^3$/day by name and capacity. The map was first published in Ref. [17,28] and updated using raw data from Ref. [5]. (For b/w version, see page 20 in the volume.)

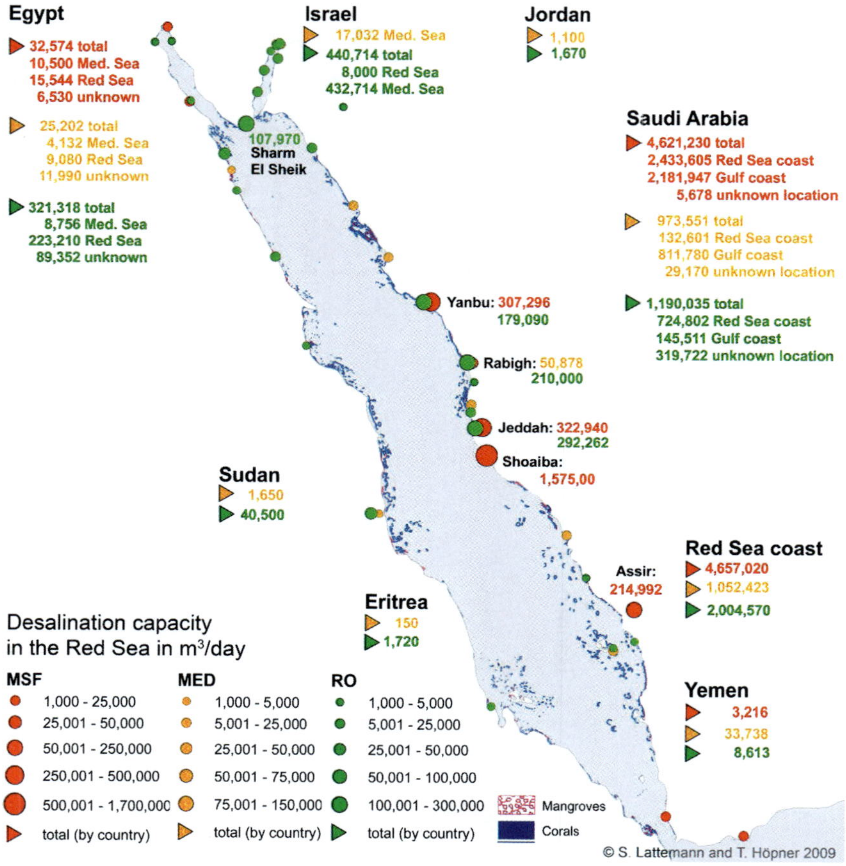

Plate 4 Cumulative MSF, MED, and RO capacities in the Red Sea in cubic meters per day by site location (dots) and by country (triangles). The map shows all sites with an installed capacity $\geq 1000\ m^3$/day and displays sites with a capacity $\geq 100{,}000\ m^3$/day by name and capacity. The map was first published in Ref. [17,28] and updated using raw data from Ref. [5]. (For b/w version, see page 21 in the volume.)

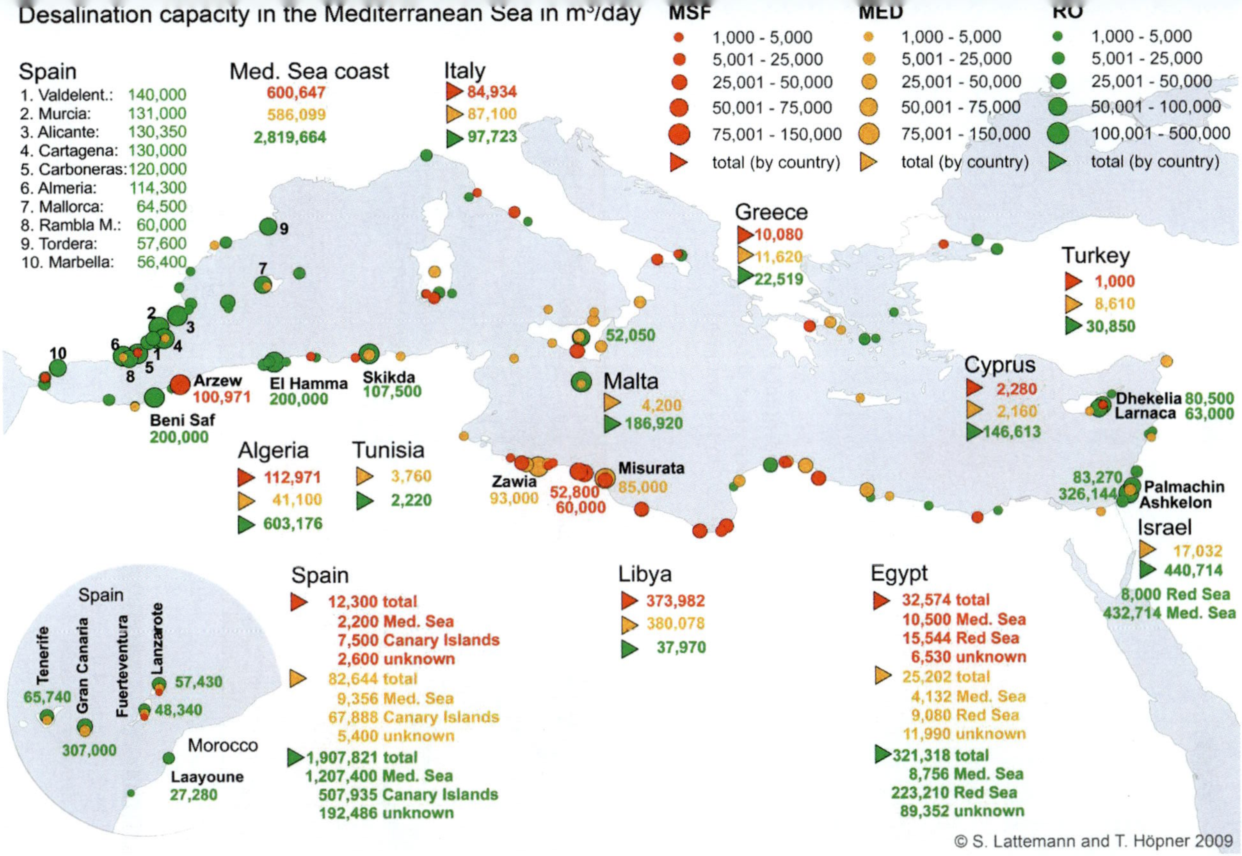

Plate 5 Cumulative MSF, MED and RO capacities in the Mediterranean Sea in cubic meters per day by site location (dots) and by country (triangles). The map shows all sites with an installed capacity $\geq 1000\,m^3$/day and displays sites with a capacity $\geq 50{,}000\,m^3$/day by name and capacity. The map was first published in Ref. [17,28] and updated using raw data from Ref. [5]. (For b/w version, see page 22 in the volume.)

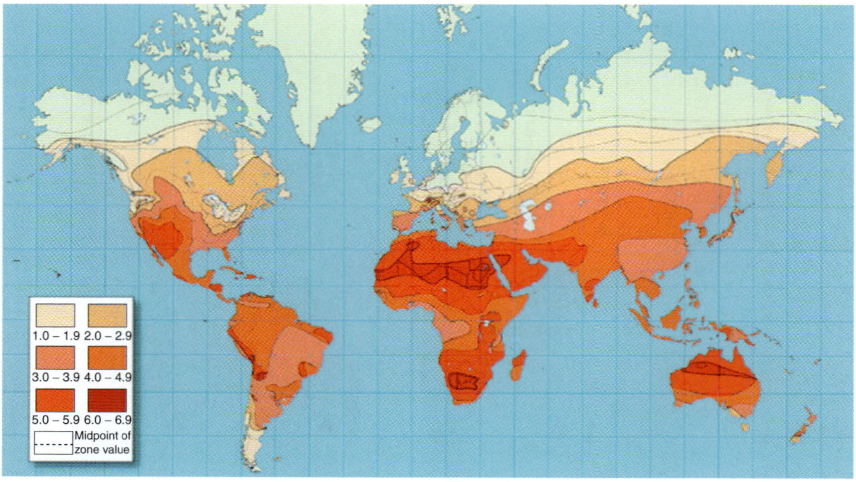

Plate 6 World solar irradiance, plotted as the number the daily number of peak sunshine hours incident on a horizontal surface at an intensity of 1 kW/m^2 (adapted from Ref. [5]). (For b/w version, see page 357 in the volume.)

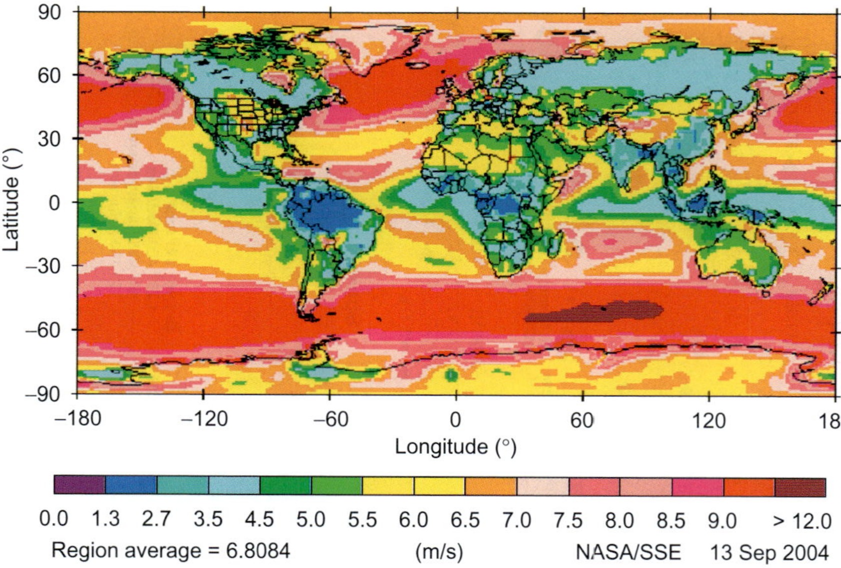

Plate 7 World map of average annual wind speed at a height of 50 m based on 10 years of data (July 1983 to June 1993). Adapted from Ref. [6]. (For b/w version, see page 358 in the volume.)

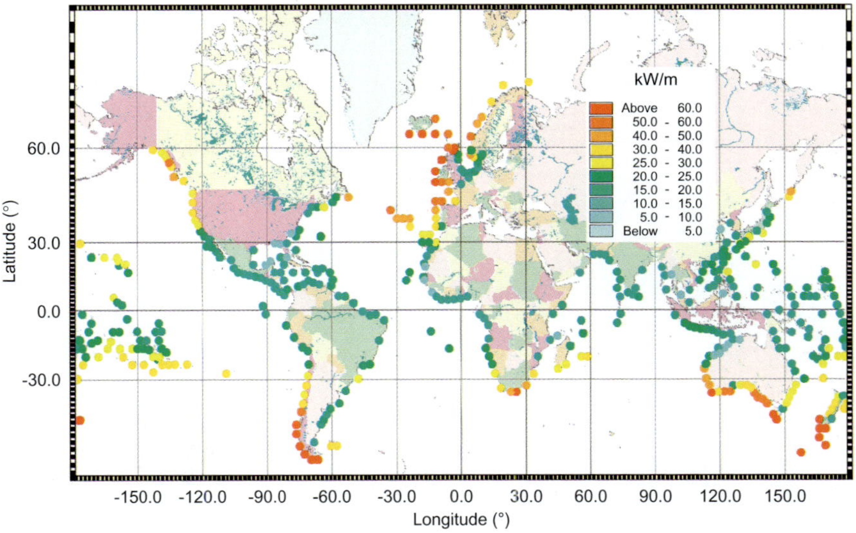

Plate 8 Wave energy estimates (in units of kW/m) along global coastlines determined via satellite data. Adapted from Ref. [7]. (For b/w version, see page 359 in the volume.)